Bio-Based Polymers' Application and Technology for Better Quality of Life

Bio-Based Polymers' Application and Technology for Better Quality of Life

Editors

Arash Moeini
Pierfrancesco Cerruti
Gabriella Santagata

 Basel • Beijing • Wuhan • Barcelona • Belgrade • Novi Sad • Cluj • Manchester

Editors

Arash Moeini
TUM School of Life Sciences
Technical University
of Munich
Munich
Germany

Pierfrancesco Cerruti
Institute of Polymers,
Composites and Biomaterials
(IPCB)
National Council of Research
of Italy (CNR)
Pozzuoli
Italy

Gabriella Santagata
Institute of Polymers,
Composites and Biomaterials
(IPCB)
National Council of Research
of Italy (CNR)
Pozzuoli
Italy

Editorial Office
MDPI
St. Alban-Anlage 66
4052 Basel, Switzerland

This is a reprint of articles from the Special Issue published online in the open access journal *Polymers* (ISSN 2073-4360) (available at: www.mdpi.com/journal/polymers/special_issues/bio_poly_appl_tech).

For citation purposes, cite each article independently as indicated on the article page online and as indicated below:

Lastname, A.A.; Lastname, B.B. Article Title. *Journal Name* **Year**, *Volume Number*, Page Range.

ISBN 978-3-7258-1372-8 (Hbk)
ISBN 978-3-7258-1371-1 (PDF)
doi.org/10.3390/books978-3-7258-1371-1

© 2024 by the authors. Articles in this book are Open Access and distributed under the Creative Commons Attribution (CC BY) license. The book as a whole is distributed by MDPI under the terms and conditions of the Creative Commons Attribution-NonCommercial-NoDerivs (CC BY-NC-ND) license.

Contents

About the Editors vii

Preface ix

Arash Moeini, Parisa Pedram, Ehsan Fattahi, Pierfrancesco Cerruti and Gabriella Santagata
Edible Polymers and Secondary Bioactive Compounds for Food Packaging Applications: Antimicrobial, Mechanical, and Gas Barrier Properties
Reprinted from: *Polymers* **2022**, *14*, 2395, doi:10.3390/polym14122395 1

Aleksandra Nesic, Sladjana Meseldzija, Antonije Onjia and Gustavo Cabrera-Barjas
Impact of Crosslinking on the Characteristics of Pectin Monolith Cryogels
Reprinted from: *Polymers* **2022**, *14*, 5252, doi:10.3390/polym14235252 22

Guillermo Conejo-Cuevas, Leire Ruiz-Rubio, Virginia Sáez-Martínez, Raul Pérez-González, Oihane Gartziandia and Amaia Huguet-Casquero et al.
Spontaneous Gelation of Adhesive Catechol Modified Hyaluronic Acid and Chitosan
Reprinted from: *Polymers* **2022**, *14*, 1209, doi:10.3390/polym14061209 33

Soňa Kontárová, Radek Přikryl, Petr Škarpa, Tomáš Kriška, Jiří Antošovský and Zuzana Gregušková et al.
Slow-Release Nitrogen Fertilizers with Biodegradable Poly(3-hydroxybutyrate) Coating: Their Effect on the Growth of Maize and the Dynamics of N Release in Soil
Reprinted from: *Polymers* **2022**, *14*, 4323, doi:10.3390/polym14204323 49

Papon Thamvasupong and Kwanchanok Viravaidya-Pasuwat
Controlled Release Mechanism of Vancomycin from Double-Layer Poly-L-Lactic Acid-Coated Implants for Prevention of Bacterial Infection
Reprinted from: *Polymers* **2022**, *14*, 3493, doi:10.3390/polym14173493 69

Ewa Rybak, Piotr Kowalczyk, Sylwia Czarnocka-Śniadała, Michał Wojasiński, Jakub Trzciński and Tomasz Ciach
Microfluidic-Assisted Formulation of -Polycaprolactone Nanoparticles and Evaluation of Their Properties and In Vitro Cell Uptake
Reprinted from: *Polymers* **2023**, *15*, 4375, doi:10.3390/polym15224375 86

Marion Ryan C. Sytu, David H. Cho and Jong-in Hahm
Self-Assembled Block Copolymers as a Facile Pathway to Create Functional Nanobiosensor and Nanobiomaterial Surfaces
Reprinted from: *Polymers* **2024**, *16*, 1267, doi:10.3390/polym16091267 106

Natalia O. Zhila, Kristina Yu. Sapozhnikova, Evgeniy G. Kiselev, Ekaterina I. Shishatskaya and Tatiana G. Volova
Biosynthesis of Polyhydroxyalkanoates in *Cupriavidus necator* B-10646 on Saturated Fatty Acids
Reprinted from: *Polymers* **2024**, *16*, 1294, doi:10.3390/polym16091294 154

Vito Gigante, Laura Aliotta, Maria-Beatrice Coltelli and Andrea Lazzeri
Upcycling of Poly(Lactic Acid) by Reactive Extrusion with Recycled Polycarbonate: Morphological and Mechanical Properties of Blends
Reprinted from: *Polymers* **2022**, *14*, 5058, doi:10.3390/polym14235058 173

Angela Marotta, Noemi Faggio and Cosimo Brondi
Curing Kinetics of Bioderived Furan-Based Epoxy Resins: Study on the Effect of the Epoxy Monomer/Hardener Ratio
Reprinted from: *Polymers* **2022**, *14*, 5322, doi:10.3390/polym14235322 187

About the Editors

Arash Moeini

Dr. Arash Moeini is a senior associate researcher at the esteemed Technical University of Munich (TUM) in Germany. With a Ph.D. in Chemical Sciences from the prestigious University of Naples Federico II, he utilizes the remarkable potential of natural compounds and biopolymers to tackle food waste through innovative food packaging and edible coatings. His creative work extends to enhancing skin regeneration with advanced wound dressings and protecting plants with eco-friendly herbicides and insecticides.

Pierfrancesco Cerruti

Dr. Pierfrancesco Cerruti is a senior researcher at IPCB-CNR, Italy. He received his Ph.D. in Materials Engineering at the University of Naples, Italy. His research fields concern bio-based and biodegradable polymers and composites, stimuli-responsive materials, polymer degradation and stabilization.

Gabriella Santagata

Dr. Gabriella Santagata is a full time permanent researcher of the Institute for Polymers, Composites and Biomaterials (IPCB) of the National Council of Research. Her research activities are mainly focused on the study and chemical and physical characterization of biodegradable thermoplastic and hydroplastic polymers obtained from renewable sources, with a focus on recovery, upgrading and exploiting waste vegetable biomass from the agro-food industry for the extraction of biomolecules, macromolecules, and ligno-cellulose fractions.

Preface

The environmental impact of petroleum-based polymers has led to extensive plastic waste globally. To address this, biopolymers with various properties like biodegradability, biocompatibility, and reasonable mechanical properties are promising alternatives in different fields such as packaging, medicine, cosmetics, and agriculture. This reprint, intended as a reference for researchers and scholars working on sustainable polymers and their application, features relevant instances of bio-based polymers that can contribute to building a better quality of life and a greener environment.

The first chapter reviews the impact of functional natural compounds on the relevant properties of edible biopolymer systems based on polysaccharides, proteins, and lipids used for food packaging materials. The following chapters showcase cutting-edge examples of applications of bio-based and biodegradable polymers. In particular, chapters 2 and 3 deal with advanced formulations to develop polysaccharide hydrogels. Chapters 4, 5, and 6 report on manufacturing polyester-based coatings and nanoparticles to develop controlled-release devices that could be applied as agricultural fertilizers and active biomaterials. Chapter 7 highlights the recent research efforts in block copolymer nanobiotechnology, particularly nanoscale surface assembly and interactions of functional biomolecules, with a specific view on biosensing. Finally, the last three chapters deal with synthesizing and modifying novel bio-based polymer formulations to pave the way for sustainable alternatives to traditional plastics. Specifically, chapter 8 explores the microbial synthesis of polymers, focusing on producing polyhydroxyalkanoates (PHA) from saturated fatty acids. Chapter 9 discloses the synergistic potential of combining poly(lactic acid) with recycled polycarbonate, shedding light on polymer compatibility. In conclusion, chapter 10 reports recent advancements in furan-based epoxy resins, offering a glimpse into their curing kinetics and providing insights into their optimization for applications ranging from coatings to adhesives.

Together, these papers underscore the pivotal role of bio-based polymers in shaping a more sustainable future for the planet. From reducing dependence on fossil fuels to mitigating plastic pollution, bio-based polymers offer tangible solutions across various industries and technologies, leading to an era of innovation that is as eco-conscious as it is impactful.

We sincerely thank all the authors for their enthusiastic and significant contributions to this reprint. We also appreciate the generous and effective assistance provided by Ms. Natalie Zhou, Section Managing Editor at MDPI, throughout the publication process.

This reprint is dedicated to my daughter Sophia, my precious one, whose presence fills my life with boundless joy and love (Yours, Arash Moeini).

This reprint is dedicated to the memory of Mario Malinconico, our beloved mentor.

Arash Moeini, Pierfrancesco Cerruti, and Gabriella Santagata
Editors

Review

Edible Polymers and Secondary Bioactive Compounds for Food Packaging Applications: Antimicrobial, Mechanical, and Gas Barrier Properties

Arash Moeini [1,*], Parisa Pedram [1], Ehsan Fattahi [1], Pierfrancesco Cerruti [2] and Gabriella Santagata [2]

1. School of Life Sciences Weihenstephan, Technical University of Munich, 85354 Freising, Germany; parisa.pedram@tum.de (P.P.); ehsan.fattahi@tum.de (E.F.)
2. Institute for Polymers, Composites and Biomaterials (IPCB CNR), Via Campi Flegrei 34, 80078 Pozzuoli, Italy; cerruti@ipcb.cnr.it (P.C.); gabriella.santagata@ipcb.cnr.it (G.S.)
* Correspondence: arash.moeini@tum.de

Abstract: Edible polymers such as polysaccharides, proteins, and lipids are biodegradable and biocompatible materials applied as a thin layer to the surface of food or inside the package. They enhance food quality by prolonging its shelf-life and avoiding the deterioration phenomena caused by oxidation, humidity, and microbial activity. In order to improve the biopolymer performance, antimicrobial agents and plasticizers are also included in the formulation of the main compounds utilized for edible coating packages. Secondary natural compounds (SC) are molecules not essential for growth produced by some plants, fungi, and microorganisms. SC derived from plants and fungi have attracted much attention in the food packaging industry because of their natural antimicrobial and antioxidant activities and their effect on the biofilm's mechanical properties. The antimicrobial and antioxidant activities inhibit pathogenic microorganism growth and protect food from oxidation. Furthermore, based on the biopolymer and SC used in the formulation, their specific mass ratio, the peculiar physical interaction occurring between their functional groups, and the experimental procedure adopted for edible coating preparation, the final properties as mechanical resistance and gas barrier properties can be opportunely modulated. This review summarizes the investigations on the antimicrobial, mechanical, and barrier properties of the secondary natural compounds employed in edible biopolymer-based systems used for food packaging materials.

Keywords: edible biopolymers; secondary compounds; antimicrobials; active food packaging; essential oils; plasticizers; gas barrier

Citation: Moeini, A.; Pedram, P.; Fattahi, E.; Cerruti, P.; Santagata, G. Edible Polymers and Secondary Bioactive Compounds for Food Packaging Applications: Antimicrobial, Mechanical, and Gas Barrier Properties. *Polymers* 2022, 14, 2395. https://doi.org/10.3390/polym14122395

Academic Editor: Sergio Torres-Giner

Received: 17 May 2022
Accepted: 10 June 2022
Published: 13 June 2022

Publisher's Note: MDPI stays neutral with regard to jurisdictional claims in published maps and institutional affiliations.

Copyright: © 2022 by the authors. Licensee MDPI, Basel, Switzerland. This article is an open access article distributed under the terms and conditions of the Creative Commons Attribution (CC BY) license (https://creativecommons.org/licenses/by/4.0/).

1. Introduction

Polymer-based food packaging materials protect food from deterioration and physical damage [1]. Currently, polyethylene terephthalate, high- and low-density polyethylene, polyvinyl chloride, polypropylene, and polystyrene are the most common polymers used in the packaging industry. However, apart from their numerous benefits in terms of performance, chemico-physical, and mechanical properties, the substantial primary concern of fossil-based packages is their end of life and their impact on environmental disposal. The statistical data show that packaging waste represents more than one-third of the overall plastic consumption [2]. These polymers are not biodegradable, and only one-fourth of them can be recycled [3]. Hence, turning to eco-friendly, biodegradable food packaging materials is absolutely unavoidableunavoidable. This is why the European Sustainable Development, by 2030, aims at the ever-wider use of bioplastics by decreasing plastic pollution [4]. Therefore, the attention of the academic and industrial world is focused on the study of bio-based and natural polymers for their potential application in food packaging. According to the European Bioplastics, biopolymers are biodegradable and compostable polymers obtained from renewable resources [5].

Based on the different sources, biodegradable polymers can be oil-derived polymers such as polycaprolactone, polybutylene succinate, and some aliphatic-aromatic copolyesters [6–9], and renewable biopolymers such as polyhydroxyalkanoates, polysaccharides, and polylactic acid coming from microorganisms, vegetables, animals, and proteins [10–13]. Generally, commercial biopolymers have excellent gas barrier properties, which are reduced by adding plasticizers and moisture [14]. Recently, there has been an ever-increasing demand for biopolymers as edible coatings (i.e., the primary packaging layer surrounding the food and consumed with it). The edible coatings preserve foods from microbial contamination by exploiting suitable gas barriers and mechanical performances. In this way, they prolong the food shelf life. Moreover, they enhance the organoleptic and sensorial properties of the coated foods. The new packaging approach is in high demand because the traditional food preservation techniques such as salting, and heat alter the food flavors, odors, colors, and textural properties [15]. Therefore, the new packaging trend has developed to keep food safe and maintain food texture and taste [16,17]. Generally, edible coatings are a challenging solution to protect perishable food. Polysaccharides, proteins, and lipids are the major edible biopolymers used in edible coatings. They are low-cost, industrially scalable, biodegradable, and biocompatible biopolymers, often used as carriers of antioxidant and antimicrobial agents to enhance the safety of the food products. Actually, the principal idea is the addition of active compounds to the packaging system to develop the so-called "active packaging" [18]. Then, this method has replaced the traditional food preservation methods to protect food from deterioration via surface interactions with food products or by modifying the headspace between the food and package surface. In the new packaging approach, biofunctional compounds are used as active agents [19]. In addition, there is a growing tendency to exploit natural products coming from plants such as the secondary metabolites that resulted in a novel research approach to use natural products as additives in active packaging [20–22].

The secondary compounds of plants and fungi such as essential oils and natural metabolites can be considered as the most promising materials for active packaging. These products can play various roles in the packaging system, for instance, acting as antioxidant/antimicrobial agents to increase the shelf-life of food and as plasticizers to improve the mechanical performance of the package. Secondary compounds are all-natural compounds derived from living organisms. Plants, fungi, and microorganisms are the primary sources of the secondary components. Given their peculiar structure, these compounds show various and unique biological properties that result in their broad range of applications in the fields of agriculture [23], medicine [24], and packaging [25,26]. Concerning active packaging applications, these products can represent a valid approach to improving packaging functionality through food protection and preservation. These products can also enhance the plasticizing effect of novel bioplastics, making them commercially attractive and comparable with some synthetic polymers. While several papers have analyzed the relevant literature on the general subject of edible coatings [27], the present review focuses on biopolymers applied for edible coatings, specifically addressing the antimicrobial, antioxidant, and plasticizing effects of secondary metabolites as active agents in edible biopolymers for food packaging applications.

2. Edible Biopolymers

One of the most common ways to apply biopolymers in the food packaging industry is represented by edible coatings. In this method, a thin layer of edible polymers covers the surface of the food and protects them from oxidation, humidity, microbial growth, and deterioration. In addition, edible coating materials have also shown gas, vapor, and oil barrier properties and could also be used as a carrier for active agents. Bioactive compounds can act as antioxidants, antimicrobials, and plasticizing agents, thus improving the food quality and extending the food shelf-life [28]. Furthermore, they can reduce the food particle clustering and maintain their surface appearance (colors and flavors) by physically strengthening the food products. Therefore, the biopolymers applied in edible coatings should pass the food approval process since they may be taken along with part of the food products [29].

On the other hand, food poisoning results from the food's microbial activity. This is why antimicrobial agents are usually formulated into the coating materials, preserve the outer food surface from bacteria activity, and protect humans from poisoning. Therefore, an edible coating is the most utilized method in active food packaging [30]. As Figure 1 shows, edible biopolymers are classified into three main sub-branches: polysaccharides, proteins, and lipids [31–34]. In addition, numerous studies have investigated edible biopolymer film-forming capabilities with potential applications in food packaging [35–39]. Therefore, different classes of edible biopolymers, along with the well-known examples of each group with film-forming capabilities, will be discussed in the following sections.

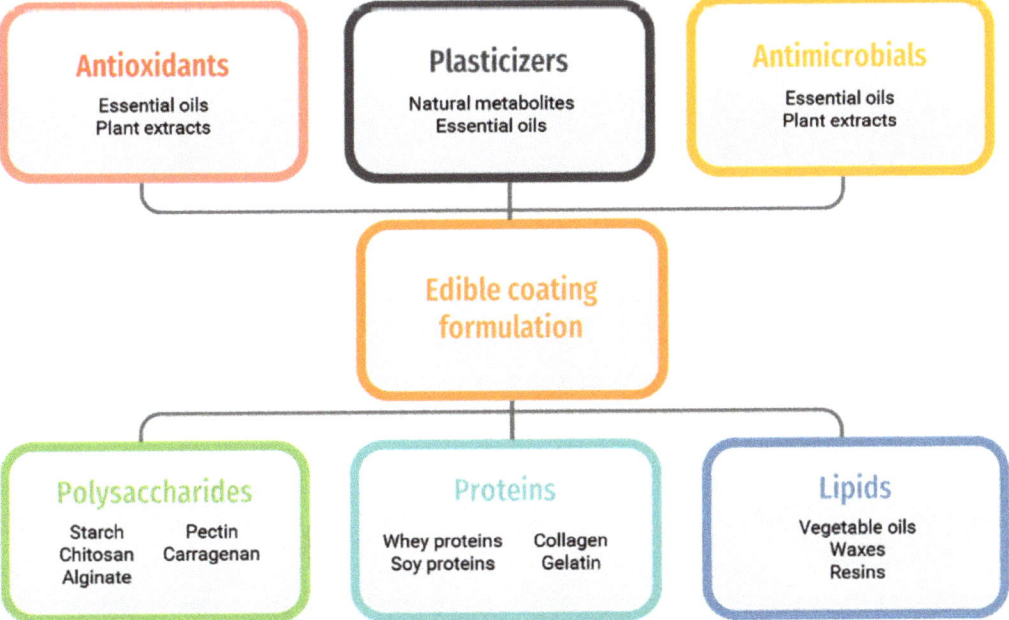

Figure 1. The main components of the edible coating formulations.

2.1. Polysaccharides

Generally, polysaccharides are constituted by saccharide units linked via glycosidic bonds. Polysaccharides for food packaging applications are widely investigated because of their non-toxic nature, low cost, fair mechanical and gas barrier properties, accessibility, biodegradability, and film-forming ability [40]. The film-forming ability of the polysaccharides is mainly due to the presence of hydroxyl groups, resulting in internal hydrogen bonding. These extensive properties make polysaccharides valid for food packaging, particularly in short-term food packaging applications [31]. The most studied polysaccharides for edible films are alginate, chitosan, cellulose, starch, pectin, alginate, and carragenan.

2.1.1. Chitosan

Chitosan consists of N-acetyl-glucosamine and N-glucosamine units derived from chitin via N-deacetylation (Figure 2). Chitin is the most abundant polysaccharide after cellulose [32]. The primary sources of chitin are shrimp shells, lobsters, crabs peritrophic membranes, and insect cocoons. Many studies have shown that the application of chitosan in different fields such as medicine [33], cosmetics [34], agriculture [41], and several other applications [42]. Of note, the chitosan application in the food industry increased after 2001 when the U.S. Food and Drug Administration (FDA) approved the edibility of chitosan [43]. Furthermore, the food packaging applications of chitosan are not only due to its film-forming ability, but also to the natural excellent antimicrobial properties against food filamentous fungi, yeast, and both Gram-negative and Gram-positive bacteria [44]. In this regard, Meng et al. proved the inhibitory effects of an edible chitosan coating against spoilage and pathogenic bacteria in pre-harvest to extend the shelf-life and quality of post-harvest fruit [45]. In another study, an edible chitosan coating was able to increase the pre- and post-harvest shelf-life of fruit before and during cold storage [46]. In this respect, better performing edible chitosan coatings require us to address the physical interaction between chitosan and the food substrate to achieve a homogenous distribution and good mechanical resistance. For this purpose, essential oils can improve the antimicrobial properties, surface adhesion, and O_2 and CO_2 gas barriers of the edible chitosan coating [47].

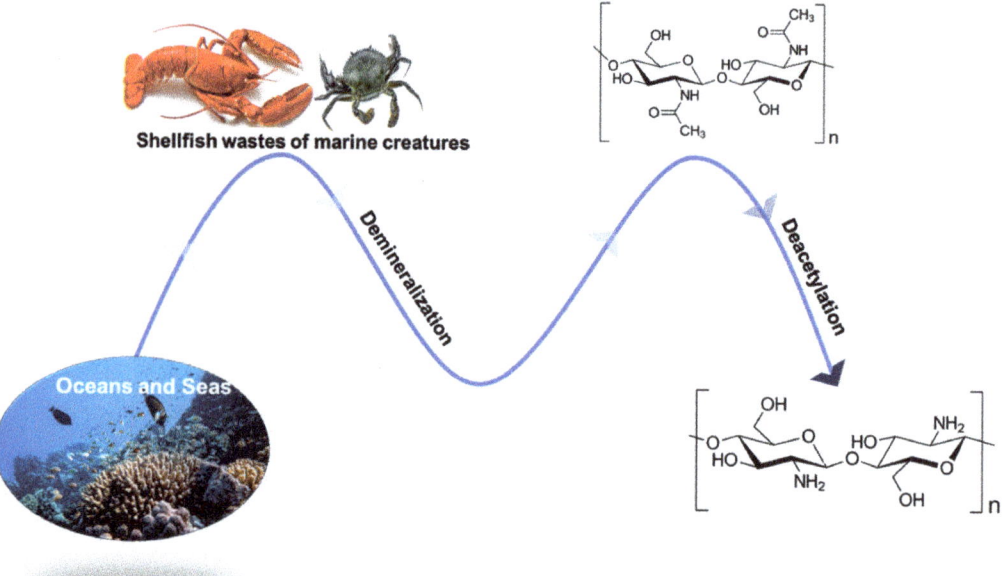

Figure 2. The chitin and chitosan manufacturing process.

2.1.2. Starch

Starch is a complex polysaccharide with (1→4)-α-D-glucopyranosyl units, arranged in linear amylose and branched amylopectin chains [48]. Starch is the most abundant renewable feedstock resources extracted from numerous plants such as maize, wheat, potato, rice, and peas [49]. Due to its biodegradability, renewability, and easy availability, starch has been extensively investigated as a low-cost component of biodegradable plastic materials. When mixed with some water and/or plasticizers such as glycerol, and following heat and shearing action, starch undergoes a spontaneous destructurization, leading to a homogeneous melt. The thermoplastic starch (TPS) is obtained, showing a typical thermoplastic behavior. Hence, TPS can be formulated and processed by means of standard equipment commonly used in the industrial manufacturing of synthetic polymers [50]. Furthermore, because of its versatility and functionality, odorless, film-forming ability, and excellent oxygen barrier properties, starch is commonly used for food preservation [49]. Moreover, several studies have investigated the exploitation of starch-based edible coatings as carriers of bioactive compounds. As an example, Wongphan et al. demonstrated that starch films could be used as carriers of specific enzymes responsible for meat tenderization [51]. Several studies have demonstrated that starch-based coatings could increase the shelf-life of fresh food products, although many efforts are necessary to improve the starch's poor water vapor barrier properties [52]. Indeed, hydrophobic additives such as lipid and essential oils are usually used for starch-based coating and films to drastically reduce the water vapor permeability [53].

2.1.3. Alginate

The other polysaccharide with an application in edible coatings for food packaging is alginate, a linear (1→4) linked polychronic acid (anionic polysaccharide) isolated from seaweed. Alginate film processing involves solvent evaporation; the interaction with divalent cations, particularly calcium salt solution, induces the development of three-dimensional networks able to resist water dissolution [52,54]. The mechanical and physicochemical properties of the alginate edible films depend on the polyvalent cations, pH, temperature, and composite ingredients, which can be elaborated using plasticizers and emulsifiers, and blended with other polymers. Alginate films present good tensile strength (TS), flexibility, tear-resistance (TR), and rigidity [53]. As with other edible polysaccharides, alginate films are tasteless, oil-resistant, glossy, and odorless [55].

2.1.4. Pectin

Pectin or poly-α-(1-4)-D-galacturonic acid is an anionic polysaccharide extracted from the cell walls of several plants and fruits such as apple, orange, lemon, and mango [56]. The solubility, gelling, and film-forming properties of pectin depend on the esterification degree (DE) [57,58]. Therefore, based on the degree of esterification, there are two categories of pectin above and below 50% of methyl ester, so-called high and low methoxy pectin, respectively. Furthermore, the cytocompatible and straight gelling mechanism widens the pectin application in various medical and tissue engineering fields [59]. Pectin has also been FDA-approved for food industry applications. Therefore, it is widely used in food and beverage industries such as in jams, yogurt drinks, fruity milk drinks, and ice creams, specifically as a gelling agent and colloidal stabilizer resulting from its gel-forming ability in the acidic condition [57,58]. Furthermore, the biodegradability, compatibility, and film-forming ability make pectin a suitable candidate for edible food packaging applications. Pectin film can be processed by solvent casting, spray coating, and knife coating. However, poor thermophysical properties and unsuitable mechanical performances are the two main significant restrictions for pectin application in the food packaging industry [60].

2.1.5. Carrageenans

Carrageenan is a high film-forming polysaccharide extracted from the cell walls of red seaweeds; it is hydrophilic, linear, and characterized by a sulfated anionic galactan including D-galactose and 3,6-anhydro-galactose linked by α-1,3 and β-1,4 glycosidic bonds [61]. Carrageenans are classified according to the amount of sulfate ester between 15–40% (degree of sulfation) and of the 3,6-anhydrous-α-D-galactopyranosyl content, influencing the degrees of negative charge and water solubility [49]. Therefore, different classes of carrageenans (λ, κ, ι, ε, μ) with a molecular weight ranging between 100 and 1000 kDa can be found. The most investigated are kappa (κ) carrageenans, with one sulfate group and one 3,6-anhydro-galactose per disaccharide, iota (ι) carrageenans with two sulfate groups and one 3,6-anhydro-galactose per disaccharide, and lambda (λ) carrageenans, with three sulfate groups and no 3,6-anhydro galactose per disaccharide [62]. Among them, κ- and ι-carrageenans show gel-forming ability in water solutions of potassium and calcium ions. At the same time, because of the high sulfate content, λ-carrageenans do not have this ability. In addition, carrageenans have a wide range of biological and antioxidant activities such as anticoagulant, antiviral, antitumor, immunomodulatory, antioxidant, and antihyperlipidemic [63,64]. All of these activities as well as their physicochemical properties have made carrageenans a suitable candidate for the food, cosmetics, and pharmaceutical industries. For example, carrageenans can act as antimicrobial and antioxidant carriers in the food packaging industry [65].

2.2. Proteins

Proteins are promising edible biopolymers with various features such as excellent mechanical, physicochemical, and optical properties with a selective and ideal fat barrier property [66–68]. Among the biopolymers, proteins are the most versatile because of the many different building blocks represented by 20 amino acids and the great variety of functional groups. Furthermore, these materials can be enzymatically, chemically, or physically modified, giving rise to biopolymers with improved physicochemical properties tailorable to each specific application. Indeed, proteins have already been exploited in the food industry to develop edible coatings that are able to preserve the food quality, thus prolonging its shelf-life [69]. Although these are advantages, the main drawbacks of protein are the great sensitivity to water, the poor oxygen barrier, and the severe structural integrity impairment, seriously compromising the protein film and coating performance and durability. On the other hand, the hydrophilic and hygroscopic nature of most of the proteins can be, in turn, positively used to develop bioactive packaging that is able to opportunely modulate the release of functional compounds willfully introduced in a protein-based matrix. Furthermore, the protein's edible coating bioactivity and functionality depends on both their specific physico-chemical properties including the size, composition, charge distribution, peculiar secondary, tertiary and quaternary structures, and on their interaction with the food substrate or nearby environment. Indeed, the mechanical performance and structural integrity are strictly correlated to the molecular flexibility/rigidity in response to the direct contact with the food substrate and to the environmental parameters (i.e., pH, temperature, and salt concentration) [70]. The most common ingredients used in order to obtain an active film or coating are antioxidant and antibacterial compounds [71]. Actually, hindering the oxidation and avoiding bacteria flora growing are the main overall aims of the active packages, also in the case of protein-based systems [72,73]. In addition, the chemical or biological changes of the protein surface can deeply tailor the release of bioactive compounds, thus ensuring prolonged and modulated food shelf-life and quality [74]. In addition, recent studies have shown that the cross-linking of proteins with transglutaminase improved the edible film barrier properties, thermal stability, and structural integrity due to the enhancement of the three-dimensional network's regularity and smoothness [75]. Actually, in the design of active biodegradable packaging in which biopolymers are the leading materials, the research on new protein sources coming from agricultural, livestock raising, fishing wastes, or from bioderived monomers is also mandatory in order to avoid any interference with the human or animal food chain and play a very crucial role in this industry [76]. Proteins based on the sources are divided into plant proteins (soy protein, corn zein, etc.) and animal proteins (casein, whey gelatin, etc.). We briefly introduce the most common proteins in the food packaging industry [77].

2.2.1. Corn Zein

Corn zein is an alcohol-soluble protein except in high- and low-pH solutions and is insoluble in anhydrous alcohol [78]. Therefore, corn zein cast films are made using aqueous ethyl alcohol or isopropanol as a solution under temperature [79]. This film formation is due to the hydrophobic, hydrogen, and disulfide bonds between the zein chains during solvent evaporation [76,80].

2.2.2. Whey Proteins

Whey proteins can be found in precipitated casein during the cheese-making process. Furthermore, commercial whey proteins are produced in two protein concentrations (25–80% and >90% protein). This protein includes hydrophobic, thiol, and disulfide groups. As a result, films with whey are transparent, flavorless, and can present varying solubility and mechanical properties [77]. In addition, the presence of disulfide bonding in whey protein films causes an increase in their stiffness, water insolubility, and stretchability [81].

2.2.3. Soy Protein

Soy protein is isolated from soybean and commercially produced in three different concentrations of soy flour: (1) 50–59%, (2) 65–72%, and (3) >90% (Gennadios, Weller, Park, Rhim & Hanna, 2002). The soy protein cast film results from disulfide bonds and hydrophobic interactions. Therefore, soy protein films can be applied to all food requiring high water permeability such as meat, fresh bakery products, vegetables, and cheese [82–84].

2.2.4. Collagen

Collagen, a hydrophilic protein, consists of glycine, hydroxyproline, and proline. Collagen films are mainly used as packaging for meat products such as steak, beef, and sausage to keep them fresh and cause a decrease in the shrinkage rate without any change in color and flavor [85].

2.2.5. Gelatin

Gelatin results from collagen hydrolysis. Gelatin film formation occurs because of cross-linking between the amino and carboxyl of the amino acid and side groups. Gelatin can be applied for coatings and used as a bioactive compound carrier [86]. Therefore, gelatin is widely applied as a soft gel capsule for oil-based foods and dietary supplements.

2.3. Lipids

In contrast to polysaccharides and proteins, lipids are hydrophobic and are usually employed for protection from moisture transfer [87]. Therefore, the main application of lipids is incorporated into edible films to provide the required moisture barrier. The lipid component can be incorporated as a coating over the polysaccharide or protein layer or be mixed in the hydrophilic component to form a dispersed-lipid phase. In general, the moisture barrier properties depend on the polarity and insaturation extent of the lipid component [88]. However, lipids can reduce food quality by changing their flavor (smooth flavor) and appearance by altering the transparency [52]. Widely used lipid materials employed in edible coatings include vegetable oils, waxes, and resins.

2.3.1. Vegetable Oils

The majority of vegetable oils are constituted of glycerol esters of fatty acids with different chain lengths and structures. Phospholipids, sterols, pigments, and polyphenols are also present as minor components. The most widely used vegetable oils are sunflower, rapeseed, and olive oils, which have been incorporated in starch and chitosan films to improve their mechanical and barrier properties [89].

2.3.2. Waxes

Other important hydrophobic substances used for lipid-based edible films and coatings include natural waxes including carnauba, candelilla, and beeswax. Unlike petroleum-based waxes such as paraffin and polyethylene wax, these natural waxes are food-approved, and are mainly used as glazing and coating agents in confections and fruit as they limit water transpiration [88].

2.3.3. Resins

Resins are a class of substances produced by plants (specially conifers) or insects, or are obtained by chemical synthesis. In food coatings, resins are used as emulsifiers, or to improve the gloss, gas barrier, or adhesion. The most representative resins are gum Arabic, shellac, and wood rosin. Gum Arabic is secreted by Acacia Senegal as protection for bark wounds. It is a complex mixture of polysaccharide with low protein content, is soluble in water, and is used as a postharvest edible dip coating on fruit to retard decay and delaying ripening during storage [90].

Shellac is secreted by scale insects on forest trees in India and Thailand. It is a natural thermoplastic with good film-forming and barrier properties, and it is readily dissolved in alcohol or in alkaline solutions. Shellac is especially employed as a coating for pharmaceutics and fruits. and it is able to extend the shelf-life of several fruits including apples, tomatoes, and pepper [91]. Wood rosins obtained from pine trees are mostly constituted of abietic acid and its derivatives and are especially used as coatings for citrus fruits [92].

3. Secondary Compounds in Food Packaging

The increase in consumer demand for healthy and safe food has resulted in the development of a new food packaging approach [93]. The original food packaging concept started to expand in the second half of the 20th century when natural and artificial additives in the package system were increasingly used to provide packaging with novel functionalities to extend the food shelf-life [94]. In recent years, many studies concerning natural additives from natural resources (secondary compounds) for food packaging applications have increased [95,96]. Secondary compounds include natural metabolites and essential oils isolated from plants and fungi [97,98], which can be incorporated into the packaging system in order to avoid food and texture changes. Recently, the number of studies concentrated on natural secondary compounds including active natural metabolites and essential oils as additives in biodegradable and edible packaging, has drastically increased [99–101]. Unlike polyester-based biopolymers and conventional plastics, edible biofilms are hydrophilic materials such as proteins and polysaccharides, commonly manufactured by casting. In this method, active agent incorporation is conducted by dissolving both the biopolymer and natural additives in a suitable solvent. The solution is then poured onto the flat surface, and the solvent is allowed to evaporate (usually at room temperature) to obtain activated biopolymer-based films [102]. This technique is frequently used to formulate essential oils and natural metabolites, particularly thermolabile ones since the process does not need heat [103,104]. In addition, the casting solution or the cast film is usually applied as a bilayer of the other films with different techniques [105]. However, the secondary compounds are mostly incorporated via emulsification or homogenization techniques and can coat the film's surface or food [106]. The advantage of the coating techniques in which active agents are formulated on the surface of biofilms is that the active agent in the inner layer of the package can protect the food without interfering with the thermal or mechanical properties of the protective biofilm and provide the maximum activity for the packaged food [58]. As with other additives, natural additives are classified based on their functionality into two main groups. The first group can act as an antimicrobial, antifungal, and/or antioxidant agent to produce active packaging and improve food shelf-life, mainly because foods are exposed to spoilage by various microbial strains that generally attack food surfaces [107]. The second group primarily affects the physicochemical properties and package functionality such as plasticity, lubricity, nucleating and blowing agents, optical brightening, ultraviolet light stabilizing, and flame retardants [108]. Therefore, active secondary compounds can be formulated into edible coatings for different purposes. However, the performance of the edible films relies on the processing techniques, preparation condition (temperature, pH, cross-linking, or enzymatic reactions, drying process), and type of interaction between the biopolymers, active ingredients, and food substrate [49].

3.1. Antimicrobial and Antioxidant Effect

Antimicrobial and antioxidant secondary compounds are among the most widely used bioadditives to manufacture functional films [106]. Antimicrobial agents can be gradually released, inhibiting bacterial growth, thereby prolonging food shelf-life, and preserving food quality [109]. In active food packaging, the active agents are formulated into biopolymers during the processing. The active agent types differ in terms of the proposed application and the physical and/or chemical interaction into polymer matrices [110]. Furthermore, the main aim of active food packaging is to protect food from microbial contamination to extend the food shelf-life. In general, active packaging inhibition mechanisms are entirely based on the migration of active compounds from the package to the food [111]. Migration tests can determine the active agent migration at a particular time and temperature based on the storage conditions and packed food types [103]. Different parameters can influence the migration rate and mechanisms such as food ingredients [112], environmental conditions such as temperature, humidity [113], package physicochemical properties, and thickness [114]. Among the secondary compounds, essential oils (EOs) include different active natural metabolites with low molecular weights (monoterpenes and sesquiterpenes) and functionality that provide a context in which EOs could show various behavior when applied as an active agent for food packaging applications [115]. The more investigated EOs for food packaging application are rosemary, cinnamon (cinnamaldehyde), tea tree, lavender, thyme oil (thymol and carvacrol), lemon, and citrus. For instance, the blueberry (*Vaccinium corymbosum* L.), grape seed, and green tea extracts showed the highest inhibition against major foodborne pathogens including (1) *Listeria ponocytogenes*; (2) *Staphylococcus*; (3) *S. enteritidis*; (4) *S. Typhimurium*; (5) *E.coli*; and (6) *Campylobacter jejun* [116,117]. In this respect, cinnamaldehyde (CAL) or (2E)-3-phenyl prop-2-enal (55–76%) isolated from cinnamon trees, camphor, and cassia showed excellent antibacterial and antifungal activity [118–120]. Mohammadi et al. showed the cinnamon EO antimicrobial activity against *E. coli*, *S. aureus*, and *S. fluorescence* by formulating a cinnamon EO into chitosan nanofiber and whey protein films for active packaging purposes [121]. The other common antioxidant and antimicrobial essential oil in active food packaging is rosemary EO, isolated from *Salvia rosmarinus* or *Rosmarinus officinalis*, which is a native Mediterranean perennial evergreen plant. In one study, rosemary was encapsulated into carboxyl methylcellulose. The antimicrobial and antioxidant activity of rosemary EO against *Pseudomonas* spp. and lactic acid bacteria has been proven for rosemary encapsulated carboxyl methylcellulose coated smoked eel [122].

In fact, the bioactivity of the functional compounds is strictly correlated to their release from the polymer matrix. This, in turn, depends on the physical interaction occurring between the food and the packaging including the functional compounds, as demonstrated by Leelaphiwat et al. [123]. Moreover, as previously stated, the water diffusion from the food surface into the biopolymer network improves the swelling of the hydrophilic macromolecular chains, which in turn enhances the macromolecular mobility and provides an easier release pattern of the bioactive compounds [124].

Like the antimicrobial effects, the antioxidant action mechanism involves releasing and spreading antioxidants in the package medium and absorbing undesirable compounds such as oxygen, food-derived chemicals, and radical oxidative species by scavengers from the food surface and package environment [125–127]. Antioxidant agents mainly apply to fresh foods such as meat, fish, fruits, and processed and raw food.

Regarding the antioxidant activity, Origanum EO extracted from the *Lamiaceae* family demonstrated excellent antioxidant and antimicrobial properties even after its formulation into whey protein and could significantly decrease the lipid peroxidation of fresh meat [128]. Fruits and vegetables are often exposed to microbial attack, mainly from phytopathogenic fungi. Antifungal compounds such as organic acids and various plant extracts or essential oils, by generating a natural obstacle against bacterial flora, could prolong the post-harvest fruits and vegetable shelf-life [129]. Aside from antimicrobial and antioxidant activity, essential oils could improve the polysaccharide-based film's hygroscopic behaviors [130]. More studies on the antimicrobial and/or antioxidant properties of other secondary compounds are listed in Table 1.

Table 1. Natural products as antioxidants and antimicrobials in active bio packaging.

Essential Oils or Metabolites	Packaging System	Role	Preparation Technique	Bioassay	Ref
Rosemary essential oil	-	Antimicrobial, antioxidant, and antibacterial	Coating	Psychrotrophics, Brochothrix thermosphacta, Pseudomonas spp., and Enterobacteriaceae	[131]
Rosemary extract	Furcellaran/gelatin hydrolysate/glycerol	Antioxidant	Casting	-	[132]
Origanum vulgare L. and Rosmarinus officinalis L.	Chitosan	Antimicrobial and antioxidant	Casting	Escherichia coli and Bacillus subtilis	[133]
Rosmarinus officinalis and Zingiber officinale	Chitosan/glycerol/montmorillonites	Antioxidant	Casting	-	[134]
Citrus sinensis	-	Antimicrobial and antifungal	Coating	E. coli, Staphylococcus aureus, and Botrytis cinerea	[135]
Citrus sinensis	Gelatin/glycerol	Antimicrobial and antioxidant	Casting	Micrococcus luteus, S. aureus, B. cereus, Pseudomonas aeruginosa, Salmonella enterica, Listeria monocytogenes, and Enterobacter sp.	[136]
S. aromaticum and C. cassia	Starch/glycerol	Antimicrobial and antioxidant	Casting	Pseudomonas, Lactobacillus, Enterobacteriaceae, Yeast and molds and Brochothrix thermosphacta	[137]
Cinnamomum cassia and Myristicafragrans	Alginate/glycerol	Antimicrobial and antioxidant	Coating	E. coli and Penicillium commune	[138]
Licorice	Carboxymethyl xylan	Antimicrobial and antioxidant	Casting	Enterococcus faecalis and L. monocytogenes	[139]
Cedrus deodara pine needle extract	Soy protein/cellulose nanofibril/lactic acid	Antimicrobial and antioxidant	Casting	E. coli, S. aureus, S. Typhimurium, and L. monocytogenes	[140]
Carvacrol	Sodium alginate	Antimicrobial	Casting	Trichoderma sp.	[141]
Cedrus deodara pine needle extract	Soy protein/cellulose nanocrystals	Antioxidant	Casting	-	[142]
Licorice essential oil	Zein-based films	Antimicrobial and antioxidant	Casting	E. faecalis and L. monocytogenes	[143]

Table 1. Cont.

Essential Oils or Metabolites	Packaging System	Role	Preparation Technique	Bioassay	Ref
Capsaicin	Chitosan/glycerol	Antimicrobial and antioxidant	Casting	S. aureus, Proteus microbilis, Proteus vulgaris, Pseudomonas aeruginosa, Enterobacter aerogenes, B. thuringiensis, S. enterica serotype typhmurium, and Streptococcus mutans	[144]
Cinnamomum cassia, Cinnamomum zeylanicum, Rosemary officinalis, Ocimum basilicum	Whey protein/glycerol	Antimicrobial, antifungal, and antioxidant	Casting	E. coli, S. aureus, and Penicillium	[103]
Mentha spicata	carboxymethyl cellulose/chitosan	Antimicrobial and antioxidant	Coating	L. monocytogenes	[145]
Mentha spicata	sodium alginate	Antibacterial and antioxidant	Coating	Pseudomonas spp.	[146]
Ziziphora clinopodioides and Mentha spicata	carboxymethyl cellulose	Antimicrobial and antioxidant	Coatings	L. monocytogenes, S. aureus, E. coli, S. Typhimurium, and Campylobacter jejuni	[147]
Thymus vulgaris	Chitosan/glycerol	Antibacterial	Casting	Aerobic Mesophilic Bacteria; Mesophilic Lactic Acid Bacteria	[148]
Citrus aurantium	Chitosan	Antioxidant	Encapsulation in nanoparticles	-	[149]
Carvacrol and citral	sago starch/guar gum	Antioxidant	Casting	B. cereus and E. coli	[150]
Cinnamon, lemon, and oregano EOs	Chitosan	Antifungal	Casting	Botrytis spp., Penicillium spp., and Pilidiella granati	[151]
Rosemary extracts	cassava starch/glycerol	Antioxidant	Casting	-	[152]
Mentha pulegium	Gelatin	Antioxidant	Coating	-	[153]
Ziziphora clinopodioides and grape seed extract	Chitosan/fish skin gelatin	Antimicrobial and antioxidant	Casting	L. monocytogenes, S. aureus and B. cereus	[154]

3.2. Plasticizing Effect

This part reviews the plasticizing effects of secondary active compounds in functional biopolymers. Food packaging is distinguished by mechanical, optical, water vapor, and gas permeability properties [155]. Biopolymers are usually fragile and brittle, hence with poor mechanical properties. These drawbacks limit biopolymer applications in the food packaging segment. Therefore, plasticizers can play a vital role in enhancing the mechanical performance of the biopolymer. The package's mechanical properties (elastic modulus, tensile strength, and percentage elongation at break) can be evaluated by tensile tests (ASTM D882, 2001) by measuring the strength versus time or distance [106]. Plasticizers are low molecular weight compounds that occupy the intermolecular space of polymer matrix chains and decrease their secondary forces [156]. The most common plasticizers are glycerol, sorbitol, and polyethylene glycol [66]. Plasticizers formed via changing the polymer chain backbone reduce the molecular interstitial movement and facilitate the formation of hydrogen bonding between the chains, which improves the polymer's mechanical performance [157]. The plasticizer's chemical properties such as molecular weight, functional groups, and chemical composition affect the degree of plasticization. The final product flexibility can differ depending on the type of plasticizer used [158,159]. A suitable plasticizer was selected based on its compatibility with the polymer matrix, the final application (packaging, medical, and others), and the processing technique [158,160,161]. However, the main parameter is physical compatibility and depends on the polarity, hydrogen bonding, dielectric constant, and solubility [160,161]. The formulated film physiochemical features result from microstructural interactions between the plasticizers and polymer matrices. This interaction depends on the amount of plasticizer content in the biopolymer, the chemical properties of the additives and biopolymers, and their functional groups [162–165]. Natural metabolites and the essential oils of plants can be used as plasticizers for biopolymers, improving its hydrophobicity and water absorption. Some of the incorporated active ingredients also facilitate the plasticization of the polymers, enhancing their hydrophobicity and modulating their water absorption [166]. However, natural compounds may have an unpredictable impact on the package structure and mechanical properties compared with conventional plasticizers because of their complex composition [110]. The number of studies addressing natural metabolites as active agents in food packaging is lower than those on essential oils. Several reasons can explain this finding; in principle, the extraction of natural metabolites often requires severe time–cost-consuming procedures than crude oil extraction; moreover, the yield of secondary metabolites is often lower when compared to the essential oil recovery. In this respect, many efforts are ongoing to improve the whole extraction procedures by exploiting more efficient green methodologies such as microwave and/or ultrasound assisted extraction [167] and supercritical fluid extraction [168]. In addition, secondary metabolites are volatile and need to be physically or chemically entrapped in protective structures to preserve their bioactivity, as successfully demonstrated by Moeini et al. [12]. The mechanical properties of active films depend on different parameters such as the polymer matrix constituents, the proportion of the samples, the preparation technique and conditions, and the type of microstructural interaction between the polymer and plasticizer functional groups [169,170].

In some cases, the tensile strength of polysaccharide-based films decreases because of the replacement of strong polymer–polymer bonds with soft polymer–secondary compounds [105,107,171,172]. In this respect, Hosseini et al. incorporated Origanum vulgare EO in fish gelatin and chitosan by the casting method [36]. The mechanical properties, namely, tensile strength and elastic modulus, significantly decreased [36]. In another study, Otoni et al. incorporated carvacrol and cinnamaldehyde into soy protein using the casting method. Tensile tests evaluated the essential oils, and the results demonstrated a reduction in the tensile strength (TS) and an increase in their elongation at break (EB) [173]. In other cases, the secondary compound additives changed the biofilm stretchability by increasing it, as shown in the study by Shojaee-Aliabadi et al. [92], who incorporated Satureja hortensis essential oil, extracted from the savory genus, into k-carrageenan biofilms.

On the other hand, some studies have proven that secondary metabolites and EOs could increase the tensile strength of the biofilms, probably due to cross-linking processes [174,175]. In this regard, essential oils with phenolic compounds can act as cross-linkers in protein-based films [176]. Likewise, when Citrus aurantifolia EO was incorporated by casting methods into starch/gelatin blends, the tensile test revealed improved mechanical properties because citric acid acted as a cross-linking agent in the blend [162]. In the following table, several examples of the plasticizing role of secondary compounds in solution cast food packaging are summarized (Table 2).

Table 2. The plasticizing effect of secondary compounds in the active biopolymer-based coating by solution casting.

Natural Metabolites and EOS	Packaging System	Effect of Additives on the Mechanical Properties	Ref.
Eugenol or ginger essential oils	Gelatin–chitosan	A significant increase in elasticity	[163]
Cinnamon oil	Corn starch/chitosan/glycerol	Tensile strength (TS) decreased and elongation at break increased	[164]
Cinnamon oil	Sodium alginate	TS and extension at break slightly increased	[165]
Cinnamon, guarana, rosemary, and boldo-do-chile ethanolic extracts	Gelatin/chitosan	A reduction in elastic modulus and tensile strength and an increase in elongation at break	[169]
Cinnamaldehyde	Soy protein	Reduction of both tensile strength and elongation at break	[170]
Rosemary acid	Rabbit skin gelatin	The elongation at break decreased, and TS significantly increased	[177]
Cymbopogon citratus and *Rosmarinus officinalis*	Banana starch	The plasticizing effect of essential oil was observed by increasing elongation at breaks in formulated films	[178]
Rosemary essential oil	Starch–carboxy methylcellulose	TS of the films decreased and elongation at break increased	[179]
Ziziphora clinopodioides and grape seed extract	Chitosan/fish skin gelatin	Decrease tensile strength and flexibility due to hydrogen bonding and hydrophobic interactions	[154]
Yerba mate extract and mango pulp	Cassava starch/glycerol	Both TS and elongation at break decreased due to heterogeneous distribution and hydrogen bonding	[174]
Carvacrol and citral	Sago starch/guar gum	The film tensile strength was significantly red, used and elongation at break increased	[150]
Citral EO	Alginate and Pectin	The tensile strength and rigidity of the active film were improved.	[171]
Lavender essential oil	Potato starch–furcellaran–gelatin	The tensile strength of the films decreased considerably with increasing concentration of oil	[172]
Oregano, lemon, and grapefruit Eos	Soy protein	the film containing grapefruit essential oil had the highest tensile strength in comparison to other samples	[175]

4. Conclusions

Active food packaging is used with promising results to extend the food shelf-life by protecting food from microbial microorganisms and oxidation. Today, synthetic polymers are still the primary materials used for packaging applications. However, the environmental impact of synthetic plastic has led to the development of bio-based food packaging materials. Furthermore recently, producers and consumers demand biopolymers that come from natural sources. As a result, biopolymer applications in the food packaging industry have grown since they are biodegradable, compostable, readily available, and have mechanical properties like those of conventional polymers. This review introduced different kinds of edible biopolymers applied in food packaging. Edible films and coatings are suitable to enhance the food quality and safety and prolong the food-shelf life. The efficiency and functionality of edible films and coatings are primarily dependent on the biopolymer matrix used such as polysaccharides, proteins, and resins, and on the plasticizers and additives. In fact, these can be used as primary packaging for different foods such as fresh-cut fruits and vegetables, cheese, and meat. The incorporation of functional molecules leads to bioactive packaging materials showing antimicrobial and antioxidant properties. In this review, particular attention was devoted to the use of secondary compounds of plants and fungi exploiting the antimicrobial, antioxidant, and plasticizing action, depending on the specific structure, functional groups, and interaction with both the polymer matrix and food substrate. Therefore, secondary natural compounds opportunely formulated into biopolymer-based films could successfully both exploit suitable activities against pathogenic microorganisms and protect food from oxidation. Additionally, the mechanical test of active biofilms showed that the effectiveness of natural additives depends on the constituents of the polymer matrix, the specific formulation, the preparation technique and conditions, and the type of microstructural interactions between the polymer and plasticizer functional groups. In this review, the drawbacks related to the hydrophilic nature of the biopolymers were highlighted, particularly concerning the barrier properties, which can sometimes be overcome by the use of lipophilic secondary compounds. At the same time, although interesting and versatile, the use of secondary compounds still requires more investigation to address the likely toxicity of these molecules, and to develop standard, green, and cost-effective procedures for their extraction. Future perspectives of more in-depth academic and industrial research should be addressed to overcome these barriers by unraveling the efficient edible film-forming mechanisms, optimizing the use of secondary compounds, improving the performance, functionality, and manufacturing of edible packaging, thus allowing for scaling-up of their production and commercial level break-through.

Author Contributions: Conceptualization, A.M.; Validation, A.M., P.C. and G.S.; Investigation, A.M. and G.S.; Writing—original draft preparation, AM., P.P., and G.S.; Writing—review, and editing, A.M., P.P., E.F., P.C. and G.S.; Visualization, P.P.; Supervision G.S., P.C. and A.M. All authors have read and agreed to the published version of the manuscript.

Funding: The current work was partially funded by the Allianz Industrie Forschung (AiF) grant (Funding code: ZF4025045AJ9 and KK5022302EB0) as part of the Central Innovation Program Initiative of the Federal Government of Germany, and funding from the EU H2020 CE-BG-06-2019 project "Developing and Implementing Sustainability-Based Solutions for Bio-Based Plastic Production and Use to Preserve Land and Sea Environmental Quality in Europe (BIO-PLASTICS EUROPE)" n. 860407.

Institutional Review Board Statement: Not applicable.

Informed Consent Statement: Not applicable.

Data Availability Statement: Not applicable.

Conflicts of Interest: The authors declare no conflict of interest.

References

1. Risch, S.J. Food packaging history and innovations. *J. Agric. Food Chem.* **2009**, *57*, 8089–8092. [CrossRef] [PubMed]
2. Schwarzböck, T.; Van Eygen, E.; Rechberger, H.; Fellner, J. Determining the amount of waste plastics in the feed of Austrian waste-to-energy facilities. *Waste Manag. Res.* **2016**, *35*, 207–216. [CrossRef]
3. European Commission. Closing the Loop-An EU Action Plan for the Circular Economy COM/2015/0614 Final. 2015. Available online: https://www.eea.europa.eu/policy-documents/com-2015-0614-final (accessed on 12 February 2015).

4. European Commission. Communication from the Commission to the European Parliament, the Council, the European Economic and Social Committee and the Committee of the Regions Closing the Loop—An EUAction Plan for the Circular Economy. 2015. Available online: https://eur-lex.europa.eu/legal-content/EN/TXT/?uri=CELEX:52015DC0614 (accessed on 12 February 2015).
5. Siracusa, V.; Rocculi, P.; Romani, S.; Rosa, M.D. Biodegradable polymers for food packaging: A review. *Trends Food Sci. Technol.* **2008**, *19*, 634–643. [CrossRef]
6. Salerno, A.; Cesarelli, G.; Pedram, P.; Netti, P.A. Modular Strategies to Build Cell-Free and Cell-Laden Scaffolds towards Bioengineered Tissues and Organs. *J. Clin. Med.* **2019**, *8*, 1816. [CrossRef] [PubMed]
7. Mallardo, S.; De Vito, V.; Malinconico, M.; Volpe, M.G.; Santagata, G.; Di Lorenzo, M.L. Poly(butylene succinate)-based composites containing β-cyclodextrin/d-limonene inclusion complex. *Eur. Polym. J.* **2016**, *79*, 82–96. [CrossRef]
8. Mitrus, M.; Wojtowicz, A.; Moscicki, L. Biodegradable Polymers and Their Practical Utility. In *Thermoplastic Starch*; Wiley-VCH Verlag GmbH & Co. KGaA: Weinheim, Germany, 2009; pp. 1–33. [CrossRef]
9. Johnson, R.M.; Mwaikambo, L.Y.; Tucker, N. Biopolymers, Rapra Technology. 2003. Available online: https://books.google.de/books?id=QQmMY_dqqCMC (accessed on 1 January 2003).
10. Moeini, A.; van Reenen, A.; Van Otterlo, W.; Cimmino, A.; Masi, M.; Lavermicocca, P.; Valerio, F.; Immirzi, B.; Santagata, G.; Malinconico, M.; et al. α-costic acid, a plant sesquiterpenoid from Dittrichia viscosa, as modifier of Poly (lactic acid) properties: A novel exploitation of the autochthone biomass metabolite for a wholly biodegradable system. *Ind. Crop. Prod.* **2020**, *146*, 112134. [CrossRef]
11. Moeini, A.; Cimmino, A.; Masi, M.; Evidente, A.; Van Reenen, A. The incorporation and release of ungeremine, an antifungal Amaryllidaceae alkaloid, in poly(lactic acid)/poly(ethylene glycol) nanofibers. *J. Appl. Polym. Sci.* **2020**, *137*, 49098. [CrossRef]
12. Moeini, A. Fungal and Plant metabolites Formulated into Biopolymers, with Anti-Mold Activity for Food Packaging. Ph.D. Thesis, University of Naples Federico II, Naples, Italy, 2020.
13. Mayer, S.; Tallawi, M.; De Luca, I.; Calarco, A.; Reinhardt, N.; Gray, L.A.; Drechsler, K.; Moeini, A.; Germann, N. Antimicrobial and physicochemical characterization of 2,3-dialdehyde cellulose-based wound dressings systems. *Carbohydr. Polym.* **2021**, *272*, 118506. [CrossRef]
14. Valdés, A.; Mellinas, A.C.; Ramos, M.; Garrigós, M.C.; Jiménez, A. Natural additives and agricultural wastes in biopolymer formulations for food packaging. *Front. Chem.* **2014**, *2*, 6. Available online: https://www.frontiersin.org/article/10.3389/fchem.2014.00006 (accessed on 26 February 2014). [CrossRef]
15. Domínguez, R.; Barba, F.J.; Gómez, B.; Putnik, P.; Kovačević, D.B.; Pateiro, M.; Santos, E.M.; Lorenzo, J.M. Active packaging films with natural antioxidants to be used in meat industry: A review. *Food Res. Int.* **2018**, *113*, 93–101. [CrossRef]
16. Battisti, R.; Fronza, N.; Júnior, V.; da Silveira, S.M.; Damas, M.S.P.; Quadri, M.G.N. Gelatin-coated paper with antimicrobial and antioxidant effect for beef packaging. *Food Packag. Shelf Life* **2017**, *11*, 115–124. [CrossRef]
17. Lorenzo, J.M.; Batlle, R.; Gómez, M. Extension of the shelf-life of foal meat with two antioxidant active packaging systems. *LWT-Food Sci. Technol.* **2014**, *59*, 181–188. [CrossRef]
18. Horita, C.N.; Baptista, R.C.; Caturla, M.Y.; Lorenzo, J.M.; Barba, F.J.; Sant'Ana, A.S. Combining reformulation, active packaging and non-thermal post-packaging decontamination technologies to increase the microbiological quality and safety of cooked ready-to-eat meat products. *Trends Food Sci. Technol.* **2018**, *72*, 45–61. [CrossRef]
19. Van Long, N.N.; Joly, C.; Dantigny, P. Active packaging with antifungal activities. *Int. J. Food Microbiol.* **2016**, *220*, 73–90. [CrossRef]
20. Granato, D.; Nunes, D.S.; Barba, F.J. An integrated strategy between food chemistry, biology, nutrition, pharmacology, and statistics in the development of functional foods: A proposal. *Trends Food Sci. Technol.* **2017**, *62*, 13–22. [CrossRef]
21. Poojary, M.M.; Putnik, P.; Kovačević, D.B.; Barba, F.J.; Lorenzo, J.M.; Dias, D.A.; Shpigelman, A. Stability and extraction of bioactive sulfur compounds from Allium genus processed by traditional and innovative technologies. *J. Food Compos. Anal.* **2017**, *61*, 28–39. [CrossRef]
22. Vincekovic, M.; Viskić, M.; Jurić, S.; Giacometti, J.; Kovačević, D.B.; Putnik, P.; Donsì, F.; Barba, F.J.; Jambrak, A.R. Innovative technologies for encapsulation of Mediterranean plants extracts. *Trends Food Sci. Technol.* **2017**, *69*, 1–12. [CrossRef]
23. Moeini, A.; Masi, M.; Zonno, M.C.; Boari, A.; Cimmino, A.; Tarallo, O.; Vurro, M.; Evidente, A. Encapsulation of inuloxin A, a plant germacrane sesquiterpene with potential herbicidal activity, in β-cyclodextrins. *Org. Biomol. Chem.* **2019**, *17*, 2508–2515. [CrossRef]
24. De Luca, I.; Pedram, P.; Moeini, A.; Cerruti, P.; Peluso, G.; Di Salle, A.; Germann, N. Nanotechnology Development for Formulating Essential Oils in Wound Dressing Materials to Promote the Wound-Healing Process: A Review. *Appl. Sci.* **2021**, *11*, 1713. [CrossRef]
25. Santagata, G.; Cimmino, A.; Poggetto, G.D.; Zannini, D.; Masi, M.; Emendato, A.; Surico, G.; Evidente, A. Polysaccharide Based Polymers Produced by Scabby Cankered Cactus Pear (Opuntia ficus-indica L.) Infected by Neofusicoccum batangarum: Composition, Structure, and Chemico-Physical Properties. *Biomolecules* **2022**, *12*, 89. [CrossRef]
26. Valerio, F.; Masi, M.; Cimmino, A.; Moeini, S.A.; Lavermicocca, P.; Evidente, A. Antimould microbial and plant metabolites with potential use in intelligent food packaging. *Nat. Prod. Res.* **2017**, *6419*, 1–6. [CrossRef]
27. Yousuf, B.; Qadri, O.S.; Srivastava, A.K. Recent developments in shelf-life extension of fresh-cut fruits and vegetables by application of different edible coatings: A review. *LWT* **2018**, *89*, 198–209. [CrossRef]
28. Garcia, C.C.; Caetano, L.C.; Silva, K.D.S.; Mauro, M.A. Influence of Edible Coating on the Drying and Quality of Papaya (*Carica papaya*). *Food Bioprocess Technol.* **2014**, *7*, 2828–2839. [CrossRef]

29. Lago-Vanzela, E.; Nascimento, P.D.; Fontes, E.; Mauro, M.; Kimura, M. Edible coatings from native and modified starches retain carotenoids in pumpkin during drying. *LWT-Food Sci. Technol.* **2013**, *50*, 420–425. [CrossRef]
30. Han, J.H. Edible Films and Coatings: A Review. In *Innovations in Food Packaging*; Academic Press: Cambridge, MA, USA, 2014. [CrossRef]
31. Falguera, V.; Quintero, J.P.; Jiménez, A.; Muñoz, J.A.; Ibarz, A. Edible films and coatings: Structures, active functions and trends in their use. *Trends Food Sci. Technol.* **2011**, *22*, 292–303. [CrossRef]
32. Moeini, A.; Pedram, P.; Makvandi, P.; Malinconico, M.; Gomez d'Ayala, G. Wound healing and antimicrobial effect of active secondary metabolites in chitosan-based wound dressings: A review. *Carbohydr. Polym.* **2020**, *233*, 115839. [CrossRef]
33. Guo, L.; Guan, Y.; Liu, P.; Gao, L.; Wang, Z.; Huang, S.; Peng, L.; Zhao, Z. Chitosan hydrogel, as a biological macromolecule-based drug delivery system for exosomes and microvesicles in regenerative medicine: A mini review. *Cellulose* **2022**, *29*, 1315–1330. [CrossRef]
34. Straccia, M.C.; Romano, I.; Oliva, A.; Santagata, G.; Laurienzo, P. Crosslinker effects on functional properties of alginate/N-succinylchitosan based hydrogels. *Carbohydr. Polym.* **2014**, *108*, 321–330. [CrossRef]
35. Bourtoom, T.; Chinnan, M.S.; Jantawat, P.; Sanguandeekul, R. Effect of select parameters on the properties of edible film from water-soluble fish proteins in surimi wash-water. Sanguandeekul, Effect of select parameters on the properties of edible film from water-soluble fish proteins in surimi wash-water. *LWT-Food Sci. Technol.* **2006**, *39*, 406–419. [CrossRef]
36. Hosseini, S.F.; Rezaei, M.; Zandi, M.; Farahmandghavi, F. Bio-based composite edible films containing *Origanum vulgare* L. essential oil. *Ind. Crop. Prod.* **2015**, *67*, 403–413. [CrossRef]
37. Moghadam, M.; Salami, M.; Mohammadian, M.; Khodadadi, M.; Emam-Djomeh, Z. Development of antioxidant edible films based on mung bean protein enriched with pomegranate peel. *Food Hydrocoll.* **2020**, *104*, 105735. [CrossRef]
38. Razzaq, H.A.; Pezzuto, M.; Santagata, G.; Silvestre, C.; Cimmino, S.; Larsen, N.; Duraccio, D. Barley β-glucan-protein based bioplastic film with enhanced physicochemical properties for packaging. *Food Hydrocoll.* **2016**, *58*, 276–283. [CrossRef]
39. Šešlija, S.; Nešić, A.; Škorić, M.L.; Krušić, M.K.; Santagata, G.; Malinconico, M. Pectin/Carboxymethylcellulose Films as a Potential Food Packaging Material. *Macromol. Symp.* **2018**, *378*, 1600163. [CrossRef]
40. Gao, C.; Pollet, E.; Avérous, L. Properties of glycerol-plasticized alginate films obtained by thermo-mechanical mixing. *Food Hydrocoll.* **2017**, *63*, 414–420. [CrossRef]
41. Malinconico, M.; Cerruti, P.; Santagata, G.; Immirzi, B. Natural Polymers and Additives in Commodity and Specialty Applications: A Challenge for the Chemistry of Future. *Macromol. Symp.* **2014**, *337*, 124–133. [CrossRef]
42. Nesic, A.; Moeini, A.; Santagata, G. 4 Marine biopolymers: Alginate and chitosan. In *Sustainability of Polymeric Materials*; Marturano, V., Ambrogi, V., Cerruti, P., Eds.; De Gruyter: Berlin, Germany, 2020; p. 73. [CrossRef]
43. No, H.; Meyers, S.; Prinyawiwatkul, W.; Xu, Z. Applications of Chitosan for Improvement of Quality and Shelf Life of Foods: A Review. *J. Food Sci.* **2007**, *72*, R87–R100. [CrossRef]
44. Cazón, P.; Velazquez, G.; Ramírez, J.A.; Vázquez, M. Polysaccharide-based films and coatings for food packaging: A review. *Food Hydrocoll.* **2017**, *68*, 136–148. [CrossRef]
45. Meng, X.; Li, B.; Liu, J.; Tian, S. Physiological responses and quality attributes of table grapefruit to chitosan preharvest spray and postharvest coating during storage. *Food Chem.* **2008**, *106*, 501–508. [CrossRef]
46. Chien, P.-J.; Sheu, F.; Yang, F.-H. Effects of edible chitosan coating on quality and shelf life of sliced mango fruit. *J. Food Eng.* **2007**, *78*, 225–229. [CrossRef]
47. Sánchez-González, L.; Cháfer, M.; Hernández, M.; Chiralt, A.; González-Martínez, C. Antimicrobial activity of polysaccharide films containing essential oils. *Food Control* **2011**, *22*, 1302–1310. [CrossRef]
48. Bertoft, E. Understanding Starch Structure: Recent Progress. *Agronomy* **2017**, *7*, 56. [CrossRef]
49. Janjarasskul, T.; Krochta, J.M. Edible Packaging Materials. *Annu. Rev. Food Sci. Technol.* **2010**, *1*, 415–448. [CrossRef]
50. Chatkitanan, T.; Harnkarnsujarit, N. Effects of nitrite incorporated active films on quality of pork. *Meat Sci.* **2020**, *172*, 108367. [CrossRef]
51. Wongphan, P.; Khowthong, M.; Supatrawiporn, T.; Harnkarnsujarit, N. Novel edible starch films incorporating papain for meat tenderization. *Food Packag. Shelf Life* **2021**, *31*, 100787. [CrossRef]
52. Tavassoli-Kafrani, E.; Shekarchizadeh, H.; Masoudpour-Behabadi, M. Development of edible films and coatings from alginates and carrageenans. *Carbohydr. Polym.* **2016**, *137*, 360–374. [CrossRef]
53. Di Donato, P.; Taurisano, V.; Poli, A.; D'Ayala, G.G.; Nicolaus, B.; Malinconinco, M.; Santagata, G. Vegetable wastes derived polysaccharides as natural eco-friendly plasticizers of sodium alginate. *Carbohydr. Polym.* **2020**, *229*, 115427. [CrossRef]
54. Straccia, M.C.; D'Ayala, G.G.; Romano, I.; Laurienzo, P. Novel zinc alginate hydrogels prepared by internal setting method with intrinsic antibacterial activity. *Carbohydr. Polym.* **2015**, *125*, 103–112. [CrossRef]
55. Theagarajan, R.; Dutta, S.; Moses, J.A.; Anandharamakrishnan, C. Alginates for Food Packaging Applications. In *Alginates*; Wiley: Hoboken, NJ, USA, 2019; pp. 205–232. [CrossRef]
56. Zannini, D.; Poggetto, G.D.; Malinconico, M.; Santagata, G.; Immirzi, B. Citrus Pomace Biomass as a Source of Pectin and Lignocellulose Fibers: From Waste to Upgraded Biocomposites for Mulching Applications. *Polymers* **2021**, *13*, 1280. [CrossRef]
57. Valerio, F.; Volpe, M.G.; Santagata, G.; Boscaino, F.; Barbarisi, C.; Di Biase, M.; Bavaro, A.R.; Lonigro, S.L.; Lavermicocca, P. The viability of probiotic Lactobacillus paracasei IMPC2.1 coating on apple slices during dehydration and simulated gastro-intestinal digestion. *Food Biosci.* **2020**, *34*, 100533. [CrossRef]

58. Santagata, G.; Mallardo, S.; Fasulo, G.; Lavermicocca, P.; Valerio, F.; Di Biase, M.; Di Stasio, M.; Malinconico, M.; Volpe, M.G. Pectin-honey coating as novel dehydrating bioactive agent for cut fruit: Enhancement of the functional properties of coated dried fruits. *Food Chem.* **2018**, *258*, 104–110. [CrossRef]
59. Nešić, A.; Onjia, A.; Davidović, S.; Dimitrijević, S.; Errico, M.E.; Santagata, G.; Malinconico, M. Design of pectin-sodium alginate based films for potential healthcare application: Study of chemico-physical interactions between the components of films and assessment of their antimicrobial activity. *Carbohydr. Polym.* **2017**, *157*, 981–990. [CrossRef] [PubMed]
60. Espitia, P.J.P.; Du, W.-X.; Avena-Bustillos, R.d.; Soares, N.d.F.; McHugh, T.H. Edible films from pectin: Physical-mechanical and antimicrobial properties-A review. *Food Hydrocoll.* **2014**, *35*, 287–296. [CrossRef]
61. Osorio, F.; Molina, P.; Matiacevich, S.; Enrione, J.; Skurtys, O. Characteristics of hydroxy propyl methyl cellulose (HPMC) based edible film developed for blueberry coatings. *Procedia Food Sci.* **2011**, *1*, 287–293. [CrossRef]
62. Necas, J.; Bartosikova, L. Carrageenan: A review. *Vet. Med.* **2013**, *58*, 187–205. [CrossRef]
63. Campo, V.L.; Kawano, D.F.; Da Silva, D.B., Jr.; Carvalho, I. Carrageenans: Biological properties, chemical modifications and structural analysis—A review. *Carbohydr. Polym.* **2009**, *77*, 167–180. [CrossRef]
64. Kim, D.; Choi, G.J.; Baek, S.; Abdullah, A.; Jang, S.; Hong, S.A.; Kim, B.G.; Lee, J.; Kang, H.; Lee, D. Characterization of anti-adhesion properties of alginate/polyethylene oxide film to reduce postsurgical peritoneal adhesions. *Sci. Adv. Mater.* **2017**, *9*, 1669–1677. [CrossRef]
65. Nieto, M.B. Structure and Function of Polysaccharide Gum-Based Edible Films and Coatings. In *Edible Films and Coatings for Food Applications*; Huber, K.C., Embuscado, M.E., Eds.; Springer: New York, NY, USA, 2009; pp. 57–112. [CrossRef]
66. Gómez-Estaca, J.; Gavara, R.; Catalá, R.; Hernandez-Munoz, P. The Potential of Proteins for Producing Food Packaging Materials: A Review. *Packag. Technol. Sci.* **2016**, *29*, 203–224. [CrossRef]
67. Kaewprachu, P.; Osako, K.; Benjakul, S.; Tongdeesoontorn, W.; Rawdkuen, S. Biodegradable Protein-based Films and Their Properties: A Comparative Study. *Packag. Technol. Sci.* **2016**, *29*, 77–90. [CrossRef]
68. Ribeiro-Santos, R.; de Melo, N.R.; Andrade, M.; Azevedo, G.; Machado, A.V.; Carvalho-Costa, D.; Sanches-Silva, A. Whey protein active films incorporated with a blend of essential oils: Characterization and effectiveness. *Packag. Technol. Sci.* **2018**, *31*, 27–40. [CrossRef]
69. Hanani, Z.A.N.; Roos, Y.H.; Kerry, J.P. Use and application of gelatin as potential biodegradable packaging materials for food products. *Int. J. Biol. Macromol.* **2014**, *71*, 94–102. [CrossRef]
70. Hauzoukim, S.S.; Mohanty, B. Functionality of protein-Based edible coating. *J. Entomol. Zool. Stud.* **2020**, *8*, 1432–1440.
71. Hanani, Z.N.; Yee, F.C.; Nor-Khaizura, M. Effect of pomegranate (*Punica granatum* L.) peel powder on the antioxidant and antimicrobial properties of fish gelatin films as active packaging. *Food Hydrocoll.* **2018**, *89*, 253–259. [CrossRef]
72. Adilah, Z.A.M.; Jamilah, B.; Hanani, Z.A.N. Functional and antioxidant properties of protein-based films incorporated with mango kernel extract for active packaging. *Food Hydrocoll.* **2018**, *74*, 207–218. [CrossRef]
73. Khaneghah, A.M.; Hashemi, S.M.B.; Limbo, S. Antimicrobial agents and packaging systems in antimicrobial active food packaging: An overview of approaches and interactions. *Food Bioprod. Process.* **2018**, *111*, 1–19. [CrossRef]
74. Romani, V.P.; Olsen, B.; Collares, M.P.; Oliveira, J.R.M.; Prentice, C.; Martins, V.G. Cold plasma and carnauba wax as strategies to produce improved bi-layer films for sustainable food packaging. *Food Hydrocoll.* **2020**, *108*, 106087. [CrossRef]
75. Restaino, O.F.; Hejazi, S.; Zannini, D.; Giosafatto, C.V.L.; Di Pierro, P.; Cassese, E.; D'Ambrosio, S.; Santagata, G.; Schiraldi, C.; Porta, R. Exploiting Potential Biotechnological Applications of Poly-γ-glutamic Acid Low Molecular Weight Fractions Obtained by Membrane-Based Ultra-Filtration. *Polymers* **2022**, *14*, 1190. [CrossRef]
76. Gennadios, A.; Weller, C.; Park, H.; Rhim, J.-W.; Hanna, M. *Protein-Based Film and Coatings*; Taylor & Francis: Abingdon, UK, 2002.
77. Brink, I.; Šipailienė, A.; Leskauskaitė, D. Antimicrobial properties of chitosan and whey protein films applied on fresh cut turkey pieces. *Int. J. Biol. Macromol.* **2019**, *130*, 810–817. [CrossRef]
78. Sahraee, S.; Milani, J.M.; Regenstein, J.M.; Kafil, H.S. Protection of foods against oxidative deterioration using edible films and coatings: A review. *Food Biosci.* **2019**, *32*, 100451. [CrossRef]
79. Hassan, B.; Chatha, S.A.S.; Hussain, A.I.; Zia, K.M.; Akhtar, N. Recent advances on polysaccharides, lipids and protein based edible films and coatings: A review. *Int. J. Biol. Macromol.* **2018**, *109*, 1095–1107. [CrossRef]
80. Padua, G.; Wang, Q. Formation And Properties Of Corn Zein Films And Coatings. In *Protein-Based and Film Coatings*; Routledge Taylor & Francis Group: Abingdon, UK, 2002; Available online: https://www.routledge.com/Protein-Based-Films-and-Coatings/author/p/book/9781420031980 (accessed on 15 May 2022).
81. Calva-Estrada, S.J.; Jiménez-Fernández, M.; Lugo-Cervantes, E. Protein-Based Films: Advances in the Development of Biomaterials Applicable to Food Packaging. *Food Eng. Rev.* **2019**, *11*, 78–92. [CrossRef]
82. Chen, H.; Cheng, R.; Zhao, X.; Zhang, Y.; Tam, A.; Yan, Y.; Shen, H.; Zhang, Y.S.; Qi, J.; Feng, Y.; et al. An injectable self-healing coordinative hydrogel with antibacterial and angiogenic properties for diabetic skin wound repair. *NPG Asia Mater.* **2019**, *11*, 3. [CrossRef]
83. González, A.; Igarzabal, C.I.A. Soy protein–Poly (lactic acid) bilayer films as biodegradable material for active food packaging. *Food Hydrocoll.* **2013**, *33*, 289–296. [CrossRef]
84. Liang, S.; Wang, L. A Natural Antibacterial-Antioxidant Film from Soy Protein Isolate Incorporated with Cortex Phellodendron Extract. *Polymers* **2018**, *10*, 71. [CrossRef] [PubMed]

85. Antoniewski, M.N.; Barringer, S.A. Meat Shelf-life and Extension using Collagen/Gelatin Coatings: A Review. *Crit. Rev. Food Sci. Nutr.* **2010**, *50*, 644–653. [CrossRef] [PubMed]
86. Adilah, A.N.; Jamilah, B.; Noranizan, M.; Hanani, Z.N. Utilization of mango peel extracts on the biodegradable films for active packaging. *Food Packag. Shelf Life* **2018**, *16*, 1–7. [CrossRef]
87. Mehyar, G.F.; Al-Ismail, K.; Han, J.H.; Chee, G.W. Characterization of Edible Coatings Consisting of Pea Starch, Whey Protein Isolate, and Carnauba Wax and their Effects on Oil Rancidity and Sensory Properties of Walnuts and Pine Nuts. *J. Food Sci.* **2012**, *77*, E52–E59. [CrossRef] [PubMed]
88. Pérez-Gago, M.B.; Rhim, J.-W. Edible Coating and Film Materials: Lipid Bilayers and Lipid Emulsions. In *Innovations in Food Packaging*, 2nd ed.; Food Science and Technology Academic Press: San Diego, CA, USA, 2014; pp. 325–350. [CrossRef]
89. Yousuf, B.; Sun, Y.; Wu, S. Lipid and Lipid-containing Composite Edible Coatings and Films. *Food Rev. Int.* **2021**, 1–24. [CrossRef]
90. Patel, S.; Goyal, A. Applications of Natural Polymer Gum Arabic: A Review. *Int. J. Food Prop.* **2015**, *18*, 986–998. [CrossRef]
91. Li, K.; Tang, B.; Zhang, W.; Tu, X.; Ma, J.; Xing, S.; Shao, Y.; Zhu, J.; Lei, F.; Zhang, H. A novel approach for authentication of shellac resin in the shellac-based edible coatings: Contain shellac or not in the fruit wax preservative coating. *Food Chem. X* **2022**, *14*, 100349. [CrossRef]
92. Dhall, R.K. Advances in Edible Coatings for Fresh Fruits and Vegetables: A Review. *Crit. Rev. Food Sci. Nutr.* **2013**, *53*, 435–450. [CrossRef]
93. Suppakul, P.; Miltz, J.; Sonneveld, K.; Bigger, S. Active Packaging Technologies with an Emphasis on Antimicrobial Packaging and its Applications. *J. Food Sci.* **2003**, *68*, 408–420. [CrossRef]
94. De Paoli, M.A.; Waldman, W.R. Bio-based additives for thermoplastics. *Polimeros* **2019**, *29*, 2. [CrossRef]
95. Coma, V. Bioactive packaging technologies for extended shelf life of meat-based products. *Meat Sci.* **2008**, *78*, 90–103. [CrossRef]
96. Moeini, A.; Santagata, G.; Antonio, E.; Malinconico, M. Natural metabolites as functional additive of biopolymers: Experimental evidence and industrial constraint. In *An Introduction to the Circular Economy*; Morganti, M.-B., Coltelli, P., Eds.; Nova Science Publishers: New York, NY, USA, 2021; pp. 309–329.
97. Ioannou, E.; Roussis, V. *Natural Products From Seaweeds in Plant-Derived Natural Products*; Springer: Berlin, Germany, 2009; pp. 51–81. [CrossRef]
98. Johnson, J.L.; Raghavan, V.; Cimmino, A.; Moeini, A.; Petrovic, A.G.; Santoro, E.; Superchi, S.; Berova, N.; Evidente, A.; Polavarapu, P.L. Absolute configurations of chiral molecules with multiple stereogenic centers without prior knowledge of the relative configurations: A case study of inuloxin C. *Chirality* **2018**, *30*, 1206–1214. [CrossRef]
99. Masi, M.; Moeini, S.A.; Boari, A.; Cimmino, A.; Vurro, M.; Evidente, A. Development of a rapid and sensitive HPLC method for the identification and quantification of cavoxin and cavoxone in Phoma cava culture filtrates. *Nat. Prod. Res.* **2017**, *6419*, 1611–1615. [CrossRef]
100. Moeini, A.; Cimmino, A.; Poggetto, G.D.; Di Biase, M.; Evidente, A.; Masi, M.; Lavermicocca, P.; Valerio, F.; Leone, A.; Santagata, G.; et al. Effect of pH and TPP concentration on chemico-physical properties, release kinetics and antifungal activity of Chitosan-TPP-Ungeremine microbeads. *Carbohydr. Polym.* **2018**, *195*, 631–641. [CrossRef]
101. Moeini, A.; Mallardo, S.; Cimmino, A.; Poggetto, G.D.; Masi, M.; Di Biase, M.; van Reenen, A.; Lavermicocca, P.; Valerio, F.; Evidente, A.; et al. Thermoplastic starch and bioactive chitosan sub-microparticle biocomposites: Antifungal and chemico-physical properties of the films. *Carbohydr. Polym.* **2020**, *230*, 115627. [CrossRef]
102. Bastarrachea, L.; Wong, D.E.; Roman, M.J.; Lin, Z.; Goddard, J.M. Active Packaging Coatings. *Coatings* **2015**, *5*, 771–791. [CrossRef]
103. dos Santos, R.R.; Silva, A.S.; Motta, J.F.G.; Andrade, M.; Neves, I.D.A.; Teofilo, R.; de Carvalho, M.G.; de Melo, N.R. Combined use of essential oils applied to protein base active food packaging: Study in vitro and in a food simulant. *Eur. Polym. J.* **2017**, *93*, 75–86. [CrossRef]
104. Shojaee-Aliabadi, S.; Hosseini, H.; Mohammadifar, M.A.; Mohammadi, A.; Ghasemlou, M.; Ojagh, S.M.; Hosseini, S.M.; Khaksar, R. Characterization of antioxidant-antimicrobial κ-carrageenan films containing Satureja hortensis essential oil. *Int. J. Biol. Macromol.* **2013**, *52*, 116–124. [CrossRef]
105. Gómez-Estaca, J.; López-de-Dicastillo, C.; Hernandez-Munoz, P.; Catalá, R.; Gavara, R. Advances in antioxidant active food packaging. *Trends Food Sci. Technol.* **2014**, *35*, 42–51. [CrossRef]
106. Atarés, L.; Chiralt, A. Essential oils as additives in biodegradable films and coatings for active food packaging. *Trends Food Sci. Technol.* **2016**, *48*, 51–62. [CrossRef]
107. Kim, Y.-T.; Kim, K.; Han, J.; Kimmel, R. Antimicrobial Active Packaging for Food. In *Smart Packaging Technologies for Fast Moving Consumer Goods*; Wiley: Hoboken, NJ, USA, 2008; pp. 99–110. [CrossRef]
108. Moeini, A.; Germann, N.; Malinconico, M.; Santagata, G. Formulation of secondary compounds as additives of biopolymer-based food packaging: A review. *Trends Food Sci. Technol.* **2021**, *114*, 342–354. [CrossRef]
109. Gutierrez, J.; Barry-Ryan, C.; Bourke, P. Antimicrobial activity of plant essential oils using food model media: Efficacy, synergistic potential and interactions with food components. *Food Microbiol.* **2009**, *26*, 142–150. [CrossRef]
110. Guo, Y.; Chen, X.; Yang, F.; Wang, T.; Ni, M.; Chen, Y.; Yang, F.; Huang, D.; Fu, C.; Wang, S. Preparation and Characterization of Chitosan-Based Ternary Blend Edible Films with Efficient Antimicrobial Activities for Food Packaging Applications. *J. Food Sci.* **2019**, *84*, 1411–1419. [CrossRef]
111. Tian, F.; Decker, E.A.; Goddard, J.M. Controlling Lipid Oxidation via a Biomimetic Iron Chelating Active Packaging Material. *J. Agric. Food Chem.* **2013**, *61*, 12397–12404. [CrossRef]

112. Zinoviadou, K.G.; Koutsoumanis, K.P.; Biliaderis, C.G. Physico-chemical properties of whey protein isolate films containing oregano oil and their antimicrobial action against spoilage flora of fresh beef. *Meat Sci.* **2009**, *82*, 338–345. [CrossRef]
113. Kuorwel, K.K.; Cran, M.J.; Sonneveld, K.; Miltz, J.; Bigger, S.W. Migration of antimicrobial agents from starch-based films into a food simulant. *LWT Food Sci. Technol.* **2013**, *50*, 432–438. [CrossRef]
114. Huang, X.-L.; Wang, Z.-W.; Hu, C.-Y.; Zhu, Y.; Wang, J. Factors Affecting Migration of Contaminants from Paper through Polymer Coating into Food Simulants. *Packag. Technol. Sci.* **2012**, *26*, 23–31. [CrossRef]
115. Seow, Y.X.; Yeo, C.R.; Chung, H.L.; Yuk, H.-G. Plant Essential Oils as Active Antimicrobial Agents. *Crit. Rev. Food Sci. Nutr.* **2013**, *54*, 625–644. [CrossRef]
116. Perumalla, A.V.S.; Hettiarachchy, N.S. Green tea and grape seed extracts—Potential applications in food safety and quality. *Food Res. Int.* **2011**, *44*, 827–839. [CrossRef]
117. Shen, X.; Sun, X.; Xie, Q.; Liu, H.; Zhao, Y.; Pan, Y.; Hwang, C.-A.; Wu, V.C. Antimicrobial effect of blueberry (*Vaccinium corymbosum* L.) extracts against the growth of Listeria monocytogenes and Salmonella Enteritidis. *Food Control* **2014**, *35*, 159–165. [CrossRef]
118. Kenawy, E.; Omer, A.M.; Tamer, T.M.; Elmeligy, M.A.; Eldin, M.S.M. Fabrication of biodegradable gelatin/chitosan/cinnamaldehyde crosslinked membranes for antibacterial wound dressing applications. *Int. J. Biol. Macromol.* **2019**, *139*, 440–448. [CrossRef]
119. Nabavi, S.M.; Di Lorenzo, A.; Izadi, M.; Sobarzo-Sánchez, E.; Daglia, M. Antibacterial Effects of Cinnamon: From Farm to Food, Cosmetic and Pharmaceutical Industries. *Nutrients* **2015**, *7*, 7729–7748. [CrossRef]
120. Wasupalli, G.K.; Verma, D. Molecular interactions in self-assembled nano-structures of chitosan-sodium alginate based polyelectrolyte complexes. *Int. J. Biol. Macromol.* **2018**, *114*, 10–17. [CrossRef]
121. Mohammadi, M.; Mirabzadeh, S.; Shahvalizadeh, R.; Hamishehkar, H. Development of novel active packaging films based on whey protein isolate incorporated with chitosan nanofiber and nano-formulated cinnamon oil. *Int. J. Biol. Macromol.* **2020**, *149*, 11–20. [CrossRef]
122. Choulitoudi, E.; Ganiari, S.; Tsironi, T.; Ntzimani, A.; Tsimogiannis, D.; Taoukis, P.; Oreopoulou, V. Edible coating enriched with rosemary extracts to enhance oxidative and microbial stability of smoked eel fillets. *Food Packag. Shelf Life* **2017**, *12*, 107–113. [CrossRef]
123. Leelaphiwat, P.; Pechprankan, C.; Siripho, P.; Bumbudsanpharoke, N.; Harnkarnsujarit, N. Effects of nisin and EDTA on morphology and properties of thermoplastic starch and PBAT biodegradable films for meat packaging. *Food Chem.* **2022**, *369*, 130956. [CrossRef]
124. Serino, N.; Boari, A.; Santagata, G.; Masi, M.; Malinconico, M.; Evidente, A.; Vurro, M. Biodegradable polymers as carriers for tuning the release and improve the herbicidal effectiveness of Dittrichia viscosa plant organic extracts. *Pest Manag. Sci.* **2021**, *77*, 646–658. [CrossRef]
125. Oun, A.; Rhim, J.-W. Characterization of carboxymethyl cellulose-based nanocomposite films reinforced with oxidized nanocellulose isolated using ammonium persulfate method. *Carbohydr. Polym.* **2017**, *174*, 484–492. [CrossRef]
126. Singh, P.; Saengerlaub, S.; Wani, A.A.; Langowski, H.-C. Role of plastics additives for food packaging. *Pigment Resin Technol.* **2012**, *41*, 368–379. [CrossRef]
127. Wyrwa, J.; Barska, A. Innovations in the food packaging market: Active packaging. *Eur. Food Res. Technol.* **2017**, *243*, 1681–1692. [CrossRef]
128. Catarino, M.D.; Alves-Silva, J.M.; Fernandes, R.P.; Gonçalves, M.J.; Salgueiro, L.; Henriques, M.H.F.; Cardoso, S.M. Development and performance of whey protein active coatings with Origanum virens essential oils in the quality and shelf life improvement of processed meat products. *Food Control* **2017**, *80*, 273–280. [CrossRef]
129. Sapper, M.; Chiralt, A. Starch-Based Coatings for Preservation of Fruits and Vegetables. *Coatings* **2018**, *8*, 152. [CrossRef]
130. Marín, A.; Atarés, L.; Chiralt, A. Improving function of biocontrol agents incorporated in antifungal fruit coatings: A review. *Biocontrol Sci. Technol.* **2017**, *27*, 1220–1241. [CrossRef]
131. Sirocchi, V.; Devlieghere, F.; Peelman, N.; Sagratini, G.; Maggi, F.; Vittori, S.; Ragaert, P. Effect of *Rosmarinus officinalis* L. essential oil combined with different packaging conditions to extend the shelf life of refrigerated beef meat. *Food Chem.* **2017**, *221*, 1069–1076. [CrossRef] [PubMed]
132. Jancikova, S.; Jamróz, E.; Kulawik, P.; Tkaczewska, J.; Dordevic, D. Furcellaran/gelatin hydrolysate/rosemary extract composite films as active and intelligent packaging materials. *Int. J. Biol. Macromol.* **2019**, *131*, 19–28. [CrossRef]
133. Ospina, J.D.; Grande, C.D.; Monsalve, L.V.; Advíncula, R.C.; Mina, J.H.; Valencia, M.E.; Fan, J.; Rodrigues, D. Evaluation of the chitosan films of essential oils from *Origanum vulgare* L. (oregano) and *Rosmarinus officinalis* L. (rosemary). *Rev. Cuba. Plantas Med.* **2018**, *24*.
134. Pires, J.R.A.; de Souza, V.G.L.; Fernando, A.L. Chitosan/montmorillonite bionanocomposites incorporated with rosemary and ginger essential oil as packaging for fresh poultry meat. *Food Packag. Shelf Life* **2018**, *17*, 142–149. [CrossRef]
135. Gao, Y.; Xu, D.; Ren, D.; Zeng, K.; Wu, X. Green synthesis of zinc oxide nanoparticles using Citrus sinensis peel extract and application to strawberry preservation: A comparison study. *LWT* **2020**, *126*, 109297. [CrossRef]
136. Jridi, M.; Boughriba, S.; Abdelhedi, O.; Nciri, H.; Nasri, R.; Kchaou, H.; Kaya, M.; Sebai, H.; Zouari, N.; Nasri, M. Investigation of physicochemical and antioxidant properties of gelatin edible film mixed with blood orange (Citrus sinensis) peel extract. *Food Packag. Shelf Life* **2019**, *21*, 100342. [CrossRef]

137. Mohan, C.C.; Krishnan, K.R.; Babuskin, S.; Sudharsan, K.; Aafrin, V.; Priya, U.L.; Mariyajenita, P.; Harini, K.; Madhushalini, D.; Sukumar, M. Active compound diffusivity of particle size reduced S. aromaticum and C. cassia fused starch edible films and the shelf life of mutton (Capra aegagrus hircus) meat. *Meat Sci.* **2017**, *128*, 47–59. [CrossRef]
138. dos Santos, S.M.; Malpass, G.R.P.; Okura, M.H.; Granato, A.C. Edible active coatings incorporated with Cinnamomum cassia and Myristica fragrans essential oils to improve shelf-life of minimally processed apples. *Ciência Rural* **2018**, *48*, 12. [CrossRef]
139. Luís, A.; Pereira, L.; Domingues, F.; Ramos, A. Development of a carboxymethyl xylan film containing licorice essential oil with antioxidant properties to inhibit the growth of foodborne pathogens. *LWT* **2019**, *111*, 218–225. [CrossRef]
140. Yu, Z.; Dhital, R.; Wang, W.; Sun, L.; Zeng, W.; Mustapha, A.; Lin, M. Development of multifunctional nanocomposites containing cellulose nanofibrils and soy proteins as food packaging materials. *Food Packag. Shelf Life* **2019**, *21*. [CrossRef]
141. Cheng, M.; Wang, J.; Zhang, R.; Kong, R.; Lu, W.; Wang, X. Characterization and application of the microencapsulated carvacrol/sodium alginate films as food packaging materials. *Int. J. Biol. Macromol.* **2019**, *141*, 259–267. [CrossRef]
142. Yu, Z.; Sun, L.; Wang, W.; Zeng, W.; Mustapha, A.; Lin, M. Soy protein-based films incorporated with cellulose nanocrystals and pine needle extract for active packaging. *Ind. Crop. Prod.* **2018**, *112*, 412–419. [CrossRef]
143. Luís, Â.; Domingues, F.; Ramos, A. Production of Hydrophobic Zein-Based Films Bioinspired by The Lotus Leaf Surface: Characterization and Bioactive Properties. *Microorganisms* **2019**, *7*, 267. [CrossRef]
144. Akyuz, L.; Kaya, M.; Mujtaba, M.; Ilk, S.; Sargin, I.; Salaberria, A.M.; Labidi, J.; Cakmak, Y.S.; Islek, C. Supplementing capsaicin with chitosan-based films enhanced the anti-quorum sensing, antimicrobial, antioxidant, transparency, elasticity and hydrophobicity. *Int. J. Biol. Macromol.* **2018**, *115*, 438–446. [CrossRef]
145. Shahbazi, Y. Application of carboxymethyl cellulose and chitosan coatings containing Mentha spicata essential oil in fresh strawberries. *Int. J. Biol. Macromol.* **2018**, *112*, 264–272. [CrossRef]
146. Shahbazi, Y.; Shavisi, N. Effects of sodium alginate coating containing Mentha spicata essential oil and cellulose nanoparticles on extending the shelf life of raw silver carp (Hypophthalmichthys molitrix) fillets. *Food Sci. Biotechnol.* **2018**, *28*, 433–440. [CrossRef] [PubMed]
147. Ala, M.A.N.; Shahbazi, Y. The effects of novel bioactive carboxymethyl cellulose coatings on food-borne pathogenic bacteria and shelf life extension of fresh and sauced chicken breast fillets. *LWT* **2019**, *111*, 602–611. [CrossRef]
148. Quesada, J.; Sendra, E.; Navarro, C.; Sayas-Barberá, E. Antimicrobial Active Packaging including Chitosan Films with Thymus vulgaris L. Essential Oil for Ready-to-Eat Meat. *Foods* **2016**, *5*, 57. [CrossRef] [PubMed]
149. Karimirad, R.; Behnamian, M.; Dezhsetan, S.; Sonnenberg, A. Chitosan nanoparticles-loadedCitrus aurantiumessential oil: A novel delivery system for preserving the postharvest quality ofAgaricus bisporus. *J. Sci. Food Agric.* **2018**, *98*, 5112–5119. [CrossRef] [PubMed]
150. Dhumal, C.V.; Ahmed, J.; Bandara, N.; Sarkar, P. Improvement of antimicrobial activity of sago starch/guar gum bi-phasic edible films by incorporating carvacrol and citral. *Food Packag. Shelf Life* **2019**, *21*, 100380. [CrossRef]
151. Munhuweyi, K.; Caleb, O.J.; Lennox, C.L.; van Reenen, A.J.; Opara, U.L. In vitro and in vivo antifungal activity of chitosan-essential oils against pomegranate fruit pathogens. *Postharvest Biol. Technol.* **2017**, *129*, 9–22. [CrossRef]
152. Piñeros-Hernandez, D.; Medina-Jaramillo, C.; López-Córdoba, A.; Goyanes, S. Edible cassava starch films carrying rosemary antioxidant extracts for potential use as active food packaging. *Food Hydrocoll.* **2017**, *63*, 488–495. [CrossRef]
153. Aitboulahsen, M.; Zantar, S.; Laglaoui, A.; Chairi, H.; Arakrak, A.; Bakkali, M.; Zerrouk, M.H. Gelatin-Based Edible Coating Combined with Mentha pulegium Essential Oil as Bioactive Packaging for Strawberries. *J. Food Qual.* **2018**, *2018*, 8408915. [CrossRef]
154. Shahbazi, Y. The properties of chitosan and gelatin films incorporated with ethanolic red grape seed extract and Ziziphora clinopodioides essential oil as biodegradable materials for active food packaging. *Int. J. Biol. Macromol.* **2017**, *99*, 746–753. [CrossRef]
155. Ribeiro-Santos, R.; Andrade, M.; de Melo, N.R.; Sanches-Silva, A. Use of essential oils in active food packaging: Recent advances and future trends. *Trends Food Sci. Technol.* **2017**, *61*, 132–140. [CrossRef]
156. Donhowe, I.G.; Fennema, O. The effects of plasticizers on crystallinity, permeability, and mechanical properties of methylcellulose films. *J. Food Process. Preserv.* **1993**, *17*, 247–257. [CrossRef]
157. Vieira, M.G.A.; da Silva, M.A.; Dos Santos, L.O.; Beppu, M.M. Natural-based plasticizers and biopolymer films: A review. *Eur. Polym. J.* **2011**, *47*, 254–263. [CrossRef]
158. Moreno, R. The Role of Slip Additives in Tape Casting Technology 2. Binders and Plasticizers. *Am. Ceram. Soc. Bull.* **1992**, *71*, 1647.
159. Cao, N.; Yang, X.; Fu, Y. Effects of various plasticizers on mechanical and water vapor barrier properties of gelatin films. *Food Hydrocoll.* **2009**, *23*, 729–735. [CrossRef]
160. Choi, J.S.; Park, W.H. Thermal and mechanical properties of poly(3-hydroxybutyrate-co-3-hydroxyvalerate) plasticized by biodegradable soybean oils. *Macromol. Symp.* **2003**, *197*, 65–76. [CrossRef]
161. van Oosterhout, J.; Gilbert, M. Interactions between PVC and binary or ternary blends of plasticizers. Part I. PVC/plasticizer compatibility. *Polymer* **2003**, *44*, 8081–8094. [CrossRef]
162. Kanatt, S.R. Irradiation as a tool for modifying tapioca starch and development of an active food packaging film with irradiated starch. *Radiat. Phys. Chem.* **2020**, *173*, 108873. [CrossRef]

163. Bonilla, J.; Poloni, T.; Lourenço, R.V.; Sobral, P.J. Antioxidant potential of eugenol and ginger essential oils with gelatin/chitosan films. *Food Biosci.* **2018**, *23*, 107–114. [CrossRef]
164. Ke, J.; Xiao, L.; Yu, G.; Wu, H.; Shen, G.; Zhang, Z. The study of diffusion kinetics of cinnamaldehyde from corn starch-based film into food simulant and physical properties of antibacterial polymer film. *Int. J. Biol. Macromol.* **2019**, *125*, 642–650. [CrossRef]
165. Chen, X.; Lu, L.-X.; Qiu, X.; Tang, Y. Controlled release mechanism of complex bio-polymeric emulsifiers made microspheres embedded in sodium alginate based films. *Food Control* **2016**, *73*, 1275–1284. [CrossRef]
166. Katekhong, W.; Wongphan, P.; Klinmalai, P.; Harnkarnsujarit, N. Thermoplastic starch blown films functionalized by plasticized nitrite blended with PBAT for superior oxygen barrier and active biodegradable meat packaging. *Food Chem.* **2022**, *374*, 131709. [CrossRef]
167. Zhang, H.-F.; Yang, X.-H.; Wang, Y. Microwave assisted extraction of secondary metabolites from plants: Current status and future directions. *Trends Food Sci. Technol.* **2011**, *22*, 672–688. [CrossRef]
168. Bader, C.D.; Neuber, M.; Panter, F.; Krug, D.; Müller, R. Supercritical Fluid Extraction Enhances Discovery of Secondary Metabolites from Myxobacteria. *Anal. Chem.* **2020**, *92*, 15403–15411. [CrossRef] [PubMed]
169. Bonilla, J.; Sobral, P.J. Investigation of the physicochemical, antimicrobial and antioxidant properties of gelatin-chitosan edible film mixed with plant ethanolic extracts. *Food Biosci.* **2016**, *16*, 17–25. [CrossRef]
170. Wu, J.; Sun, Q.; Huang, H.; Duan, Y.; Xiao, G.; Le, T. Enhanced physico-mechanical, barrier and antifungal properties of soy protein isolate film by incorporating both plant-sourced cinnamaldehyde and facile synthesized zinc oxide nanosheets. *Colloids Surf. B Biointerfaces* **2019**, *180*, 31–38. [CrossRef] [PubMed]
171. Siracusa, V.; Romani, S.; Gigli, M.; Mannozzi, C.; Cecchini, J.P.; Tylewicz, U.; Lotti, N. Characterization of Active Edible Films based on Citral Essential Oil, Alginate and Pectin. *Materials* **2018**, *11*, 1980. [CrossRef]
172. Jamróz, E.; Juszczak, L.; Kucharek, M. Investigation of the physical properties, antioxidant and antimicrobial activity of ternary potato starch-furcellaran-gelatin films incorporated with lavender essential oil. *Int. J. Biol. Macromol.* **2018**, *114*, 1094–1101. [CrossRef]
173. Otoni, C.G.; Avena-Bustillos, R.J.; Olsen, C.W.; Bilbao-Sáinz, C.; McHugh, T.H. Mechanical and water barrier properties of isolated soy protein composite edible films as affected by carvacrol and cinnamaldehyde micro and nanoemulsions. *Food Hydrocoll.* **2016**, *57*, 72–79. [CrossRef]
174. Reis, L.C.B.; de Souza, C.O.; da Silva, J.B.A.; Martins, A.C.; Nunes, I.L.; Druzian, J.I. Active biocomposites of cassava starch: The effect of yerba mate extract and mango pulp as antioxidant additives on the properties and the stability of a packaged product. *Food Bioprod. Process.* **2015**, *94*, 382–391. [CrossRef]
175. Xue, F.; Gu, Y.; Wang, Y.; Li, C.; Adhikari, B. Encapsulation of essential oil in emulsion based edible films prepared by soy protein isolate-gum acacia conjugates. *Food Hydrocoll.* **2019**, *96*, 178–189. [CrossRef]
176. Tongnuanchan, P.; Benjakul, S.; Prodpran, T. Properties and antioxidant activity of fish skin gelatin film incorporated with citrus essential oils. *Food Chem.* **2012**, *134*, 1571–1579. [CrossRef]
177. Zhang, X.; Ma, L.; Yu, Y.; Zhou, H.; Guo, T.; Dai, H.; Zhang, Y. Physico-mechanical and antioxidant properties of gelatin film from rabbit skin incorporated with rosemary acid. *Food Packag. Shelf Life* **2019**, *19*, 121–130. [CrossRef]
178. E Restrepo, A.; Rojas, J.D.; García, O.R.; Sánchez, L.T.; I Pinzón, M.; Villa, C.C. Mechanical, barrier, and color properties of banana starch edible films incorporated with nanoemulsions of lemongrass (*Cymbopogon citratus*) and rosemary (*Rosmarinus officinalis*) essential oils. *Food Sci. Technol. Int.* **2018**, *24*, 705–712. [CrossRef]
179. Mohsenabadi, N.; Rajaei, A.; Tabatabaei, M.; Mohsenifar, A. Physical and antimicrobial properties of starch-carboxy methyl cellulose film containing rosemary essential oils encapsulated in chitosan nanogel. *Int. J. Biol. Macromol.* **2018**, *112*, 148–155. [CrossRef]

Article

Impact of Crosslinking on the Characteristics of Pectin Monolith Cryogels

Aleksandra Nesic [1,*], Sladjana Meseldzija [1], Antonije Onjia [2] and Gustavo Cabrera-Barjas [3]

[1] Department of Chemical Dynamics and Permanent Education, Vinca Institute of Nuclear Sciences—National Institute of the Republic of Serbia, University of Belgrade, Mike Petrovica-Alasa 12-14, 11000 Belgrade, Serbia
[2] Department of Analytical Chemistry and Quality Control, Faculty of Technology and Metallurgy, University of Belgrade, 11000 Belgrade, Serbia
[3] Unidad de Desarrollo Tecnológico (UDT), Universidad de Concepción, Avda. Cordillera 2634, Parque Industrial Coronel, Coronel 4190000, Chile
* Correspondence: anesic@vin.bg.ac.rs

Abstract: In this research, the pectin monoliths were prepared via the sol-gel process through different routes of crosslinking and additional freeze-drying. The crosslinking reaction was induced by the use of calcium ions in aqueous solutions and in alcohol/water solutions. The resulting pectin monoliths obtained by freeze-drying were macroporous with open cells, limited specific surface area, moderate mechanical stability and moderate biodegradation rate. The presence of alcohol in crosslinking solution significantly changed the morphology of final pectin monoliths, which was evidenced by the reduction of their pore size for one order. The specific surface area of pectin monoliths obtained through the calcium-water-alcohol route was 25.7 m^2/g, the Young compressive modulus was 0.52 MPa, and the biodegradation rate was 45% after 30 days of immersion in compost media. Considering that pectin can be obtained from food waste, and its physical properties could be tailored by different crosslinking routes, the pectin monoliths could find wide application in the pharmaceutical, agricultural, medical and food industries, providing sustainable development concepts.

Keywords: pectin; cryogels; alcohol gelation; biodegradation

1. Introduction

Pectin belongs to the group of heteropolysaccharides and is one of the most abundant cellular components of the plant wall. It consists of homogalacturonans (HG), xylogalacturonans (XGA), rhamnogalacturonans type I (RG-I), rhamnogalacturonans type II (RG-II), arabinans and arabinogalactans [1]. The largest part of the pectin backbone (60–70%) consists of homogalacturonan, i.e., of homopolymers of D-galacturonic acid, connected by α (1→4) glycosidic bonds, where some of the carboxyl groups are methyl-esterified, while some O–2 and O–3 positions might be O–acetylated [2]. Pectin can be found in the pulp and peel of lemons, oranges, limes, sunflowers, apples, potatoes, tomatoes, and carrots. Commercially, pectin is mostly obtained from citrus fruits, as well as from apple pomace, as secondary products in the production of juices [3]. Depending on the source, pectins can differ in molecular weight, degree of acetylation and methylation, as well as in the content of galacturonic acid and neutral sugars. One of the unique properties of pectin is that it can be physically and/or ionically crosslinked by external stimuli, thus easily forming a three-dimensional hydrogel network. The degree of esterification affects pectin gelation and processing conditions, so pectin with a high degree of methyl-esterification is sensitive in an acidic environment and gels at low pH values and in the presence of high concentrations of sugars. On the other side, pectin, with a low degree of methyl-esterification, has the ability to form gels in the presence of divalent cations, and the crosslinking reaction is described by the egg-box model [4]. Moreover, in both cases (high-methylated and low-methylated pectin), it is confirmed that higher temperature improves the gelation ability. Hence, hydrogen bonds, as well as hydrophobic interactions, have an important role in the gelation of pectin [5].

The ability of pectin to undergo gelation by different external stimuli using sol-gel technology allows the production of three-dimensional networks in the form of hydrogels (wet gels), xerogels (vacuum drying), aerogels (supercritical CO_2 drying) and cryogels (freeze-drying). The main characteristic of wet gels is their ability to bind substantial amounts of water, thus increasing their volume. The drying route of wet gels dictates the textural (porosity, pore size distribution) and functional properties of final porous networks and their application potential. During the drying process under vacuum comes pore collapse, so final materials usually possess high density and low porosity [6]. On the other side, it has been demonstrated that supercritical drying preserves the morphology of wet gels, giving mesoporous materials with low density [6–8]. Furthermore, by freeze-drying, it is possible to tailor the macropore and mesopore space within the final material [9]. These properties, along with the non-toxicity, biodegradability and biocompatibility, make pectin-based materials adequate for a wide range of applications such as agriculture, pharmacy, food and biomedicine. Up to date, pectin-based three-dimensional networks have been investigated for tissue engineering [10], drug delivery [11], wound healing [12,13], food technology [14] and as food packaging material [15].

The aim of this work is to study the influence of different crosslinking routes of pectin on the physical properties of final materials. Namely, three different crosslinker solutions were used: (a) aqueous calcium bath, (b) tert-butanol bath, and (c) calcium-tert butanol/water bath. The obtained hydrogels were freeze-dried, and their textural, morphological, mechanical, and thermal properties were evaluated. It was previously demonstrated that pectin and alginate could gel in the presence of different alcohols, such as ethanol, methanol, propanol and 1-butanol and that obtained hydrogels can easily be turned into mesoporous aerogels by supercritical drying [16–18]. In this work, tert-butanol was chosen as a precursor for gelation since it is widely used as a green solvent and is proven as an efficient crosslinker for various polysaccharides, providing the cryogels with different levels of hierarchy, i.e., tailoring the macro and mesopores in the system by different used concentrations [9]. Up to date, there are several papers reported in the literature that deal with the characterization of pectin-calcium crosslinked aerogels or cryogels [6–8,19,20], but to the best of our knowledge, there is no published data related to the influence of tert-butanol and calcium ions combined together on physical properties of final materials. Moreover, there is no comparison of the impact of different crosslinking routes (including ionic and alcohol-induced crosslinking, separately and combined) on the final properties of the pectin of the same origin. Finally, since pectin is a natural polymer derived from renewable resources, the aerobic degradation rate of all obtained pectin-based cryogels was assessed in order to evaluate if these materials can contribute to the sustainable development goals.

2. Materials and Methods

2.1. Chemicals

Citrus pectin, with a degree of methylation of 50%, was obtained from Herbstreith & Fox KG, Pektin-Fabriken (Neuenburg, Germany). Tert-butanol and $CaCl_2 \times 2H_2O$ were purchased from Sigma Aldrich (St. Louis, MO, USA).

2.2. Preparation of Cryogels

Pectin-based cryogel monoliths were prepared through three different crosslinking routes in a two-step process. In all cases, the primary step was the dissolution of pectin in an aqueous solution (2.0 wt%) at room temperature. The first gelation route was by use of $CaCl_2$ aqueous solution (P1 sample), the second gelation route was by use of tert-butanol (P2 sample), and the third gelation route by use of tert-butanol/water solution containing $CaCl_2$ (P3 sample). The final concentration of calcium ions in samples P1 and P3 was 2 wt%, with respect to the mass of pectin, whereas the final concentration of tert-butanol in sample P3 was 40 wt%. The crosslinking reaction occurs through slow diffusion of crosslinking solution into pectin solution. The mixture was left until complete gelation, which took place after 4 h. The obtained pectin gel samples were frozen in liquid nitrogen and freeze-dried for at least 24 h on a VirTis SP Scientific Sentry 2.0 freeze drier (New Life Scientific Inc., Woonsocket, RI, USA). The drying conditions were as follows: vacuum set to 100 m Torr and a condenser temperature of −80.0 °C.

2.3. Morphology and Texture of Cryogels

The morphology of the prepared samples was analyzed by FEI Quanta 200 FEG Scanning Electron Microscope (SEM) (FEI Company, Hillsboro, OR, USA) at an accelerating voltage of 5–10 kV. Prior to the SEM analysis, the cryogels were sputtered with gold (~7 nm). The bulk density of the samples was calculated as the mass-to-volume ratio, and the presented results are an average of three measurements. The specific surface area was analyzed by nitrogen adsorption-desorption test (Nova 3000e surface area analyzer, Quantachrome Instruments, Bointon Beach, FL, USA) and mercury porosimetry test (Carlo Erba Porosimeter 2000 equipped with the software Milestone 200, Carlo Erba, Washington, DC, USA). Before the measurements, the samples were degassed at 75 °C for 24 h. The presented results are the average values of 3 measurements; the standard deviation was within 10%.

2.4. Calcium Content in Cryogels

An accurately weighed amount of cryogel was digested with hydrogen peroxide (30%, w/w) and nitric acid (70%, w/w) in a microwave oven (working conditions: power = 800 W; temperature = 200 °C, ramp 10 min, working time = 20 min). Upon cooling, the solution was diluted up to 50 mL in a volumetric flask using deionized water. Appropriate dilutions were carried out, and the calcium ion content in the cryogel was determined by the use of Inductively coupled plasma mass spectrometry (ICP–MS) with a quadrupole detector at ^{44}Ca. The presented results are the average values of 3 measurements; the standard deviation was within 5%.

2.5. Water Uptake and Solubility of Cryogels

The water uptake (WU, %) of cryogels was determined by the gravimetric method. Firstly, the cryogels were weighed (m_0, g) and placed in 100 mL of distilled water at room temperature (25 °C). The cryogels were taken out from the water after 24 h; the excess water from the surface was removed by filter paper, and the weight of the swollen cryogels (m_{t1}, g) was measured. The water uptake was calculated by the following equation:

$$WU\ (\%) = \frac{(m_{t1} - m_0) \times 100}{m_0} \quad (1)$$

The solubility test was determined as the content of dry matter solubilized after 24 h in distilled water. The swollen cryogels were taken out after 24 h and dried until constant weight (m_{t2}, g) in an oven at 105 °C. The solubility degree (SLD) was calculated according to the following equation:

$$SLD(\%) = \frac{(m_0 - m_{t2})}{m_0} \times 100 \quad (2)$$

The presented results are the average values of 3 measurements; the standard deviation was within 10%.

2.6. Mechanical Analysis

The uniaxial compression test was performed according to the ASTM D695 standard. Cylindrical specimens with a height-to-diameter ratio of approximately 5 were used. A load of 10 kN was applied. The samples were compressed upon the breaking point at a compression rate of 2 mm/min. The force-time curves were converted to the compressive stress (σ_c, MPa)-compressive strain (ε_c, %) curves and Young's modulus (E_c, MPa) was obtained from the linear part of the curve. The results are presented as the average of five experiments, and values of compressive strength and Young's modulus are within ±15%.

2.7. Thermal Analysis

Thermogravimetric analysis (TG) was performed using a TGA Q500 (TA Instruments New Castle, DE, USA) instrument under dynamic nitrogen flow in the temperature range from 25 to 600 °C. The nitrogen flow rate was 60 cm^3/min while the heating rate was 10 °C/min. The weight of the samples was approximately 10 mg.

2.8. Biodegradation

The aerobic biodegradation of pectin monolith cryogels was determined according to the ISO 14855-2:2018 standard. Each sample (~0.2 g) was ground, mixed with compost soil and stored in a 250 mL sample chamber at room temperature. The amount of released CO_2 within the time was monitored and measured by Micro-Oxymax Respirometer (Columbus Instruments, Columbus, OH, USA). The degree of biodegradation was calculated according to the following equation:

$$D = \frac{V_1 - V_b}{V_2} \times 100 \quad (3)$$

where V_1 and V_b are the amount of CO_2 released within the sample reactor and the blank reactor, respectively, and V_2 is the theoretical amount of CO_2 available from the samples. The presented results are the average values of 3 measurements; the standard deviation was within 10%.

3. Results

3.1. Morphology and Texture of Cryogels

Pectin monolith cryogels were obtained by gelation in (a) an aqueous solution of calcium ions (2 wt%), (b) tert-butanol, and (c) a solution that combines calcium ions and tert-butanol/water solution. The morphology of obtained cryogels is presented in Figure 1. The pore size distribution of all samples was evaluated by scanning 3 pieces of each sample at different magnifications and presented in Table 1. It can be seen that all pectin monolith samples have dense sheet-like morphology with some level of interconnected voids. This type of structure is common when the slow growth of crystals occurs, caused by slow freezing and slow drying process. A similar structure is reported for other pectin cryogel systems in literature [6,19] and for other biopolymer-based cryogels, such as alginate [21,22] and starch [23,24]. It is interesting to note that the structure is dense in samples that contain tert-butanol (P2 and P3). Namely, the range of diameter of voids for samples crosslinked only in calcium ion solution (P1) is between 150 and 400 µm, whereas in the case of samples crosslinked in tert-butanol (P2) and Ca/tert-butanol-water (P3), the range is between 10 and 50 µm. Furthermore, the bulk density of the P2 and P3 (0.1–0.12 g cm^{-3}) samples is higher than for sample P1 (0.06 g cm^{-3}). It appears that the presence of tert-butanol in crosslinking solution increases the hydrophobic interactions of pectin chains and leads to interpolymer bridging, thus changing the morphology and texture of final cryogels.

Table 1. Characterization of pectin-based cryogels by nitrogen adsorption.

| Sample | Specific Surface Area, m^2 g^{-1} | | Average Pore Size, µm | Density, g cm^{-3} |
	Hg Porosity	BET	SEM	
P1	17	8	150–400	0.06
P2	23	14	10–50	0.10
P3	26	19	10–50	0.12

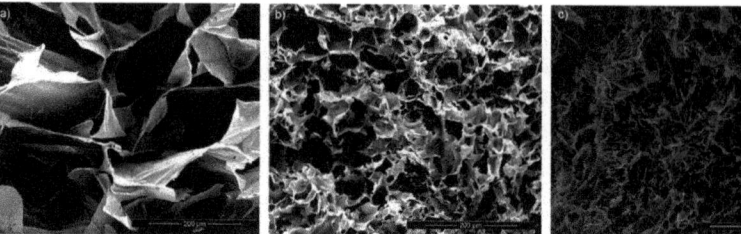

Figure 1. SEM morphology of cryogels: (**a**) P1, (**b**) P2 and (**c**) P3.

In order to get an idea about the specific surface area of samples, a mercury porosimetry test and nitrogen adsorption test was performed, and these results are presented in Table 1. Mercury porosimetry is a generally used method to characterize the texture of materials with large macropores. The pressure of mercury causes the contraction of biobased cryogels, which can cause the destruction of pores, so valuable information can be obtained sometimes only at smaller applied pressures [25]. However, during the measurement, not only the compression of samples occurs but also the partial crushing of cryogels, so the pore size distribution can not be determined by this technique. On the other side, the nitrogen adsorption test is not precise enough for the evaluation of pore volume and pore size distribution of materials that are macroporous. Hence, both techniques can give just an idea about the specific surface area of obtained samples and its relationship between the samples, i.e., similarities or differences among the samples.

It is demonstrated that gelation in the presence of tert-butanol (sample P2 and P3) provides materials with higher surface areas in comparison to the pectin crosslinked in an aqueous solution of calcium ions. The obtained specific surface area of pectin cryogels is slightly higher by Hg porosimetry than by nitrogen adsorption test, and it ranges between 17 and 26 $m^2\ g^{-1}$. These values are higher than for the starch freeze-dried cryogels (7.7 $m^2\ g^{-1}$) [23] and have a similar range for the other reported pectin cryogel systems (10–20 $m^2\ g^{-1}$) [6]. Apparently, the specific surface area originates from the porosity in the walls of voids. Similar results are obtained in literature for the starch cryogels and resorcinol-formaldehyde cryogels, where measurable, specific surface area is obtained by nitrogen adsorption test, but SEM micrographs displayed macroporous systems with large macropores of the 200 μm [23,26]. According to the SEM and Hg porosimetry/nitrogen adsorption test results, the highest specific surface area and lowest pore size distribution, but higher density exhibits the pectin-based cryogel crosslinked by the solution of Ca/tert-butanol-water. The presence of alcohol induces the re-organization of macromolecular chains, with competitive interactions of; (a) self-association, i.e., interpolymer bridging through intermolecular hydrogen bonding and hydrophobic interactions, and (b) ionic crosslinking with the Ca-ions. It has been demonstrated before that the crosslinking of alginate in the alcoholic bath (methanol) forms the association structures through interpolymer bridging [18]. Moreover, Rodriguez-Dorado et al. have reported that alginate aerogels crosslinked in Calcium-ethanolic solution exhibit higher specific surface area than alginate microbead aerogels crosslinked in Calcium-aqueous solution [27].

3.2. Water Uptake, Solubility and Calcium Content in Cryogels

The water uptake and solubility of pectin-based monolith cryogels after 24 h of immersion in distilled water, as well as the content of calcium ions in them, are presented in Table 2. As it is shown, the pectin sample crosslinked only by alcohol dissolves after 24 h, confirming that occurred crosslinking reaction is only physical, through intermolecular hydrogen bonding, thus making these cryogels less stable in aqueous solutions, in comparison to the ionically crosslinked cryogels. On the other side, the solubility of the P1 and P3 samples is only 5% after 24 h of immersion in distilled water. The water uptake for these two samples is in the range between 346 and 983%, proving that these samples have the ability for large water uptake, which makes them suitable for the range of applications in the food, agriculture and biomedical sector. It is interesting to note that water uptake is significantly higher for the P1 sample than the P3 sample. The presence of tert-butanol and Ca in cryogel induces lower uptake of water due to the formation of a denser network. As it is confirmed by SEM, in the presence of Ca/tert-butanol-water crosslinker, it comes to more dense morphology of pectin cryogels, with less interconnected voids, when compared to only Ca-crosslinker. In addition, this result can be related to calcium ion content entrapped into pectin cryogel. Namely, it is shown that the P3 sample contains more calcium ions than the P1 sample. Apparently, the competitive interaction of tert-butanol and Ca-ions with pectin chains induces the better physical re-organization of chains, allowing better entrapment of calcium ions in comparison to the samples crosslinked only by calcium ions.

Table 2. Water-related properties of pectin-based cryogels.

Sample	WU, %	SLD, %	Ca Content, ppm
P1	983	5	70
P2	dissolved	100	–
P3	346	5	103

The main ionic crosslinking process of pectin is described by the physical entrapment of divalent cations (in this case, calcium ions) between non-methyl esterified galacturonate units from pectin, where the junction zones are formed in so-called "egg box" model [28]. Further research has confirmed that ionic crosslinking does not involve only physical entrapment but also the interaction of calcium ions with the oxygen atom from the carboxylate group of pectin, making the stable three-dimensional and thermos-irreversible network [29]. It has been also demonstrated that, at some level, hydrophobic interactions also contribute to network development [30,31]. On the other side, Oakenfull et al. have shown that alcohol presence can stabilize the junction zones to form three-dimensional pectin networks through hydrogen bonds and hydrophobic interactions of ester methyl groups from pectin [5]. In both crosslinking processes, the stabilization of hydrogel, i.e., the gelation process, depends on the degree of methylation and distribution pattern of these groups. Taking into consideration the results presented in Section 3.1 and in Table 2, it can be concluded that the presence of tert-butanol and calcium ions in crosslinking bath has a synergic effect on developing a more stable three-dimensional network.

3.3. Mechanical Properties

Mechanical properties of pectin-based cryogels, in terms of compressive Young Modulus and compressive strength, were studied by uniaxial compression test and presented in Figure 2. The Young compressive modulus represents the level of stiffness or resistance to deforming until applied load (compression). In this work, the obtained Young modulus values for samples P1, P2 and P3 are 0.18 MPa, 0.33 MPa and 0.52 MPa, respectively. This trend is predictable due to pore size distribution and density results obtained for pectin-based cryogels (see Section 3.1). Generally, the materials with higher porosity and lower density negatively impact the Young modulus in comparison to the materials with a more dense structure and less porosity. Moreover, it is demonstrated in the literature that ethanol, used as a co-solvent in calcium crosslinking solution, improves the mechanical strength of alginate films in comparison to the film crosslinked with an aqueous solution of calcium ions [32]. The maximum compressive strength represents the maximum compressive stress that the material can withstand until it breaks. The lowest compressive strength (0.14 MPa) is obtained for the P1 sample, whereas the highest compressive strength exhibits the sample P3 (0.3 MPa). Also, it is important to note that all samples plastically deform until strain between 67% and 82%. Overall, the presence of tert-butanol and calcium ions presents a synergic effect for the formation of more mechanically stable three-dimensional networks. Chen et al. have reported the Young compressive modulus of neat pectin cryogels between 0.04 (2.5 wt% content of pectin in cryogel) and 48 MPa

(15 wt% content of pectin in cryogel) [19]. Moreover, the addition of clay significantly improved the mechanical performances and Young modulus of pectin cryogels, reaching a range between 0.16 and 114 MPa. Furthermore, Yang et al. have obtained a Young modulus of neat pectin cryogels of 0.09 MPa, and enhanced by 119% with the inclusion of boron nitride nanosheets into the pectin matrix [33]. The difference in Young modulus value presented in this work and literature is due to different types of used pectin and different degree of methylation of pectin. Namely, it is proved that degree of methylation, but also the distribution pattern of methylated and non-methylated galacturonic units, influences the pectin hydrogel strength [4,34–36]. In addition, the pectin concentration used to make hydrogels and cryogels give contribution too to the mechanical strength of final materials. In this work, the cryogels are made of 2 wt% of pectin because the focus was to evaluate the lowest possible concentration of raw material. i.e., pectin to obtain cryogels without any visible surface damage and to characterize them. Our preliminary results (not presented here) showed that cryogels with pectin content of 1 wt% and 1.5 wt% had visible cracks and "peeling off" surfaces. On the other side, cryogels reported in literature usually contained 2.5–6 wt% of pectin, which provided better mechanical stability. Nevertheless, the influence of different pectin concentrations on the physical-chemical properties of cryogels obtained by the most efficient crosslinking route established in this work will be assessed in the future and part of the forthcoming paper.

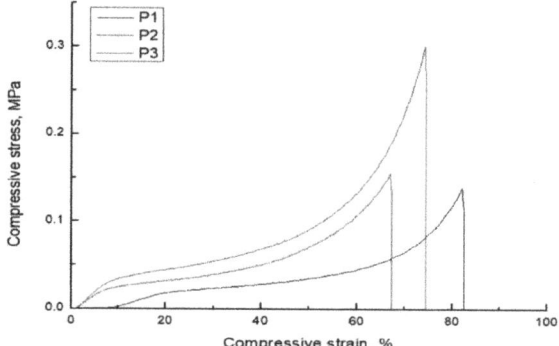

Figure 2. Compressive stress–strain behavior for pectin-based cryogels.

3.4. Thermal Analysis

There are three types of absorbed water in hydrophilic materials: free unbound water, freezing bound and non-freezing water or bound water. Free water does not interact, via hydrogen bonding, with the polymeric chains, and it is released from samples up to 100 °C. The freezing-bound water interacts only weakly with the polymeric chain, while the non-freezing water is represented by molecules of water bound to the polymeric chains through hydrogen bonds [37]. Generally, these two types of water are released from samples between 100 °C and 200 °C. On the basis of the above claims, it can be concluded by the analysis of the pectin-based cryogels thermograms (Figure 3 and Table 3) that the mass loss in the first region up to 100 °C is associated with the evaporation of free water (or free absorbed moisture), while the second region up to 200 °C represents a loss of freezing and non-freezing bound water. All tested samples contain free water because they are macroporous materials subjected to the fast absorption of moisture. The highest content of bound water is in sample P1, according to the weight loss percentage at 200 °C (see Table 3, W_{L200}, %). This result is expected since it is already confirmed by SEM that pectin crosslinked only by calcium ions provides the materials with the highest pore size distribution and less dense structure, thus being more subjected to moisture absorption in comparison to the samples that contain tert-butanol. In all cases, the decomposition process of pectin chains starts above 200 °C, which includes the random split of the glycosidic bonds, vaporization and elimination of volatile products [1,15]. It is interesting to note that there is a significant shift of Tonset (the onset temperature at which starts degradation) to lower values for samples P2 and P3 when compared to the P1 sample, indicating that hydrogen bonds formed between pectin chains are easier to break at increasing temperature, than bonds formed ionically with calcium ions. However, it is important to highlight that the second and third step of degradation of the P1 sample is a continuous process, where it is not possible precisely to say where one degradation step ends, and another starts. In the case of samples P2 and P3, the onset temperature of the third degradation step is clear since these two samples do not contain too much-bound freezing and non-freezing water; hence no continual steps are occurring. Although the main degradation of pectin chains starts earlier for cryogels that contain tert-butanol, the Tdeg is shifted to a higher value for the P3 sample and to a lower value for the P2 sample, in comparison to the P1. This result confirms once again that the self-associated hydrogen bonds of pectin chains are less thermally stable than bonds occurred by ionic crosslinking of calcium ions with pectin chains and bonds formed in combination with ionic (Ca^{2+}) and physical (tert-butanol).

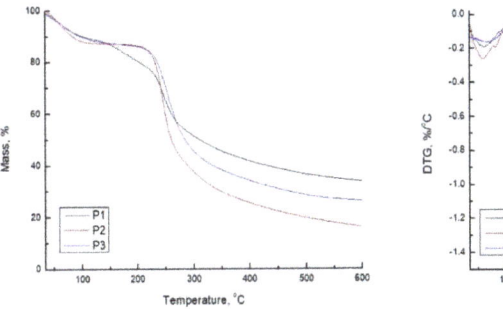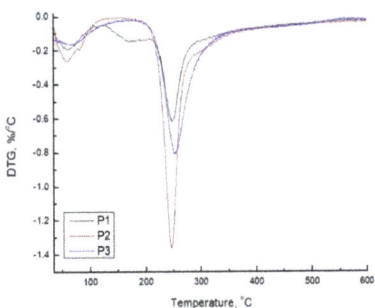

Figure 3. TGA/DTG of pectin-based cryogels.

Table 3. Thermal properties of pectin monolith cryogels, where W_{L100} and W_{L180} present the weight loss at 100 °C and 200 °C, respectively, Tonset is the temperature at which starts degradation, and Tdeg is the temperature of maximum degradation rate.

Sample	W_{L100}, %	W_{L200}, %	Tonset, °C	Tdeg, °C	Char Residue at 600 °C, %
P1	10	20	216	248	34
P2	12	13	212	245	16
P3	11	14	209	252	26

3.5. Biodegradation

Biodegradation of materials is essential for the balance of nature. Although there is increased awareness about the sustainable development of biodegradable materials and promotion of new approaches to replace plastic materials or upgrade them to be biodegradable, there are still not enough data in the literature related to this topic, where is studied on which conditions polysaccharides degrade, and how to speed their process of degradation. Hence, in this work, the aerobic biodegradation of pectin-based cryogels was performed by a respiratory method and monitored within the period of 30 days, and results are presented in Figure 4. As can be seen, the aerobic biodegradation rate is in the range of 43 and 62%, and the equilibrium plateau is reached within 15 days. The highest rate of aerobic biodegradation has sample P1, whereas there is no significant difference in the biodegradation rate between samples P2 and P3. It appears that the presence of tert-butanol decreases the aerobic biodegradation of pectin cryogels. This result is expected since it is well known that samples with higher moisture/water content promote the propagation of microorganisms, which as a result, increase the respiration rate, as well as biodegradation. A similar biodegradation rate determined by the respiratory method is obtained for the polysaccharide cryogels. Chen et al. have demonstrated that control pectin cryogel reaches a biodegradation rate of 60%. However, the equilibrium plateau is reached within 10 days, which is faster than in this work. Also, Praglowska et al. obtained a biodegradation rate of 80% for chitosan cryogels within 11 days of immersion in compost media [38]. On the other side, the alginate-calcium xerogel beads degrade only 32% after 2 months [39], ammonium alginate-phytic acid cryogel 92% after 45 days [40], cellulose powder degrades 83% after 45 days [41].

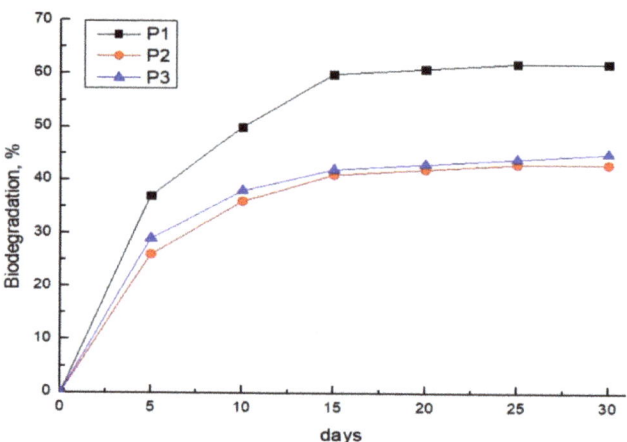

Figure 4. Aerobic biodegradation of pectin-based cryogels.

4. Conclusions

In this work, calcium ions and tert-butanol, separately and in combination, were used as the precursors for crosslinking of pectin. The hydrogels were freeze-dried and subjected to several different characterization techniques in order to evaluate their textural, morphological, mechanical, thermal and biodegradable property. It was shown that a combination of tert-butanol and calcium ions provides materials with higher density but also higher specific surface area and smaller pore size distribution. The highest Young modulus was obtained for this sample, too and reached a value of 0.55 MPa. On the other side, tert-butanol had a negative influence on the thermal stability of pectin-based cryogels, causing the shift of the onset degradation temperature to lower values of 7 °C. Moreover, the biodegradation rate of pectin-based samples that contain tert-butanol was 45% after 30 days of immersion in compost media, whereas the pectin sample crosslinked by calcium ions had a biodegradation rate of 62%. The results obtained in this work demonstrate that different crosslinking routes can significantly impact the morphology and also final properties of materials, dictating the direction of application. Pectin, as a natural polymer from renewable sources, presents a suitable feedstock for the processing of sustainable and biodegradable materials with tailored macroporosity through different crosslinking routes.

Author Contributions: Conceptualization, A.N.; methodology, A.N. and G.C.-B.; formal analysis, S.M. and A.O.; investigation, S.M. and G.C.-B.; data curation, S.M. and A.O.; writing—original draft preparation, A.N. and G.C.-B.; writing—review and editing, A.N. All authors have read and agreed to the published version of the manuscript.

Funding: This research received no external funding.

Data Availability Statement: Not applicable.

Acknowledgments: This work was supported by the Ministry of Education, Science and Technological Development of the Republic of Serbia (Contract number 451-03-68/2022-14/200017). This work has been realized in the frame of AERoGELS COST Action CA18125–Advanced Engineering and Research of aeroGels for Environment and Life Sciences.

Conflicts of Interest: The authors declare no conflict of interest.

References

1. Nesic, A.R.; Trifunovic, S.S.; Grujic, A.S.; Velickovic, S.J.; Antonovic, D.G. Complexation of amidated pectin with poly (itaconic acid) as a polycarboxylic polymer model compound. *Carbohydr. Res.* **2011**, *346*, 2463–2468. [CrossRef] [PubMed]
2. Nunes, C.; Silva, L.; Fernandes, A.P.; Guiné, R.P.F.; Domingues, M.R.M.; Coimbra, M.A. Occurrence of cellobiose residues directly linked to galacturonic acid in pectic polysaccharides. *Carbohydr. Polym.* **2012**, *87*, 620–626. [CrossRef] [PubMed]
3. Meseldzija, S.; Petrovic, J.; Onjia, A.; Volkov-Husovic, T.; Nesic, A.; Vukelic, N. Utilization of agro-industrial waste for removal of copper ions from aqueous solutions and mining-wastewater. *J. Ind. Eng. Chem.* **2019**, *75*, 246–252. [CrossRef]
4. Gawkowska, D.; Cybulska, J.; Zdunek, A. Structure-Related Gelling of Pectins and Linking with Other Natural Compounds: A Review. *Polymers* **2018**, *10*, 762. [CrossRef]
5. Oakenfull, D.; Scott, A.; Chai, E. The mechanism of formation of mixed gels by high methoxyl pectins and alginates. In *Gums and Stabilisers for the Food Industry*, 2nd ed.; Williams, P.A., Phillips, G.O., Eds.; Woodhead Publishing: Cambridge, UK, 1990; Volume 10, pp. 243–264.
6. Groult, S.; Buwalda, S.; Budtova, T. Pectin hydrogels, aerogels, cryogels and xerogels: Influence of drying on structural and release properties. *Eur. Polym. J.* **2021**, *149*, 110386. [CrossRef]
7. Rudaz, C.; Courson, R.; Bonnet, L.; Calas-Etienne, S.; Sallée, R.; Budtova, T. Aeropectin: Fully biomass-based mechanically strong and thermal superinsulating aerogel. *Biomacromolecules* **2014**, *15*, 2188–2195. [CrossRef]
8. Veronovski, A.; Tkalec, G.; Knez, Z.; Novak, Z. Characterisation of biodegradable pectin aerogels and their potential use as drug carriers. *Carbohydr. Polym.* **2014**, *113*, 272–278. [CrossRef]
9. Borisova, A.; De Bruyn, M.; Budarin, V.L.; Shuttleworth, P.S.; Dodson, J.R.; Segatto, M.L.; Clark, J.H. A Sustainable Freeze-Drying Route to Porous Polysaccharides with Tailored Hierarchical Meso- and Macroporosity. *Macromol. Rapid Commun.* **2015**, *36*, 774–779. [CrossRef]
10. Lapomarda, A.; De Acutis, A.; De Maria, C.; Vozzi, G. Pectin-Based Scaffolds for Tissue Engineering Applications. In *Pectins—The New-Old Polysaccharides*, 2nd ed.; Masuelli, M.A., Ed.; IntechOpen: London, UK, 2021; Volume 8, pp. 146–216.
11. Li, D.; Li, J.; Dong, H.; Li, X.; Zhang, J.; Ramaswamy, S.; Xu, F. Pectin in biomedical and drug delivery applications: A review. *Int. J. Biol. Macromol.* **2021**, *185*, 49–65. [CrossRef]
12. Nordin, N.N.; Aziz, N.K.; Naharudin, I.; Anuar, N.K. Effects of Drug-Free Pectin Hydrogel Films on Thermal Burn Wounds in Streptozotocin-Induced Diabetic Rats. *Polymers* **2022**, *14*, 2873. [CrossRef]
13. Giusto, G.; Vercelli, C.; Comino, F.; Caramello, V.; Tursi, M.; Gandini, M. A new, easy-to-make pectin-honey hydrogel enhances wound healing in rats. *BMC Complement. Altern. Med.* **2017**, *17*, 266. [CrossRef]
14. Vanitha, T.; Khan, M. Role of Pectin in Food Processing and Food Packaging. In *Pectins–Extraction, Purification, Characterization and Applications*, 2nd ed.; Masuelli, M., Ed.; IntechOpen: London, UK, 2019; Volume 7, pp. 85–106.
15. Nešić, A.; Gordić, M.; Davidović, S.; Radovanović, Ž.; Nedeljković, J.; Smirnova, I.; Gurikov, P. Pectin-based nanocomposite aerogels for potential insulated food packaging application. *Carbohydr. Polym.* **2018**, *195*, 128–135. [CrossRef] [PubMed]
16. Tkalec, G.; Knez, Ž.; Novak, Z. Fast production of high-methoxyl pectin aerogels for enhancing the bioavailability of low-soluble drugs. *J. Supercrit. Fluids* **2015**, *106*, 16–22. [CrossRef]
17. Tkalec, G.; Knez, Z.; Novak, Z. Formation of polysaccharide aerogels in ethanol. *Carbohydr. Polym.* **2015**, *5*, 1593–1599. [CrossRef]
18. Tkalec, G.; Kranvogl, R.; Perva Uzunali, A.; Knez, Ž.; Novak, Z. Optimisation of critical parameters during alginate aerogels' production. *J. Non. Cryst. Solids* **2016**, *443*, 112–117. [CrossRef]
19. Chen, H.-B.; Chiou, B.-S.; Wang, Y.-Z.; Schiraldi, D.A. Biodegradable Pectin/Clay Aerogels. *ACS Appl. Mater. Interfaces* **2013**, *5*, 1715–1721. [CrossRef] [PubMed]
20. Subrahmanyam, R.; Gurikov, P.; Meissner, I.; Smirnova, I. Preparation of biopolymer aerogels using green solvents. *J. Vis. Exp.* **2016**, *2016*, 3–7. [CrossRef] [PubMed]
21. Barros, A.; Quraishi, S.; Martins, M.; Gurikov, P.; Subrahmanyam, R.; Smirnova, I.; Duarte, A.R.C.; Reis, R.L. Hybrid Alginate-Based Cryogels for Life Science Applications. *Chem. Ing. Tech.* **2016**, *88*, 1770–1778. [CrossRef]
22. Chen, H.-B.; Ao, Y.-Y.; Liu, D.; Song, H.-T.; Shen, P. Novel Neutron Shielding Alginate Based Aerogel with Extremely Low Flammability. *Ind. Eng. Chem. Res.* **2017**, *56*, 8563–8567. [CrossRef]
23. Baudron, V.; Gurikov, P.; Smirnova, I.; Whitehouse, S. Porous Starch Materials via Supercritical- and Freeze-Drying. *Gels* **2019**, *5*, 12. [CrossRef]
24. Ago, M.; Ferrer, A.; Rojas, O.J. Starch-Based Biofoams Reinforced with Lignocellulose Nanofibrils from Residual Palm Empty Fruit Bunches: Water Sorption and Mechanical Strength. *ACS Sustain. Chem. Eng.* **2016**, *4*, 5546–5552. [CrossRef]
25. Horvat, G.; Pantić, M.; Knez, Ž.; Novak, Z. A Brief Evaluation of Pore Structure Determination for Bioaerogels. *Gels* **2022**, *8*, 438. [CrossRef] [PubMed]
26. Job, N.; Théry, A.; Pirard, R.; Marien, J.; Kocon, L.; Rouzaud, J.-N.; Béguin, F.; Pirard, J.-P. Carbon aerogels, cryogels and xerogels: Influence of the drying method on the textural properties of porous carbon materials. *Carbon* **2005**, *43*, 2481–2494. [CrossRef]
27. Rodríguez-Dorado, R.; López-Iglesias, C.; García-González, C.; Auriemma, G.; Aquino, R.; Del Gaudio, P. Design of Aerogels, Cryogels and Xerogels of Alginate: Effect of Molecular Weight, Gelation Conditions and Drying Method on Particles' Micromeritics. *Molecules* **2019**, *24*, 1049. [CrossRef]

28. Grant, G.T.; Morris, E.R.; Rees, D.A.; Smith, P.J.C.; Thom, D. Biological interactions between polysaccharides and divalent cations: The egg-box model. *FEBS Lett.* **1973**, *32*, 195–198. [CrossRef]
29. Wellner, N.; Kačuráková, M.; Malovíková, A.; Wilson, R.H.; Belton, P.S. FT-IR study of pectate and pectinate gels formed by divalent cations. *Carbohydr. Res.* **1998**, *308*, 123–131. [CrossRef]
30. Cardoso, S.M.; Coimbra, M.A.; Lopes da Silva, J.A. Temperature dependence of the formation and melting of pectin–Ca^{2+} networks: A rheological study. *Food Hydrocoll.* **2003**, *17*, 801–807. [CrossRef]
31. Racape, E.; Thibault, J.F.; Reitsma, J.C.E.; Pilnik, W. Properties of amidated pectins. II. Polyelectrolyte behavior and calcium binding of amidated pectins and amidated pectic acids. *Biopolymers* **1989**, *28*, 1435–1448. [CrossRef]
32. Li, J.; He, J.; Huang, Y.; Li, D.; Chen, X. Improving surface and mechanical properties of alginate films by using ethanol as a co-solvent during external gelation. *Carbohydr. Polym.* **2015**, *123*, 208–216. [CrossRef] [PubMed]
33. Yang, W.; Yuen, A.C.Y.; Ping, P.; Wei, R.-C.; Hua, L.; Zhu, Z.; Li, A.; Zhu, S.-E.; Wang, L.-L.; Liang, J.; et al. Pectin-assisted dispersion of exfoliated boron nitride nanosheets for assembled bio-composite aerogels. *Compos. Part A Appl. Sci. Manuf.* **2019**, *119*, 196–205. [CrossRef]
34. Fraeye, I.; Duvetter, T.; Doungla, E.; Van Loey, A.; Hendrickx, M. Fine-tuning the properties of pectin–calcium gels by control of pectin fine structure, gel composition and environmental conditions. *Trends Food Sci. Technol.* **2010**, *21*, 219–228. [CrossRef]
35. Fraeye, I.; Doungla, E.; Duvetter, T.; Moldenaers, P.; Van Loey, A.; Hendrickx, M. Influence of intrinsic and extrinsic factors on rheology of pectin–calcium gels. *Food Hydrocoll.* **2009**, *23*, 2069–2077. [CrossRef]
36. Ngouémazong, D.E.; Tengweh, F.F.; Fraeye, I.; Duvetter, T.; Cardinaels, R.; Van Loey, A.; Moldenaers, P.; Hendrickx, M. Effect of de-methylesterification on network development and nature of Ca^{2+}-pectin gels: Towards understanding structure–function relations of pectin. *Food Hydrocoll.* **2012**, *26*, 89–98. [CrossRef]
37. Russo, R.; Malinconico, M.; Santagata, G. Effect of Cross-Linking with Calcium Ions on the Physical Properties of Alginate Films. *Biomacromolecules* **2007**, *8*, 3193–3197. [CrossRef]
38. Radwan-Pragłowska, J.; Piątkowski, M.; Janus, Ł.; Bogdał, D.; Matysek, D. Biodegradable, pH-responsive chitosan aerogels for biomedical applications. *RSC Adv.* **2017**, *7*, 32960–32965. [CrossRef]
39. Achmon, Y.; Dowdy, F.R.; Simmons, C.W.; Zohar-Perez, C.; Rabinovitz, Z.; Nussinovitch, A. Degradation and bioavailability of dried alginate hydrocolloid capsules in simulated soil system. *J. Appl. Polym. Sci.* **2019**, *136*, 48142. [CrossRef]
40. Cao, M.; Liu, B.-W.; Zhang, L.; Peng, Z.-C.; Zhang, Y.-Y.; Wang, H.; Zhao, H.-B.; Wang, Y.-Z. Fully biomass-based aerogels with ultrahigh mechanical modulus, enhanced flame retardancy, and great thermal insulation applications. *Compos. Part B Eng.* **2021**, *225*, 109309. [CrossRef]
41. Way, C.; Wu, D.Y.; Dean, K.; Palombo, E. Design considerations for high-temperature respirometric biodegradation of polymers in compost. *Polym. Test.* **2010**, *29*, 147–157. [CrossRef]

Article

Spontaneous Gelation of Adhesive Catechol Modified Hyaluronic Acid and Chitosan

Guillermo Conejo-Cuevas [1], Leire Ruiz-Rubio [1,2], Virginia Sáez-Martínez [3], Raul Pérez-González [3], Oihane Gartziandia [3], Amaia Huguet-Casquero [3] and Leyre Pérez-Álvarez [1,2,*]

1. Macromolecular Chemistry Group (LABQUIMAC), Department of Physical Chemistry, Faculty of Science and Technology, University of the Basque Country, UPV/EHU, Barrio Sarriena, s/n, 48940 Leioa, Spain; gconejo001@ikasle.ehu.eus (G.C.-C.); leire.ruiz@ehu.eus (L.R.-R.)
2. BCMaterials, Basque Center for Materials, Applications and Nanostructures, UPV/EHU Science Park, 48940 Leioa, Spain
3. i+Med S. Coop. Parque Tecnológico de Álava, Albert Einstein 15, nave 15, 01510 Vitoria-Gasteiz, Spain; v.saez@imasmed.com (V.S.-M.); r.perez@imasmed.com (R.P.-G.); o.gartziandia@imasmed.com (O.G.); amaya.huguet@imasmed.com (A.H.-C.)
* Correspondence: leyre.perez@ehu.eus

Citation: Conejo-Cuevas, G.; Ruiz-Rubio, L.; Sáez-Martínez, V.; Pérez-González, R.; Gartziandia, O.; Huguet-Casquero, A.; Pérez-Álvarez, L. Spontaneous Gelation of Adhesive Catechol Modified Hyaluronic Acid and Chitosan. *Polymers* 2022, 14, 1209. https://doi.org/10.3390/polym14061209

Academic Editor: Luminita Marin

Received: 28 January 2022
Accepted: 14 March 2022
Published: 17 March 2022

Publisher's Note: MDPI stays neutral with regard to jurisdictional claims in published maps and institutional affiliations.

Copyright: © 2022 by the authors. Licensee MDPI, Basel, Switzerland. This article is an open access article distributed under the terms and conditions of the Creative Commons Attribution (CC BY) license (https://creativecommons.org/licenses/by/4.0/).

Abstract: Spontaneously formed hydrogels are attracting increasing interest as injectable or wound dressing materials because they do not require additional reactions or toxic crosslinking reagents. Highly valuable properties such as low viscosity before external application, adequate filmogenic capacity, rapid gelation and tissue adhesion are required in order to use them for those therapeutic applications. In addition, biocompatibility and biodegradability are also mandatory. Accordingly, biopolymers, such as hyaluronic acid (HA) and chitosan (CHI), that have shown great potential for wound healing applications are excellent candidates due to their unique physiochemical and biological properties, such as moisturizing and antimicrobial ability, respectively. In this study, both biopolymers were modified by covalent anchoring of catechol groups, and the obtained hydrogels were characterized by studying, in particular, their tissue adhesiveness and film forming capacity for potential skin wound healing applications. Tissue adhesiveness was related to o-quinone formation over time and monitored by visible spectroscopy. Consequently, an opposite effect was observed for both polysaccharides. As gelation advances for HA-CA, it becomes more adhesive, while competitive reactions of quinone in CHI-CA slow down tissue adhesiveness and induce a detriment of the filmogenic properties.

Keywords: hyaluronic acid; chitosan; catechol; tissue adhesive

1. Introduction

Hydrogels are polymers based on three-dimensional networks capable of retaining large amounts of water due to their hydrophilic nature, while remaining insoluble due to polymer chains crosslinking [1]. Crosslinking by covalent bonds results in covalently crosslinked networks, and when polymers are joined by non-covalent interactions such as hydrogen bonds and hydrophobic or dipole–dipole interactions, physical hydrogels are formed [2]. Dried hydrogels behave similarly to a hard solid, but in an aqueous medium, water penetration between the polymer chains causes the swelling of the network [3,4]. Water content affects dramatically the mechanical properties of these gels [1], leading to soft, elastic and permeable materials for which its properties are similar to those of biological tissues [5]. For this reason, hydrogels are well known as interesting materials in biomedical applications [6].

Hydrogel characteristics make them interesting candidates for wound healing applications. On the one hand, their hydrophilic nature allows the required moist environment in the wound for extracellular matrix formation and re-epithelialization and provides protection against infections. On the other hand, the incorporation of therapeutic agents into hydrogels acting as wound dressings provides their topical release in the wound that has been shown to be more effective than systemic treatment [7]. Indeed, the promotion of an effective wound healing or regeneration, which consists of a series of complex biochemical reactions that aims to repair the wound, is highly demanded. This process occurs in three stages that can take place simultaneously. Firstly, the inflammation phase takes place, which can be summarized as the elimination of bacteria and the migration of cells that act in the second stage. Secondly, the proliferation phase comprises an increase in collagen with the aim of forming new tissues and blood vessels, as well as the contraction of the wound. Lastly, in the last phase called maturation, the elimination of the excess cells and the repositioning of the collagen occur. This entire process is complex and highly susceptible to be interrupted or fail [8]. Due to the latter, this process can be supported by healing species, which can help by functioning as antibacterial barriers or by acting as cellular scaffolds that enhance wound closure [6].

Hydrophilic polymers, due to their ability to mimic physical and biological properties of tissues, can promote damaged tissue regeneration. In this sense, it is worth highlighting that hydrogels are derived from natural polymers, especially polysaccharides, which have been widely investigated and exploited in recent years due to their abundance, biocompatible, filmogenic, and beneficial biological properties that make them interesting candidates for wound dressing applications [2]. The most studied polysaccharides include alginates, chondroitin, chitosan and chitin, cellulose, dextran, hyaluronic acid and heparin.

Hyaluronic acid (HA) is a natural polysaccharide based on a D-glucuronic acid and N-acetyl-D-glucosamine (Figure 1b) that is used in various biomedical applications, such as wound healing, visco-supplementation for wrinkle fillers, drug delivery carriers and tissue scaffolds. Furthermore, this polysaccharide is completely degraded in the body by hyaluronidase, in which the velocity of biodegradation is influenced by its molecular weight [9]. HA can interact intra/intermolecularly thanks to hydrogen bonds or ionic interactions by its carboxylic groups and their deprotonated form, carboxylates. It also can be easily modified by its carboxyl or hydroxyl groups. There are many examples of HA modified hydrogels for healing, such as hyaluronic acid modified with bromo acetate or those modified with polyhydrazides [10].

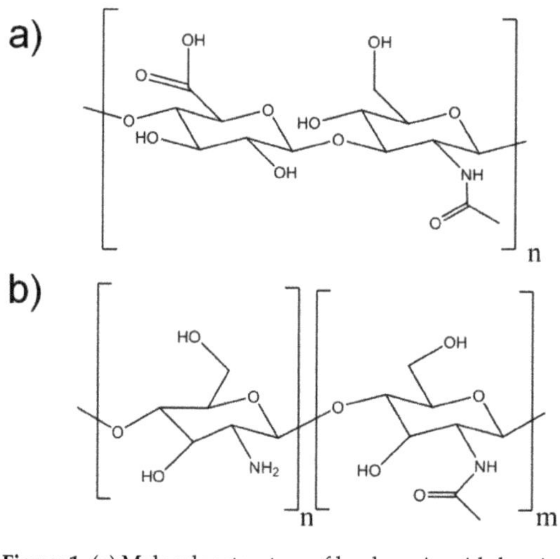

Figure 1. (a) Molecular structure of hyaluronic acid showing D-glucuronic acid (left) and a N-acetyl-D-glucosamine (right) units. (b) Molecular structure of chitosan monomers possessing D-glucosamine (left) and N-acetylglucosamine (right).

Chitosan (CHI) is also a natural polymer that comes from the partial deacetylation of chitin, a natural polymer synthesized by some arthropods, fungi and insects [11,12]. Thus, chitosan has a *D*-glucosamine structure mixed with *N*-acetylglucosamine structures for the acetylated monomer, as observed in Figure 1a. Chitosan is degraded in the human body by the action of lysozyme and colonic bacterial enzymes and its biodegradation strongly depends on its deacetylation degree and molecular weight [2]. It is considered one of the most promising materials in the fields of pharmacy, chemistry and the food industry due to its highly reactive hydroxyl and amino groups, as well as being a biocompatible, antibacterial and nontoxic polymer [13]. It is also noteworthy its ability to form films, which it is able to cause the suppression of essential nutrients for microbial growth, in other words protecting the open wound from the outside due to its good barrier properties [14].

Due to the presence of cited chemical groups in its structure, CHI is very good at interacting through hydrogen bonds, and by electrostatic interactions with negative charges at the appropriate pH due to the protonation of its amino groups ($-NH_3^+$) [2].

Since wound healing treatment requires prolonged time periods, the development of filmogenic materials with tissue adhesiveness, such as adhesive hydrogels, is crucial for a suitable performance on the skin. For this, acrylate derived hydrogels have been typically developed in the last decades based on their adhesive properties [11,15].

Tissue adhesion (Figure 2) is promoted by the interaction of tissues with many functional groups that are present along hydrogels polymeric chains through covalent bonds, such as imine formations or Schiff bases and Michael additions, among others. Moreover, tissue adhesion promoted by physical interactions such as hydrogen bonding is the most frequent. However, these interactions are reversible, which causes a decrease in the ability of the hydrogel to remain attached to tissues [16].

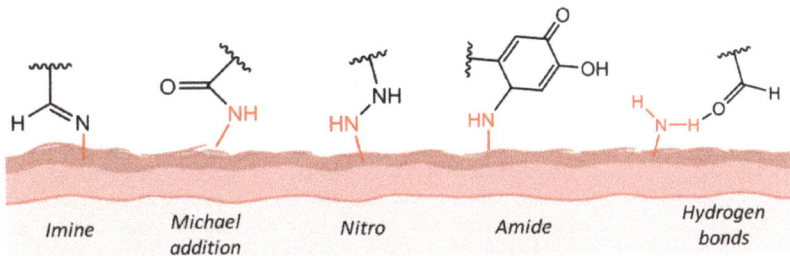

Figure 2. Intermolecular interactions between polymer chains and tissue.

In nature, the adhesive ability of mussels has been ascribed to the presence of an amino acid: L-3,4-dihydroxyphenylalanine (DOPA), which is responsible of their adhesion to both inorganic and organic surfaces, especially in humid conditions. This dihydroxy group is called catechol (CA) [17,18].

Taking the inspiration of these natural organisms, the strategy of modifying polymers with catechol groups has recently been developed to discover new materials, such as hydrogels with adhesive properties [12]. The derivatives of catechol are particularly interesting, as they are also natural and, therefore, biodegradable and biocompatible. The main advantage of natural polysaccharides chitosan and hyaluronic acid relies on the fact that they are easily modifiable through chemical reactions by their amine or carboxylic acid groups, respectively [9,19]. Moreover, the CA group can be oxidized at basic pH or even in the presence of the oxygen of the atmosphere [9], and it is transformed to o-quinone (Figure 3a). It can also be oxidized intentionally and at a higher rate with sodium periodate [20]. This spontaneously formed group behaves as a Michael acceptor and reacts with specific substrates, such as amines, thiols, alcohols, etc. [21,22].

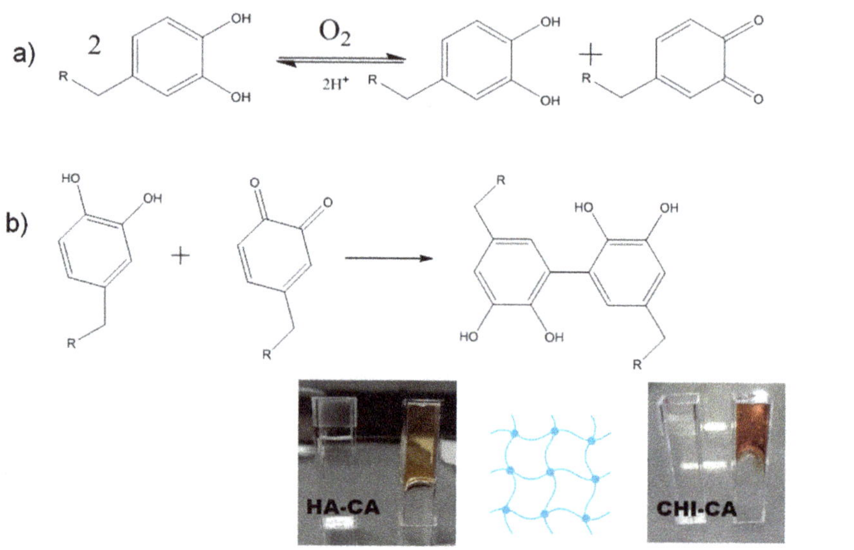

Figure 3. (**a**) Catechol group oxidation to quinone by atmospheric oxygen. (**b**) Crosslinking between catechol and quinone groups. R equals to CHI or HA.

In the case of organic surfaces that contain electron donor groups such as alcohols, thiols or amines, quinone reacts irreversibly, making more resistant covalent bonds than physical interactions of catechol [23]. Once CA is oxidized to quinone, the polymer that carries this substituent begins to react with itself (Figure 4) [22,23], causing its self-crosslinking, increasing viscosity, while a change of colour takes places and becomes brownish. This reaction results in a rapid hardening of the product (Figure 3b) that can be seen as an advantage, because it is a method for spontaneously promoting the gelation of the polymer, which proceeds from a viscous liquid state to a gelled state without the addition of any external crosslinking agents. However, to the best of our knowledge, the effect of this spontaneous gelation on the tissue adhesiveness and filmogenic properties of these polysaccharides has not been explored and comparatively analyzed.

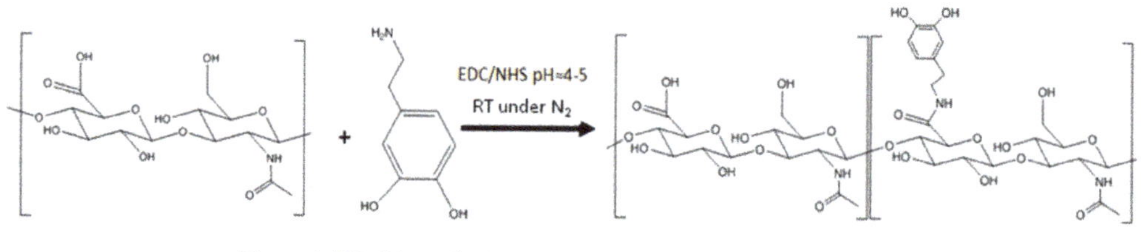

Figure 4. HA-CA synthesis reaction at pH = 4–5 at room temperature and under nitrogen atmosphere.

Taking all this into account, this work aims to explore the formation of hydrogels of catechol derivatives obtained by the chemical modification of chitosan and hyaluronic acid as tissue adhesive and filmogenic materials for potential wound healing purposes.

2. Materials and Methods

2.1. Materials

For the synthesis of the hydrogels, chitosan (CHI, $1.2 \times 10^6 \pm 153.9$ g/mol, Sigma-Aldrich, St. Louis, MO, USA; DD = 80%), hyaluronic acid (HA, $1.9–2.2 \times 10^6$ g/mol, Contripo, Dolní Dobrouč, Czech Republic), hydrochloric acid (HCl, 37%, Panreac, Barcelona,

Spain) and ethanol (EtOH, 99.8%, Panreac, Barcelona, Spain) as solvent were used. 3,4-Dihydroxycinnamic acid or hydrocaffeic acid (HCF, 98%, Sigma-Aldrich, St. Louis, MO, USA) and dopamine hydrochloride (DOPA, 98%, Sigma-Aldrich, St. Louis, MO, USA) were employed to introduce the catechol group. To carry out the conjugation of the catechol to the polymer, N-hydroxysuccinimide (NHS, 98%, Sigma-Aldrich), St. Louis, MO, USA and N-(3-Dimethylaminopropyl)-N'-ethylcarbodiimide hydrochloride (EDC, 98%, Sigma-Aldrich, St. Louis, MO, USA) were used. Subsequently, to clean the modified polymer, a dialysis was carried out with 12,000 Da membranes (Medicell Membranes Ltd., London, UK). The magnesium chloride salt ($MgCl_2$, 98%, Sigma-Aldrich) was used to control humidity in a closed atmosphere. Sodium metaperiodate ($NaIO_4$, 99%, Sigma-Aldrich, St. Louis, MO, USA) was used for catechol oxidation in spectroscopy calibration. In order to prepare a phosphate buffer saline or PBS, monobasic sodium phosphate (NaH_2PO_4, 99%, Sigma-Aldrich, St. Louis, MO, USA) and sodium hydroxide (NaOH, 99%, Panreac) were used. PET films (75 μm) were supplied by HIFI Film Industria (Stevenage, UK).

2.2. Experimental Synthesis

2.2.1. Synthesis of Hyaluronic Acid-Catechol (HA-CA)

The synthesis of hyaluronic acid with catechol was carried out following the described method [24]. Briefly, high molecular weight hyaluronic acid (1 g, 2.5 mmol) was dissolved in distilled water (200 mL) for 12 h and under a nitrogen atmosphere. EDC (959 mg and 5 mmol) and NHS (575 mg and 5 mmol) (Figure 4) were then slowly added to the reaction flask. After 20 min under stirring, dopamine hydrochloride (948 mg, 5 mmol) was added at pH 4–5 for 4 h. It was left to react overnight and dialyzed in 12,000–14,000 Da dyalisis membranes against acidified deionized water (pH 5) for 3 days. Finally, the product was lyophilized and stored in a vacuum desiccator at 3 °C.

2.2.2. Chitosan-Catechol Synthesis

The synthesis of chitosan modified with the catechol group was carried out following the described method [17]. Briefly, high molecular weight chitosan (591 mg and 1.6 mmol) was dissolved in 22.5 mL of water together with 2.5 mL of 1 M HCl overnight under a nitrogen atmosphere. The next day, hydrocaffeic acid (600 mg and 3.25 mmol), previously dissolved in 1.5 mL in distilled water, was added. Then, EDC (930 mg and 4.75 mmol) and NHS (558 mg and 4.75 mmol) (Figure 5), dissolved in 50 mL of an ethanol/water solution (1:1, v/v), were added. The reaction was left overnight, and the pH value was between 4 and 5. The product was dialyzed on 12,000–14,000 Da membranes in acidified deionized water (pH 5) for 3 days. Finally, the product was lyophilized (Benchtop Freeze Dryer operating at −50 °C, 0.1 mBar) and stored in a vacuum desiccator at 3 °C.

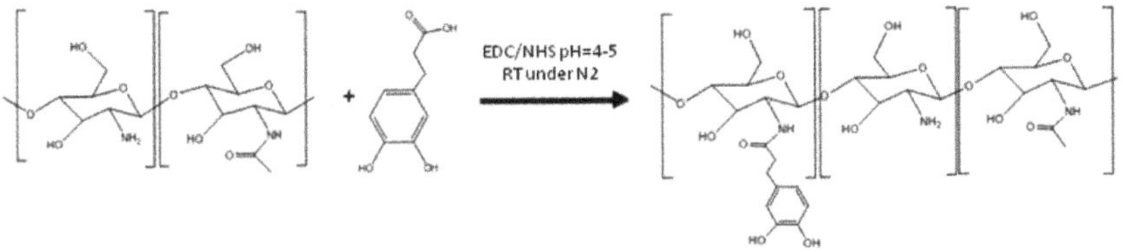

Figure 5. CHI-CA synthesis reaction at pH = 4–5 at room temperature under nitrogen atmosphere.

2.2.3. Films

HA-CA and CHI-CA films were prepared by using a doctor blade technique to form wet films with well-defined thicknesses from solutions at a concentration of 7 g/L in water at room temperature. Films that were 1-millimeter-thick were obtained onto the PET sheet.

2.3. Characterization Techniques

Proton nuclear magnetic resonance (^{1}H NMR) spectra were performed at room temperature on a Bruker AV-500 spectrometer (500 MHz for 1H), using deuterated acetic acid and water as solvents. Chemical shifts (δ) are expressed in parts per million with respect to deuterated water. The concentration of quinone group was determined by ultraviolet and visible spectroscopy (UV-VIS) measuring the absorbance at 414 nm, respectively, in the Double beam Cintra303 GBC equipment. The gelation time of prepared hydrogels was determined at different polysaccharide concentrations by the known inverted tube test [9], in which it is considered that the gelation point corresponds to the moment in which the solution stops flowing once inverting the tube. An inverted optical microscope Olympus IX71 from Japan was used as a non-destructive technique. Photographs were obtained in order to study the stability of the films during and after drying. A Hitachi S-4800 brand scanning electron microscope (FEG-SEM) from Japan was used in order to obtain high resolution images of the films at micron scale. In the case of polymers, a layer of gold was applied to allow the mobility of the electrons because they are not conductive. The adhesion of the synthesized hydrogels was determined by measuring the force necessary to detach gels from a piece of tissue with mechanical test equipment (Metrotec, MTEf), using a 20 N load cell. For this purpose, porcine skin without external fat was cut into circular sections of 196 mm^2 and kept for 4–5 h in a PBS solution (pH $\approx$ 7.4) at 37 °C to simulate physiological conditions [15,25]. Then, skin was fixed with cyanoacrylate (Loctite®) [26] to a test tube and placed on the surface of the gel sample. Finally, the force per area required to detach it from the sample was measured. The stress–displacement curves were obtained for each sample. All measurements were conducted with the following parameters: test speed: 2 mm/min; skin/sample contact time: 1 min; contact area: 196 mm^2; preload: 0 N; drop: 100%.

3. Results

3.1. Catechol Conjugation

CHI and HA were chemically modified, as described in the Experimental Section, in order to introduce catechol groups along polysaccharides chains to promote gelation and enhance adhesiveness to biological tissue. This conjugation was confirmed and quantified by ^{1}H NMR and UV analyses. Figure 6 compares the ^{1}H-NMR spectra of initial HA, dopamine hydrochloride reagent and the finally modified HA-CA.

^{1}H-NMR Hyaluronic acid (D$_2$O, 500 MHz, 20 °C): δ (ppm) = 4.30 (s, 2H, anomeric CH), 3.00–4.00 (m, 10H, ring CH and CH$_2$), 1.99 (s, 3H, acetamide).

^{1}H-NMR Dopamine Hydrochloride (D$_2$O, 500 MHz, 20 °C): δ (ppm) = 6.70 (m, 3H, CH aromatic ring), 3.15 (d, 2H, -CH$_2$N), 2.75 (d, 2H, -CH$_2$Ar).

^{1}H-NMR Hyaluronic acid-catechol (D$_2$O, 500 MHz, 20 °C): δ (ppm) = 6.75–7.30 (m, 3H, CH aromatic ring), 4.30 (s, 4H, anomeric CH), 3.00–4.00 (m, 20H, ring CH and CH$_2$), 2.82 (d, 2H, CH$_2$N), 2.80 (d, 2H, -CH$_2$Ar), 1.99 (s, 6H, acetamide).

The appearance of new peaks corresponding to the phenyl hydrogens of catechol moieties (Figure 6c) at 6.5–6.75 ppm and those appearing at 2.75 ppm ascribed to the aliphatic carbons of catechol [9] demonstrates the successful conjugation of HA with catechol functionality. In addition, the integration of the peaks at 6.5–6.75 ppm with respect to 1.99 ppm peak corresponding to the methyl protons of the acetamide group of HA allows the quantification of the percentage of introduced catechol groups, obtaining average substitution values of 38 $\pm$ 8%.

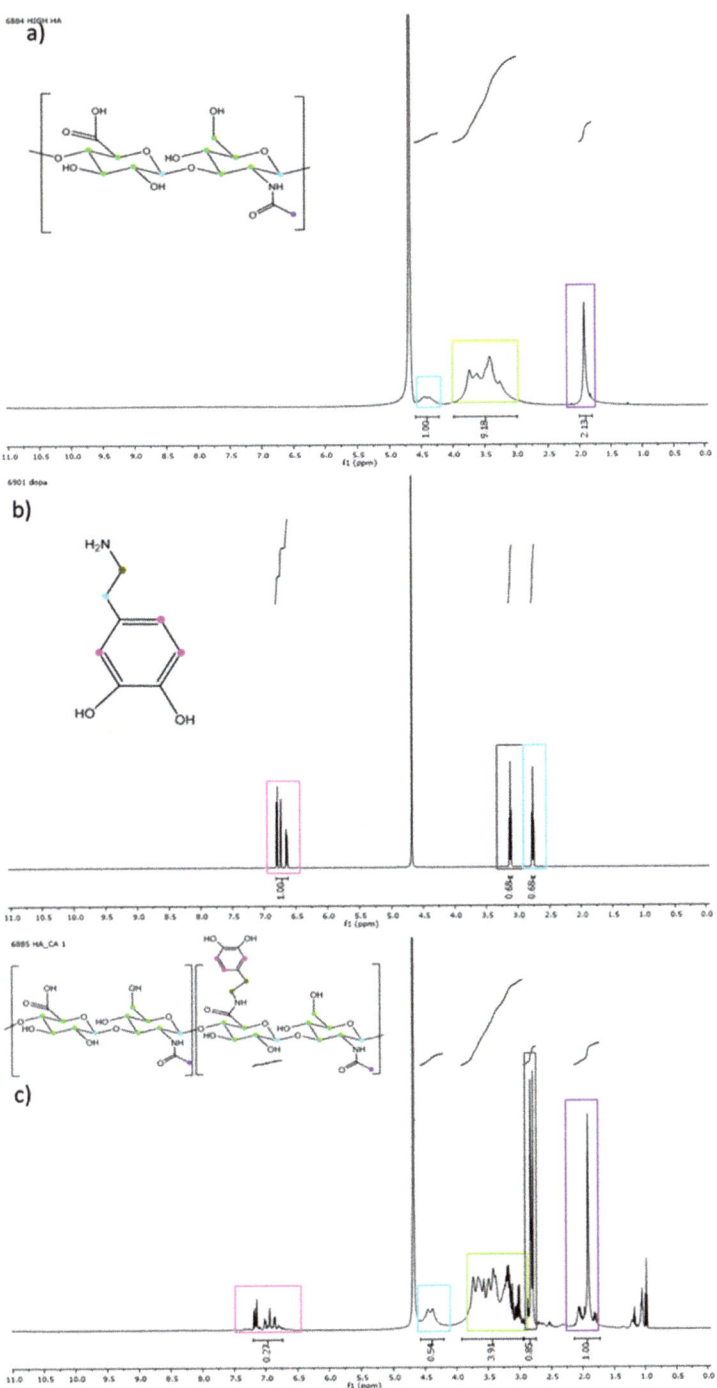

Figure 6. (**a**) High molecular weight hyaluronic acid spectrum. (**b**) Dopamine hydrochloride spectrum, a reagent that adds catechol to the product. (**c**) Hyaluronic acid modified with catechol (HA-CA) spectrum.

In the CH-CA spectrum (Figure 7c), the appearance of the characteristic peaks of catechol groups (6.5–6.75 ppm) can be observed, indicating that the reaction takes place successfully. In addition, the peaks at 2.5 ppm corresponding to the aliphatic carbons of hydrocaffeic acid and the appearance of the signal at 4 ppm, which corresponds to the hydrogen of the C2 of the glucosamine unit of the chitosan bound to catechol, were also observed. The percentage of substitution of catechol was calculated by the integration of the peak at 6.5–6.75 ppm with respect to that of chitosan appearing at 1.99 ppm, knowing the degree of deacetylation. Different synthesis conditions were explored in CHI modification. On the one hand, the following results were obtained: CHI-CA 1 with 24 h of reaction and 1:2 CHI:HCF feed ratio, CHI-CA 2 with 8 h of reaction and 1:2 CHI:HCF feed ratio and, finally, CHI-CA 3 with 12 h of reaction and 1:1 CHI:HC feed ratio. The resulting percentage of catechol varied according to these synthetic conditions (Table 1). Indeed, lower reagent equivalents (CHI-CA 3) and reaction time (CHI-CA 2) resulted in a significant decrease in the conjugation with catechol.

Table 1. Substitution percentages of catechol in the samples.

Sample	Catechol % (^{1}H NMR) [a]
HA-CA	38 ± 8
CHI-CA 1	82 ± 10
CHI-CA 2	8 ± 2
CHI-CA 3	2 ± 2

[a] $n = 3$.

1H-NMR Chitosan (D2O, 500 MHz, 20 °C): δ (ppm) = 4.50 (s, 2H, anomeric CH), 3.30–4.00 (m, 10H, ring CH and CH2), 3.10 (s, 2H, CH -N ring), 1.99 (s, 3H, acetamide).

1H-NMR Hydrocaffeic acid (D2O, 500 MHz, 20 °C): δ (ppm) = 6.5–6.75 (m, 3H, CH of the aromatic ring), 2.60 (d, 2H, CH2COOH), 2.50 (d, 2H, CH2Ar).

1H-NMR Chitosan-catechol (D2O, 500 MHz, 20 °C): δ (ppm) = 6.5–6.75 (m, 3H, CH of the aromatic ring), 4.50 (s, 3H, anomeric CH), 4.20 (dd, 1H, CH-N (catechol)), 3.25–3.80 (m, 15H, ring CH and CH2), 3.10 (s, 2H, CH-N), 2.60 (d, 2H, CH2COOH), 2.30 (d, 2H, CH2Ar), 1.99 (s, 3H, acetamide).

3.2. Hydrogel Formation of Catechol Derivatives

The spontaneous oxidation of the catechol group leads to their transformation to the quinone group that presents an absorption in the visible spectrum at λ = 380–480 nm (depending on the degree of oxidation) [27]. Accordingly, the quantification of quinone moiety was carried out by using the calibration curve obtained with a standard solution of 1 mM dopamine hydrochloride previously oxidized with sodium periodate (1:1 Dopa/Periodate). When periodate was added to the dopamine solution, it immediately took on a yellow hue, and after 10 min, it became reddish and brown, since the absorption spectrum varies with oxidation time. For this reason, the calibration was carried out at the isosbestic point, at which absorption is not a function of time [27] (λ = 413.6 nm, Abs = 1112C + 0.009R^2 = 0.991). This color change allows monitoring the oxidation of catechol-modified polysaccharides.

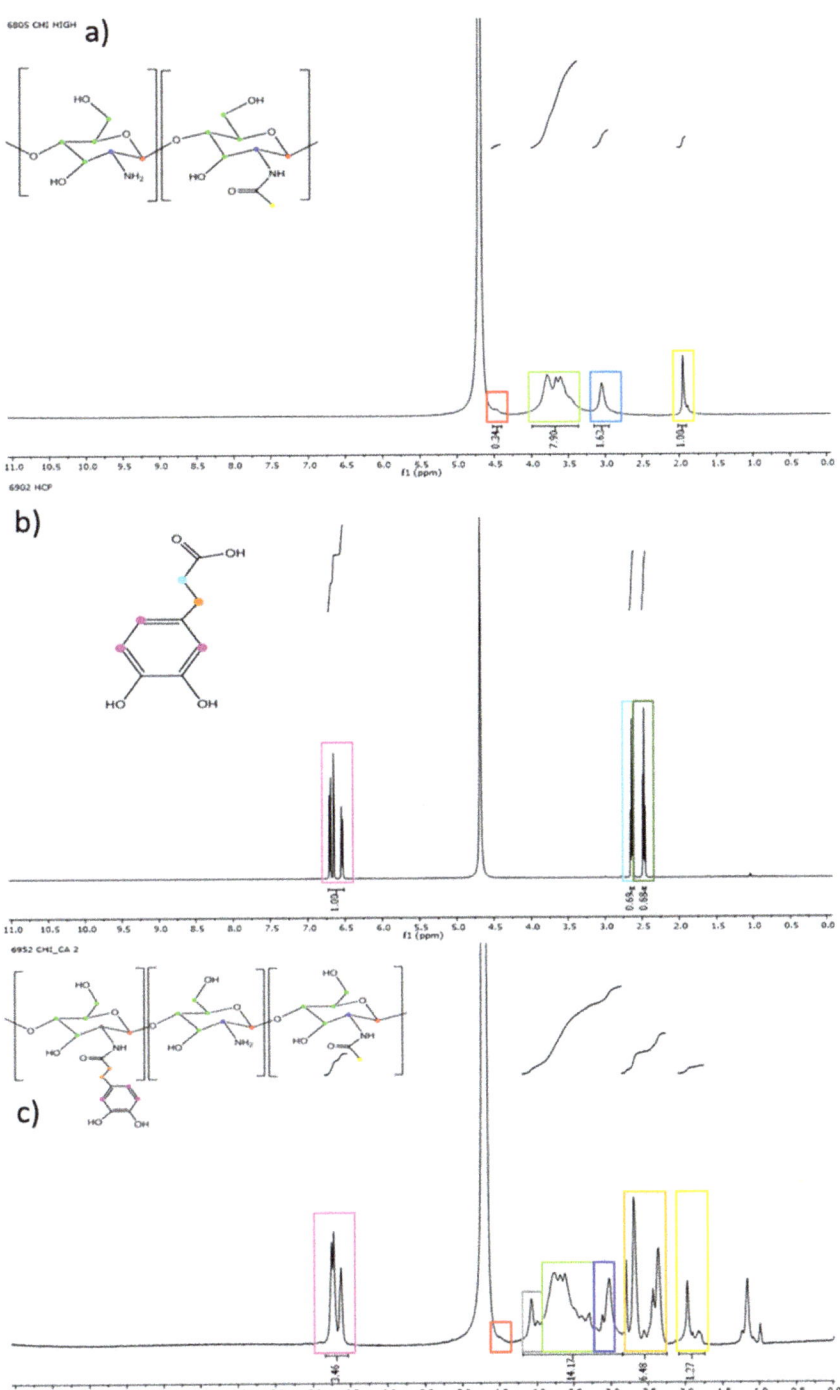

Figure 7. (**a**) High molecular weight chitosan spectrum. (**b**) Hydrocaffeic acid spectrum. (**c**) Chitosan modified with catechol (CHI-CA) spectrum.

It is known that this oxidation of catecholized polymers solution with air [9,12,23] causes sol–gel transition that changes adhesion to skin. Taking all this into account, the oxidation of HA-CA and CHI-CA was analyzed along the time for different polymer concentrations in terms of the variation of tissue adhesion and quinone concentration (Figures 8 and 9). As expected, as the time of exposure to air increases, as a consequence of the appearance of the oxidized species, o-quinone, the increase in coloration took place (Figure 8) and the characteristic absorption band at 413.6 nm was observed for both polymers. It is also shown that higher concentration of catechol-derived polymer in the solution leads to a greater quinone concentration. For instance, in the case of 11 g/L of HA-CA, the presence of quinone during the first 30 min is four times higher than that of 4.6 g/L solution. In addition, an increase in viscosity was observed, caused by the gelling of the polymers through the formation of covalent bonds between quinones, originating a covalent three-dimensional network that results in the formation of the hydrogel [22]. This spontaneous formation of the gels can be considered a great advantage since, as mentioned above, it allows obtaining cross-linked systems without additional reactions or reagents. The gelation time, determined by the vial inversion method [9], is indicated in Figure 8 (yellow star) for all studied concentrations.

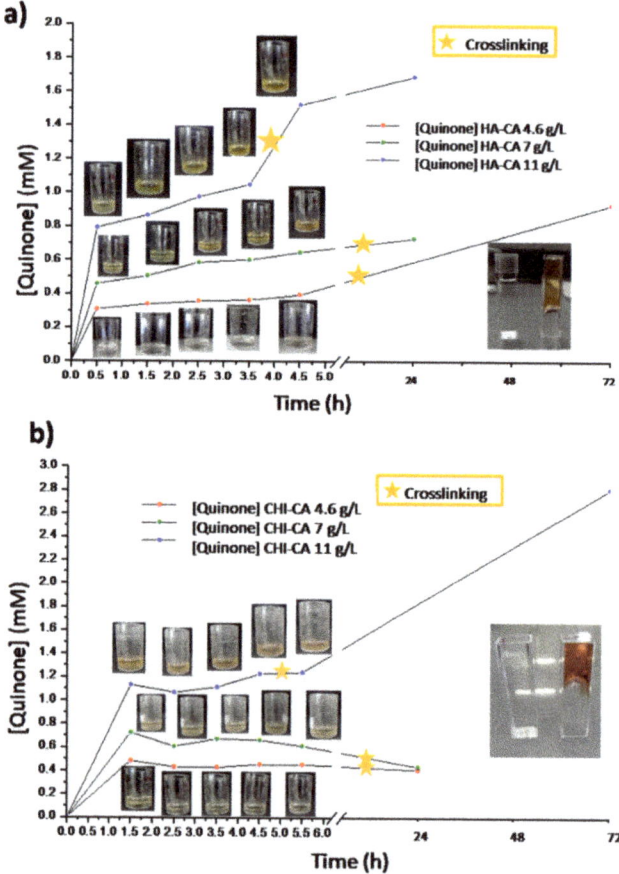

Figure 8. Quinone concentration over the time in (**a**) HA-CA and (**b**) CHI-CA samples calculated from calibration. In addition, a yellow star indicates the gelation time determined by the vial inversion method.

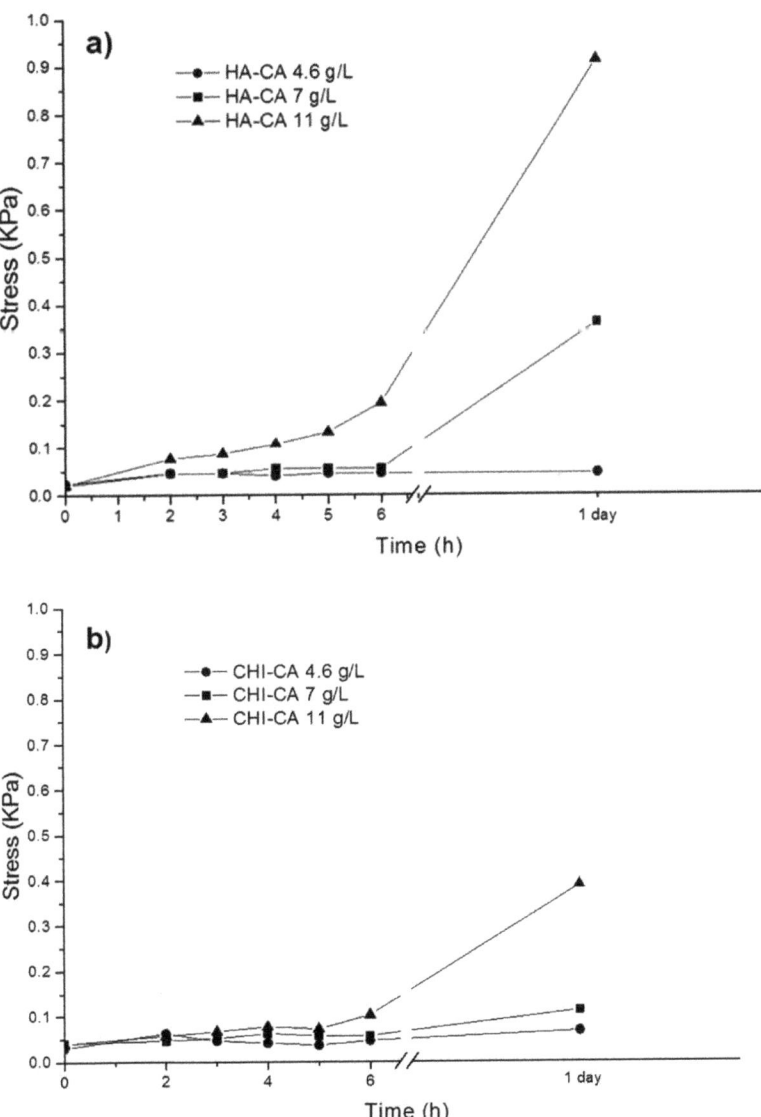

Figure 9. Skin adhesion graph of (**a**) HA-CA and (**b**) CHI-CA over time at different concentrations.

Regarding CHI-CA, it is worth highlighting that the quinone concentration is higher than that for HA-CA for all studied concentrations. This difference can be ascribed to the higher degree of catechol substitution in the CHI-CA (82%) in comparison with HA-CA (38%). Tissue adhesion tests were also carried out at different modified polysaccharide concentrations (4.6, 7, 11 g/L). The stress–displacement curves were obtained, and the maximum stress points required for tissue-polymer detachment were analyzed (Figure 9).

It could be observed (Figure 9) for both polymers that as the concentration of initial catechol groups and oxidation time increases, where quinone content increases, a greater detachment force is measured. As it is known, in organic surfaces such as porcine skin, which contains thiol and amino groups [12], quinone is more adhesive than catechol [17] due to the covalent nature of the formed quinone bonds compared with the weaker physical interactions established by catechol groups [9,12,21]. When HA-CA and CHI-CA are compared, it can be highlighted that a higher adhesion was measured for HA-CA than CHI-CA samples, despite the higher catechol content of CHI-CA samples. This is because CHI-CA undergoes faster and additional self-crosslinking [12]. Moreover, in addition to intramolecular reactions between quinones, the amine group from the deacetylated CHI segments also reacts through Michael additions with quinone [22]. Thus, there is a lower content of quinone moieties available to tissue interaction than in the case of the HA-CA system. Additionally, hydration is a key factor in adhesion, as it enhances the mobility of the polymer chains that promotes tissue adhesion [28]. In this sense, HA is one of the most hydrating polymers known [29,30], which can interestingly enhance the tissue adhesion of HA-CA gels.

Figure 10 shows specifically the stress applied at the detachment (Figure 10a) and the displacement produced by the gels before breakage (Figure 10b) for both catechol modified polysaccharides after 4 and 24 h of oxidation by air exposure. As it is observed in the stress graphs (Figure 10a), the adhesion of HA-CA is greater than that of CHI-CA even though a greater content of quinone was determined by VIS spectroscopy.

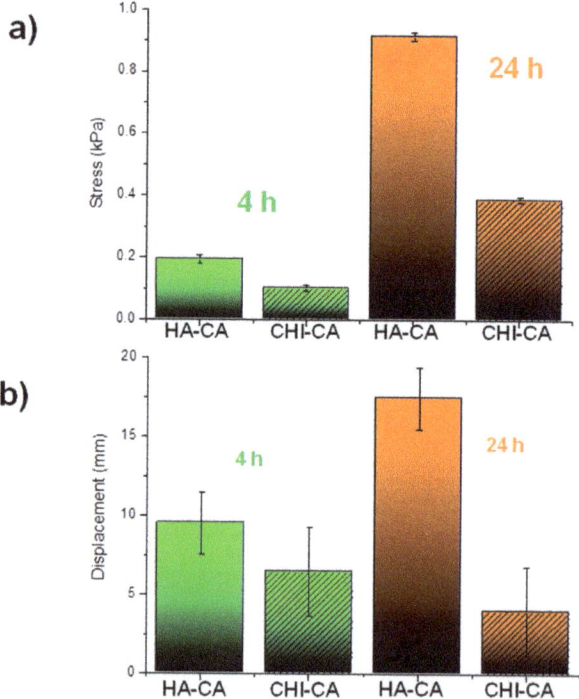

Figure 10. Comparative graph between the studied systems that gathers (**a**) the maximum stress applied to detach the sample from the skin and (**b**) the maximum elongations until rupture of the systems studied after 4 h (green) and after 24 h (orange) for HA-CA and CHI-CA hydrogels. $n = 5$.

As it is observed in the stress graphs (Figure 10a), the adhesion of HA-CA is greater than that of CHI-CA even though there is a greater content of quinone determined by VIS spectroscopy. In addition to this, it is known that CHI hydrates more slowly than HA [29,31], which has a negative influence on adhesion, as we have already mentioned. Indeed, hydration, which has the function of a lubricant, seems to increases the mobility of HA chains in contrast to CHI, which improves adhesion properties.

3.3. Microscopy Analysis of the Film

The filmogenic ability of wound dressing materials is a valuable property due to the fact that it promotes wound protection and reduces bacterial growth [14]. For this reason, the films of the studied systems were developed by casting CA-modified polysaccharide solution, as described in the experimental section. After 4 h (Figure 11a,b) and 24 h (Figure 11c,d) of drying, photographs were taken under the light microscope to HA-CA and CHI-CA hydrogel films in order to study the possible formation of fractures and the homogeneity of the films as a non-destructive technique. High resolution images were also obtained by scanning electron microscopy (SEM) for HA-CA (Figure 11e) and CHI-CA (Figure 11f) after 24 h of drying for more information on the micron scale [32].

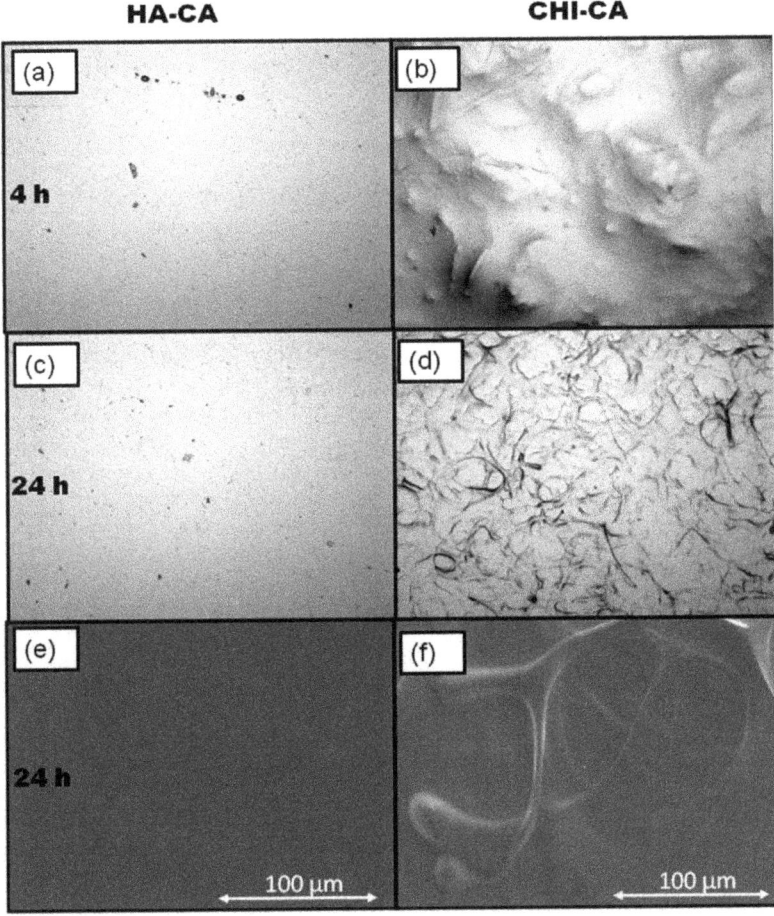

Figure 11. (a) HA-CA photographs taken with microscope after 4 h with (×4 lens and 1.62 µm/pixel) and (b) CHI-CA photographs taken with microscope after 4 h with (×4 lens and 1.62 µm/pixel) (c) HA-CA photographs taken with microscope after 24 h (×4 lens and 1.62 µm/pixel) (d) CHI-CA photographs taken with microscope after 24 h of drying with ×4 lens (1.62 µm/pixel) (e) HA-CA SEM images after 24 h of drying and (f) CHI-CA SEM images after 24 h of drying.

After 4 h of drying, noticeable differences between some systems can be appreciated. In the case of HA-CA films (Figure 11a), photographs indicate high homogeneity of the film without cracks or wrinkles. However, for CHI-CA films (Figure 11b), high roughness can be observed, probably due to lower solubility in the water of polymers. In addition, its poor filmogenic capacity is observed as a consequence of its high drop forming ability during the casting process that is indicative of the existence of important cohesive forces [32], which corresponds to those derived from highly crosslinked CHI-CA (quinone-NH2) [12,22].

After 24 h of drying, the HA-CA hydrogels do not display difference with 4 h of drying (Figure 11c), while signs of increased crosslinking were observed as roughness or irregularities due to the formation of small aggregates in the case of CHI-CA. Indeed, CHI-CA (Figure 11d) shows fiber-like structures as a consequence of solvent evaporation from its previous gelled aggregates. The rapid crosslinking of CHI-CA seems to decrease the homogeneity of the film, and high cohesion forces do not allow smooth surface formation.

Finally, SEM images taken after 24 h of drying confirmed that highly stable films are obtained for HA-CA gels (Figure 11e). However, regarding CHI-CA (Figure 11f), fibers remained after the total elimination of the solvent in high resolution due to its poor ability to spread, high cohesion and crosslinking forces [22].

4. Conclusions

The conjugation of CHI and HA with catechol groups was successfully developed and quantified by ^{1}H-NMR spectroscopy. The spontaneous oxidation of introduced groups to o-quinone by the action of air resulted in hydrogel formation that could be easily observed by the change of color along the polymeric chains where crosslinking took place. Higher catechol content, derived from higher polymer concentration as well as higher modification degree, resulted in greater quinone concentrations that resulted in faster gelation for both polymers. In addition, it was demonstrated that gelation induces a clear variation on tissue adhesiveness of these catecholized polymers. As HA-CA and CHI-CA air induced gelation, an increase in tissue detachment force was measured in stress–strain curves due to the stronger interactions of quinone groups with tissue in comparison with initial catechol moiety. CHI-CA spontaneous hydrogels showed reduced adhesiveness in comparison with HA-CA, even though its initial catechol content was greater higher due to the additional reaction of quinone groups with free amine groups present along CHI polymer. In addition, the analysis of the comparative filmogenic ability of CHI-CA and HA-CA points out a lower homogeneity of CHI-CA, possibly ascribed to the lower solubility of this conjugate, as well as for its additional crosslinking and consequent viscosity. The spontaneous gelation of HA-CA and CHI-CA allows obtaining highly cross-linked systems without additional reactions or reagents, which in the case of HA-CA interestingly show also excellent filmogenic properties and n enhanced tissue-adhesive ability.

Author Contributions: Conceptualization, R.P.-G., O.G. and L.P.-Á.; methodology, G.C.-C., L.R.-R. and L.P.-Á.; formal analysis, G.C.-C. and L.P.-Á.; investigation, G.C.-C.; writing—original draft preparation, G.C.-C. and L.P.-Á.; writing—review and editing, V.S.-M. and L.R.-R.; supervision, L.P.-Á.; funding acquisition, A.H.-C. and V.S.-M. All authors have read and agreed to the published version of the manuscript.

Funding: This research was funding by Basque Government (ELKARTEK program, Department of Development and Infrastructures of the Basque Country, KK-2021-00040), University of the Basque Country UPV/EHU (GIU 207075), Ministry of Economy, Industry and Competitiveness (grant MAT2017-89553-P), CDTI of the Ministry of Science and Innovation (Spain) (GAMMAREGEN INNO-20182003) and i+Med S. Coop.

Data Availability Statement: The data presented in this study are available upon request from the corresponding author.

Acknowledgments: Technical and human support provided by SGIker (UPV/EHU, MICINN, GV/EJ, EGEF and ESF) is gratefully acknowledged.

Conflicts of Interest: The authors declare no conflict of interest.

References

1. Ullah, F.; Othman, M.B.H.; Javed, F.; Ahmad, Z.; Akil, H.M. Classification, processing and application of hydrogels: A review. *Mater. Sci. Eng. C* **2015**, *57*, 414–433. [CrossRef] [PubMed]
2. Hamedi, H.; Moradi, S.; Hudson, S.M.; Tonelli, A.E. Chitosan based hydrogels and their applications for drug delivery in wound dressings: A review. *Carbohydr. Polym.* **2018**, *199*, 445–460. [CrossRef]
3. Geng, H. A one-step approach to make cellulose-based hydrogels of various transparency and swelling degrees. *Carbohydr. Polym.* **2018**, *186*, 208–216. [CrossRef] [PubMed]
4. Horkay, F.; Tasaki, I.; Basser, P.J. Osmotic swelling of polyacrylate hydrogels in physiological salt solutions. *Biomacromolecules* **2000**, *1*, 84–90. [CrossRef] [PubMed]
5. Yang, Z.; Peng, H.; Wang, W.; Liu, T. Crystallization behavior of poly(ε-caprolactone)/layered double hydroxide nanocomposites. *J. Appl. Polym. Sci.* **2010**, *116*, 2658–2667. [CrossRef]
6. Hoare, T.R.; Kohane, D.S. Hydrogels in drug delivery: Progress and challenges. *Polymer* **2008**, *49*, 1993–2007. [CrossRef]
7. Min, J.G.; Sanchez Rangel, U.J.; Franklin, A.; Oda, H.; Wang, Z.; Chang, J.; Foxa, P.M. Topical antibiotic elution in a collagen-rich hydrogel successfully inhibits bacterial growth and biofilm formation in vitro. *Antimicrob. Agents Chemother.* **2020**, *64*, e00136-20. [CrossRef]
8. Midwood, K.S.; Williams, L.V.; Schwarzbauer, J.E. Tissue repair and the dynamics of the extracellular matrix. *Int. J. Biochem. Cell Biol.* **2004**, *36*, 1031–1037. [CrossRef]
9. Kim, J.; Lee, C.; Ryu, J.H. Adhesive catechol-conjugated hyaluronic acid for biomedical applications: A mini review. *Appl. Sci.* **2021**, *11*, 21. [CrossRef]
10. Dovedytis, M.; Liu, Z.J.; Bartlett, S. Hyaluronic acid and its biomedical applications: A review. *Eng. Regen.* **2020**, *1*, 102–113. [CrossRef]
11. Borzacchiello, A.; Ambrosio, L.; Netti, P.A.; Nicolais, L.; Peniche, C.; Gallardo, A.; San Roman, J. Chitosan-based hydrogels: Synthesis and characterization. *J. Mater. Sci. Mater. Med.* **2001**, *12*, 861–864. [CrossRef] [PubMed]
12. Ryu, J.H.; Hong, S.; Lee, H. Bio-inspired adhesive catechol-conjugated chitosan for biomedical applications: A mini review. *Acta Biomater.* **2015**, *27*, 101–115. [CrossRef]
13. Guo, Z.; Ni, K.; Wei, D.; Ren, Y. Fe^{3+}-induced oxidation and coordination cross-linking in catechol-chitosan hydrogels under acidic pH conditions. *RSC Adv.* **2015**, *5*, 37377–37384. [CrossRef]
14. Fernández-de Castro, L.; Mengíbar, M.; Sánchez, Á.; Arroyo, L.; Villarán, M.C.; Díaz de Apodaca, E.; Heras, Á. Films of chitosan and chitosan-oligosaccharide neutralized and thermally treated: Effects on its antibacterial and other activities. *LWT—Food Sci. Technol.* **2016**, *73*, 368–374. [CrossRef]
15. Cook, M.T.; Khutoryanskiy, V.V. Mucoadhesion and mucosa-mimetic materials—A mini-review. *Int. J. Pharm.* **2015**, *495*, 991–998. [CrossRef]
16. Guo, Z.; Mi, S.; Sun, W. A Facile Strategy for Preparing Tough, Self-Healing Double-Network Hyaluronic Acid Hydrogels Inspired by Mussel Cuticles. *Macromol. Mater. Eng.* **2019**, *304*, 1800715. [CrossRef]
17. Kim, K.; Kim, K.; Ryu, J.H.; Lee, H. Chitosan-catechol: A polymer with long-lasting mucoadhesive properties. *Biomaterials* **2015**, *52*, 161–170. [CrossRef] [PubMed]
18. Gao, Z.; Li, Y.; Shang, X.; Hu, W.; Gao, G.; Duan, L. Bio-inspired adhesive and self-healing hydrogels as flexible strain sensors for monitoring human activities. *Mater. Sci. Eng. C* **2020**, *106*, 110168. [CrossRef]
19. Almeida, A.C.; Vale, A.C.; Reis, R.L.; Alves, N.M. Bioactive and adhesive properties of multilayered coatings based on catechol-functionalized chitosan/hyaluronic acid and bioactive glass nanoparticles. *Int. J. Biol. Macromol.* **2020**, *157*, 119–134. [CrossRef] [PubMed]
20. Peng, X.; Peng, Y.; Han, B.; Liu, W.; Zhang, F.; Linhardt, R.J. IO_4^--stimulated crosslinking of catechol-conjugated hydroxyethyl chitosan as a tissue adhesive. *J. Biomed. Mater. Res.—Part B Appl. Biomater.* **2019**, *107*, 582–593. [CrossRef] [PubMed]
21. Park, M.K.; Li, M.X.; Yeo, I.; Jung, J.; Yoon, B.I.L.; Joung, Y.K. Balanced adhesion and cohesion of chitosan matrices by conjugation and oxidation of catechol for high-performance surgical adhesives. *Carbohydr. Polym.* **2020**, *248*, 116760. [CrossRef] [PubMed]
22. Yang, J.; Cohen Stuart, M.A.; Kamperman, M. Jack of all trades: Versatile catechol crosslinking mechanisms. *Chem. Soc. Rev.* **2014**, *43*, 8271–8298. [CrossRef] [PubMed]
23. Narkar, A.R.; Barker, B.; Clisch, M.; Jiang, J.; Lee, B.P. PH Responsive and Oxidation Resistant Wet Adhesive based on Reversible Catechol-Boronate Complexation. *Chem. Mater.* **2016**, *28*, 5432–5439. [CrossRef] [PubMed]
24. Joo, H.; Byun, E.; Lee, M.; Hong, Y.; Lee, H.; Kim, P. Journal of Industrial and Engineering Chemistry Biofunctionalization via flow shear stress resistant adhesive polysaccharide, hyaluronic acid-catechol, for enhanced in vitro endothelialization. *J. Ind. Eng. Chem.* **2016**, *34*, 14–20. [CrossRef]
25. Xu, J.; Strandman, S.; Zhu, J.X.X.; Barralet, J.; Cerruti, M. Genipin-crosslinked catechol-chitosan mucoadhesive hydrogels for buccal drug delivery. *Biomaterials* **2015**, *37*, 395–404. [CrossRef]
26. Woertz, C.; Preis, M.; Breitkreutz, J.; Kleinebudde, P. Assessment of test methods evaluating mucoadhesive polymers and dosage forms: An overview. *Eur. J. Pharm. Biopharm.* **2013**, *85*, 843–853. [CrossRef] [PubMed]
27. Muñoz, J.L.; García-Molina, F.; Varón, R.; Rodriguez-Lopez, J.N.; García-Cánovas, F.; Tudela, J. Calculating molar absorptivities for quinones: Application to the measurement of tyrosinase activity. *Anal. Biochem.* **2006**, *351*, 128–138. [CrossRef] [PubMed]

28. Vorvolakos, K.; Isayeva, I.S.; do Luu, H.M.; Patwardhan, D.V.; Pollack, S.K. Ionically cross-linked hyaluronic acid: Wetting, lubrication, and viscoelasticity of a modified adhesion barrier gel. *Med. Devices Evid. Res.* **2011**, *4*, 1. [CrossRef] [PubMed]
29. Smejkalova, D.; Huerta-Angeles, G.; Ehlova, T. Hyaluronan (Hyaluronic Acid): A Natural Moisturizer for Skin Care. In *Harry's Cosmeticology*, 9th ed.; Chemical Publishing Company: Gloucester, MA, USA, 2015; Volume 2, pp. 605–622.
30. Olejnik, A.; Goscianska, J.; Zielinska, A.; Nowak, I. Stability determination of the formulations containing hyaluronic acid. *Int. J. Cosmet. Sci.* **2015**, *37*, 401–407. [CrossRef]
31. Lim, S.T.; Martin, G.P.; Berry, D.J.; Brown, M.B. Preparation and evaluation of the in vitro drug release properties and mucoadhesion of novel microspheres of hyaluronic acid and chitosan. *J. Control. Release* **2000**, *66*, 281–292. [CrossRef]
32. Cao, W.; Yan, J.; Liu, C.; Zhang, J.; Wang, H.; Gao, X.; Yan, H.; Niu, B.; Li, W. Preparation and characterization of catechol-grafted chitosan/gelatin/modified chitosan-AgNP blend films. *Carbohydr. Polym.* **2020**, *247*, 116643. [CrossRef] [PubMed]

Article

Slow-Release Nitrogen Fertilizers with Biodegradable Poly(3-hydroxybutyrate) Coating: Their Effect on the Growth of Maize and the Dynamics of N Release in Soil

Soňa Kontárová [1,*], Radek Přikryl [1], Petr Škarpa [2], Tomáš Kriška [2], Jiří Antošovský [2], Zuzana Gregušková [1], Silvestr Figalla [1], Vojtěch Jašek [1], Marek Sedlmajer [1], Přemysl Menčík [1] and Mária Mikolajová [3]

1. Institute of Materials Chemistry, Faculty of Chemistry, Brno University of Technology, 61200 Brno, Czech Republic
2. Department of Agrochemistry, Soil Science, Microbiology and Plant Nutrition, Mendel University in Brno, 61200 Brno, Czech Republic
3. Institute of Natural and Synthetic Polymers, Faculty of Chemical and Food Technology, Slovak University of Technology in Bratislava, Radlinského 9, 812 37 Bratislava, Slovakia
* Correspondence: kontarova@fch.vut.cz

Citation: Kontárová, S.; Přikryl, R.; Škarpa, P.; Kriška, T.; Antošovský, J.; Gregušková, Z.; Figalla, S.; Jašek, V.; Sedlmajer, M.; Menčík, P.; et al. Slow-Release Nitrogen Fertilizers with Biodegradable Poly(3-hydroxybutyrate) Coating: Their Effect on the Growth of Maize and the Dynamics of N Release in Soil. *Polymers* 2022, 14, 4323. https://doi.org/10.3390/polym14204323

Academic Editors: Gabriella Santagata, Arash Moeini and Pierfrancesco Cerruti

Received: 22 September 2022
Accepted: 12 October 2022
Published: 14 October 2022

Publisher's Note: MDPI stays neutral with regard to jurisdictional claims in published maps and institutional affiliations.

Copyright: © 2022 by the authors. Licensee MDPI, Basel, Switzerland. This article is an open access article distributed under the terms and conditions of the Creative Commons Attribution (CC BY) license (https://creativecommons.org/licenses/by/4.0/).

Abstract: Fertilizers play an essential role in agriculture due to the rising food demand. However, high input fertilizer concentration and the non-controlled leaching of nutrients cause an unwanted increase in reactive, unassimilated nitrogen and induce environmental pollution. This paper investigates the preparation and properties of slow-release fertilizer with fully biodegradable poly(3-hydroxybutyrate) coating that releases nitrogen gradually and is not a pollutant for soil. Nitrogen fertilizer (calcium ammonium nitrate) was pelletized with selected filler materials (poly(3-hydroxybutyrate), struvite, dried biomass). Pellets were coated with a solution of poly(3-hydroxybutyrate) in dioxolane that formed a high-quality and thin polymer coating. Coated pellets were tested in aqueous and soil environments. Some coated pellets showed excellent resistance even after 76 days in water, where only 20% of the ammonium nitrate was released. Pot experiments in Mitscherlich vegetation vessels monitored the effect of the application of coated fertilizers on the development and growth of maize and the dynamics of N release in the soil. We found that the use of our coated fertilizers in maize nutrition is a suitable way to supply nutrients to plants concerning their needs and that the poly(3-hydroxybutyrate) that was used for the coating does not adversely affect the growth of maize plants.

Keywords: slow-release; nitrogen fertilizers; coating; poly(3-hydroxybutyrate); ammonium nitrate; biodegradable; dioxolane; biomass; control-release; maize

1. Introduction

The importance of fertilizers for the human population is indisputable. Without the Haber–Bosch synthesis of ammonia (commercialized in 1913), and the subsequent synthetic nitrogen compounds and their applications, we would not be able to produce more than half of today's world food [1]. Due to the rapid growth of the population by 2–3 million people in the next 80 years, more crop production will be needed, which will also increase the need for fertilizers [2,3]. Fertilizers provide plants with fundamental macronutrients (nitrogen/phosphorus/potassium), where nitrogen is essential to plant growth [4]. However, crops' poor absorption of nutrients from fertilizers and the related losses of nitrogen in the soil, water, and air caused a cascade of environmental problems [5]. Despite the fact that the use of nitrogen fertilizers plays an essential role in meeting the demand for crop production, nitrogen use efficiency is relatively low due to their excessive use (in general between 25 and 50%), which often leads to losses of redundant nitrogen from agroecosystems [6]. Environmental problems include eutrophication and the expansion of dead zones in coastal ocean waters resulting from the leaching of nitrates into rivers, lakes, ponds, and ground waters. The atmospheric deposition of ammonia and nitrates affects natural ecosystems; nitrous oxide (N_2O) is now the third most crucial greenhouse gas, following CO_2 and CH_4 [1,4]. The excessive use of fertilizers can cause soil acidification, depletion of soil cations, reduction of carbon uptake, and contamination with heavy metals. All this negatively influences animal and human health, the environment, biodiversity, and the climate [7,8]. Even more significant losses of nitrogen in the environment (that will occur with a growing population) could be reduced by more moderate meat consumption, low-protein animal feeding, better agronomic management, and a portfolio that includes split applications and balanced use of fertilizers, precision farming, optimized crop rotations, and the use of expensive slow-release compounds [1].

Controlled- and slow-release fertilizers (CRFs and SRFs) provide a more efficient, economical, and safe way to deliver nutrients to plants. They are able to retain nutrients in the soil for a longer period, as they are available for the plants at the desired rate or concentration level. Thus, nutrient utilization efficiency (NUE) improves due to less frequent dosing and reduced nutrient removal from the soil by rain or irrigation, which also reduces environmental hazards [4,9–12]. Slow- or controlled-release fertilizers are a good alternative to conventional mineral fertilizers, and their use allows for a reduction of the fertilizer rate by 20% or 30% of the recommended value to achieve the same yield [13]. Slow-release fertilizers are described as "low solubility compounds with a complex/high molecular weight chemical structure that releases nutrients through either microbial or chemically decomposable compound" [4,14]. Controlled-release fertilizers (CRFs) can be described as "products containing sources of water-soluble nutrients, the release of which in the soil is controlled by a coating applied to the fertilizer" [4,15]. SRFs are usually classified into "condensation products of urea-aldehydes, fertilizers with a physical barrier (coated or incorporated into the matrix), and super granules. CRF is a subset of SRF, which falls under the category of fertilizer with a physical barrier" [4].

Current commercial CRFs with polymer coatings are often made of a thermoplastic resin such as polyolefin, polyvinylidene chloride, and copolymers, which cannot degrade easily in soil and accumulate over time. Generally, these CRFs containing synthetic, non-biodegradable polymers accumulate up to 50 kg/ha per year in the soil after releasing their nutrients, remaining and causing white pollution [4,16]. With the agreement of the European Green Deal, EU Fertilizing Products Regulation (FPR), and Circular Economy Action Plan, manufacturers are faced with significant challenges; they will have to adapt their current practices to the new FPR requirements, including biodegradability criteria for polymer coatings of controlled-release fertilizers [4,17].

The present study aimed to prepare CRFs with a coating of a biodegradable nature. The pelletization of powder fertilizer ammonium nitrate with dolomite (CAN) with selected additive materials (poly(3-hydroxybutyrate) P3HB, struvite, dried biomass containing P3HB) has become the most technologically feasible. Using struvite and biomass from P3HB extraction meets the requirements of the circular economy. P3HB can be made from waste cooking oil and a variety of byproducts [18–21]. However, pellets alone without a non-defective coating would not meet the requirements of slow release. A 6% solution of fully biodegradable P3HB with green solvent dioxolane [22] was applied to the pellets by 6-fold immersion [23] and produced a good quality coating.

The kinetics of fertilizer release from coated pellets in an aqueous environment was monitored by the conductometry method. The results showed the high quality of the prepared water-barrier polymer coating. The dynamics of mineral nitrogen release of these CRFs were studied under laboratory conditions, followed by an evaluation of the effectiveness of the newly developed coated fertilizers on maize growth and an assessment of their effects on soil mineral nitrogen content. We assumed that the coated CAN-based fertilizers would reduce the environmental load in the soil without simultaneously limiting plant growth.

2. Materials and Methods

2.1. Preparation of Coated and Encapsulated Fertilizers

2.1.1. The Fertilizer

The fertilizer used in pellets was calcium ammonium nitrate (CAN) supplied by Lovochemie, a.s. (Lovosice, Czech Republic) [24]. CAN is a nitrogen fertilizer containing 27% of nitrogen (up to 13.5% $N-NH_4^+$ and 13.5% $N-NO_3^-$) and 21% of dolomite (CAS 16389-88-1). It is a mixture of ammonium nitrate with finely ground dolomite in the form of whitish to light brown granules. Most particles (up to 90%) of CAN had a size in the range of 2–5 mm. Before pelletization, it was necessary to grind CAN into a powder on a grain mill (Sana Products Ltd., České Budějovice, Czech Republic) and sieve (Sana mesh sieve, pore size 0.5 × 0.5 (mm × mm)). Thus, the CAN was finally below 0.5 mm.

2.1.2. Filling Material in Pellets

The primary filler material (together with the fertilizer) in the pellets prepared manually using a hydraulic press was poly-3-hydroxybutyrate (P3HB, ρ = 1.23 g·cm^{-3}, M_w = 450,000 g·mol^{-1}, purity 98–99%) supplied by TianAn Biopolymer (Ningbo, People's republic of China) [25] and commercially purified in acetone as a white powder (Nafigate Corporation, Prague, Czech Republic) [26].

Another filler material used in some fertilizer mixtures was a struvite ($NH_4MgPO_4 \cdot 6H_2O$). A homogenized mixture of struvite from Nafigate and Aldrich (Merck KGaA, St. Louis, Missouri, USA) [27], in the form of a white powder, free of lumps by sieving, was used.

Dried biomass was also used in some mixtures with fertilizer as a filler material. The biomass was supplied by Nafigate Corporation and was produced from waste cooking oil by fermentation with a bacterial culture of *Cupriavidus necator* H16. The specific procedure for biomass production is described in [18]. After cultivation, the biomass was concentrated by centrifugation to 50% dry matter and stored frozen at −20 °C. It was dried at 105 °C to constant weight before use. Before pelletization, the lumps were removed by sieving. The dried biomass contained about 60% P3HB.

2.1.3. Preparation of Fertilizer Pellets

Pelletization was performed by compressing powdered materials of fertilizer CAN and other filler materials (P3HB, struvite, dried biomass containing P3HB) into pellets using a manual hydraulic press SPECAC (Orpington, Great Britain) [28] with a compressive force equivalent to 2 tons. Each resulting pellet had weight = 0.5 g, height = 4 mm, and diameter = 10 mm (approx.). Before the powder mixture preparation and homogenization, the fertilizer powder and P3HB powder were dried in an oven at 60 °C overnight and freed of lumps with a sieve (other filler materials: struvite and biomass were also sieved, and the sieve pore size was 0.5 × 0.5 (mm × mm)). Several series of cold-pressed pellets were prepared and composed of: 50% fertilizer CAN + 50% P3HB, 50% CAN + 50% biomass, 50% CAN + 25% P3HB + 25% struvite and 100% CAN serving as reference.

2.1.4. The Material for the Coating Solution

Coatings were prepared with a 6% solution of P3HB dissolved in chloroform with amylene (($CHCl_3$, M_r = 119.38; 99.93%, stabilized with amylene max. 55 ppm, supplied by Lach-Ner Ltd. (Neratovice, Czech Republic) [29]), then a 7% solution of P3HB in chloroform with ethanol ((99.8%, stabilized with approx. 1% ethanol p.a., supplied by PENTA Ltd. (Chrudim, Czech Republic) [30]) and 6% solution of P3HB in 1,3-dioxolane (($C_3H_6O_2$, M_w = 74.08 g·mol^{-1}, density 1.07 kg/L, stabilized with 0.03% BHT (2,6-Di-Tert-Butyl-4-Methylphenol), supplied by VWR International Ltd. (Stříbrná Skalice, Czech Republic) [31]). The solutions were prepared for preliminary tests of coating solutions and for dip coating of pelletized fertilizers purposes.

2.1.5. Coating of Fertilizer Pellets and Encapsulation

Finally, after preliminary tests of coating solutions, dioxolane was chosen as the best solvent for preparing the coating solution. The prepared pellets were coated manually by dipping them six times in a beaker with a 6% coating solution of P3HB in dioxolane.

The preparation of 6% P3HB in dioxolane for manual coating was first carried out by dissolving in an oven at a temperature of 90–95 °C approx. 1.5–2 h. However, new and faster ways of preparing larger volumes of solutions (at least 2000 mL) for later coating in a coating drum were sought (will be part of another article).

The resulting coating solution had suitable viscosity, and the final coating weight was only 12–17% of the original pellet weight (depending on the composition of the pellet to which the coating was applied).

The prepared pellets of 50% CAN + 50% P3HB were also encapsulated in an experimental biodegradable film based on P3HB, thermoplastic starch TPS, and PLA instead of coating. The biodegradable film was prepared by Panara Nitra (Slovakia, [32]) with Slovak University of Technology in Bratislava (Bratislava, Slovakia, [33]). The intention was to use biodegradable films based on P3HB and auxiliary polymers that were developed for complete decomposition in environmental components. These foils were prepared to use fully degradable agricultural foils, both in compost and in soils. An annular welding head was tailor-made for encapsulation that was electrically heated to a suitable temperature close to the melting point of the film (approx. 160 °C). By applying the appropriate pressure and time, a defect-free connection around the entire pellet was achieved. These encapsulated pellets were also tested in aqueous and soil environments.

2.2. Testing Nitrogen Release from Coated Fertilizer

The conductometric method monitored the release of ammonium and nitrate nitrogen in the aqueous medium from our six times hand-coated and encapsulated pellets prepared on a hydraulic press.

Ammonium nitrate is an inorganic, highly soluble fertilizer that dissociates into individual ions in aqueous environments: ammonium (NH_4^+) and nitrate (NO_3^-). The solubility of ammonium nitrate is 213 g/100 g H_2O [34], and conductometry can be used to determine the solvated fraction. Conductometry is a method based on measuring the conductivity of the second type of conductor: a solution whose conductivity is ensured by the ions present. The specific conductivity of the solution (κ) is related to the conductance (G), which can be determined by measuring the electric current flowing between the two electrodes of the conductometric probe through free charge carriers (ions). The conductance (G) is a quantity depending on the geometry of the measuring probe, and therefore it is used only to obtain a specific conductance (κ). Since, in the case of fertilizer analysis, there is a fundamental relationship between ion concentration (c_{Ion}) and specific conductance (κ), we describe the dependence of conductivity on ion concentration by the molar conductivity quantity (Λ_m):

$$G = \kappa \cdot \frac{A}{l} \quad (1)$$

$G \ldots$ conductance (S)
$\kappa \ldots$ specific conductivity ($S \cdot cm^{-1}$)
$A \ldots$ probe electrode area (cm^2)
$l \ldots$ probe electrode distance (cm)

$$\kappa = \Lambda_m \cdot c_{Ion} \quad (2)$$

$\Lambda_m \ldots$ molar conductivity ($S \cdot cm^2 \cdot mol^{-1}$)
$c_{Ion} \ldots$ ion concentration ($mol \cdot cm^{-3}$) [35]

The measurements of nutrient loss from the prepared pellets were performed in closable plastic test tubes with a volume of 50 mL. For all tests, one pellet (approx. 0.5 g) of fertilizer compressed into the form of a carrier with additives was placed in test tubes with a volume of distilled water of 40 mL. Ten pellets were measured for each experimental formulation, and the results were averaged. The measurements were performed at laboratory conditions (temperature 22 °C). The current signal was electronically converted to a frequency. The dependence of the change in the frequency of the alternating current on time was measured using a METEX® M-3850D multimeter (Metex Corporation, Toronto, Canada) [36]. Values were recorded initially after 24 h, and then the frequency of monitoring was reduced as necessary, and the time interval between measurements was usually 48 h. The method of calibration dependence between the frequency of alternating current and electrolyte concentration was used to monitor the mass transport of the nutrients from controlled release fertilizer. The calibration dependences were fitted using the Origin program (OriginLab, Northampton, MA, USA) [37] with two exponential fits to obtain the calibration dependence equation. Finally, the weight loss of the fertilizer was determined from this calibration dependence equation. We counted 21% dolomite content in the CAN fertilizer in the calculations of ammonium and nitrate loss from the pellets.

2.3. The Greenhouse Pot Experiment

The pot greenhouse experiment aimed to verify the effect of the application of coated fertilizers on the development and growth of maize and the dynamics of N release in the soil. A pot experiment was established in the vegetation hall of Mendel University in Brno (Brno, Czech Republic), evaluating the effect of fertilizing of coated fertilizers applied during the sowing period of maize.

2.3.1. Experimental Design

The experiment was performed in Mitscherlich vegetation pots (on 6.5 kg of soil). Five kg of soil was weighed into each pot. Maize (6 seeds per pot) was sown on the soil surface followed by uniform coverage with soil (750 g). Fertilizers were manually spread on the surface of this layer and evenly covered with another 750 g of soil (Figure 1). The basic agrochemical properties of the soil used in the pot experiment are presented in Table 1. SY Orpheus maize variety from Syngenta (Oseva, a.s., Bzenec, Czech Republic) [38] was used in the experiment. The pot experiment was performed under semi-natural conditions (rain shelter) in the vegetation hall. An identical controlled watering regime was used for all pots during the experiment. Plants were watered to 70% of maximum water holding capacity throughout the growing season. The pots were hand-watered with demineralized water on the soil surface.

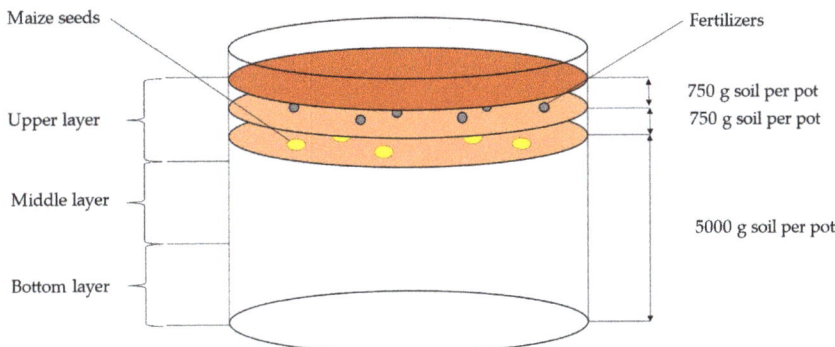

Figure 1. Scheme of Mitscherlich vegetation pot.

Table 1. Agrochemical properties of used soil in the pot experiment.

Soil Parameters	Value	Ref.
Clay	20%	
Dust	27%	[39]
Sand	53%	
Oxidizable C content (Cox)	0.80%	[40]
pH (CaCl$_2$)	6.09	
Cation exchange capacity	164 mmol/kg	
N overall	0.19%	
NH$_4^+$ (K$_2$SO$_4$)	1.48 mg/kg	
NO$_3^-$ (K$_2$SO$_4$)	17.2 mg/kg	[41]
P (Mehlich 3)	36.4 mg/kg	
K (Mehlich 3)	400 mg/kg	
Ca (Mehlich 3)	2720 mg/kg	
Mg (Mehlich 3)	214 mg/kg	

The fertilization treatments used in the experiment are presented in Table 2.

Table 2. Scheme of the pot experiment with maize.

Treatment	Composition of Pellets Used in Vegetation Tests	Number of Pellets (pcs/Pot)	N Dose (g/Pot)
CAN-c	100% CAN with P3HB coating	4	0.54
CAN/P-c	50% CAN + 50% P3HB with P3HB coating	8	0.54
CAN/P/S-c	50% CAN + 25% P3HB + 25% struvite with P3HB coating	8	0.54
CAN/P/B-c	50% CAN + 50% biomass with P3HB coating	8	0.54
CAN/P-bf	50% CAN + 50% P3HB encapsulated in biodegradable film	8	0.54
CAN	100% CAN (positive reference)	4	0.54
CON-c	without fertilizer (negative reference)	0	0

Each of these treatments was based on 12 replicates (pots) distributed in the vegetation hall at random. Maize was sown on 18 March 2021. The influence was monitored of the applied fertilizers on plant growth and nitrogen content in the soil profile after the establishment of the experiment (sowing) in regular 3-week intervals (t_1 = 8 April; t_2 = 29 April; and t_3 = 20 May 2021). After emergence (29 March 2021), the maize plants were adjusted to a final number of three plants per pot. Soil and plant analyses were performed during the vegetation.

2.3.2. Soil and Plant Sampling and Analyses

Soil sampling was carried out at the observed intervals (t_1–t_3) for each variant (each variant from four pots in each term). Soil samples were collected from each pot from three soil profile depths (upper, middle, and bottom layer, Figure 1). The contents of mineral nitrogen (Nmin)—ammonium (NH$_4^+$) and nitrate (NO$_3^-$) were determined in the soil samples [42].

The effects of the coated fertilizers on the growth of maize plants were monitored during the experiment at identical intervals (terms t_1–t_3). Chlorophyll content (N tester value), vegetative index (NDVI), basic photosynthetic parameters based on the action and measurement of light signal in PS II, dry weight of maize aboveground biomass (AGB), nitrogen content in AGB and capacitance of the root system were evaluated in plants (Table 3).

Table 3. Observed plant growth parameters.

Plant Parameter	Device Used	Terms	Ref.
Chlorophyll content (N-tester value)	Yara N-Tester chlorophyll meter (Yara International ASA, Oslo, Norway)	t1–t3	[43]
Vegetation index (NDVI)	PlanPen NDVI310 device (Photon Systems Instruments, Drásov, Czech Republic)	t1–t3	
Quantum yield of the PSII (Φ_{PSII})	PAR-Fluorpen FP110-LM/D device (Photon Systems Instruments, Drásov, Czech Republic)	t1–t3	[44]
Dry weight of AGB	Laboratory-scale PCB Kern (KERN & Sohn GmbH, Balingen, Germany)	t1–t3	
N content in AGB	Kjeltec 2300 device (Foss Analytical, Hillerød, Denmark)	t1–t3	[45]
Root electrical capacitance (C_R)	VOLTCRAFT LCR 4080 (Conrad Electronic GmbH, Wels, Austria)	t3	[44]

2.3.3. Statistical Data Analysis

The soils and plants were analyzed in STATISTICA 12 software [46] using Tukey's analysis of variance followed by testing at the 95% level of significance ($p \leq 0.05$). Normality and homogeneity of variances were checked using the Shapiro–Wilk test and Levene's test. In the case of the greenhouse pot experiment, the results are expressed as arithmetic mean ± standard error (SE).

3. Results

Several mixtures (formulations) of fertilizer and filler materials were successfully prepared by compressing powdered fertilizer (CAN) with filler materials (P3HB, struvite, dried biomass containing P3HB) into pellets using a manual hydraulic press. The coating was performed manually by dipping them six times in a beaker with a 6% coating solution of P3HB in dioxolane.

3.1. Preliminary Tests of Coating Solutions

Dioxolane was finally chosen as a solvent for P3HB in the coating solution based on preliminary tests that compared coating solutions of 6% P3HB in dioxolane, 7% P3HB in chloroform with ethanol, and 6% P3HB in chloroform with amylene. A solution of 6% P3HB in dioxolane was prepared for this test in an oven at 90–95 °C for 2 h (however, now, we are able to reduce the time required to prepare the solution to only 25 min using stirred autoclave with a heating jacket). Coating solutions of P3HB with chloroform were prepared under reverse reflux at 65 °C for 1 h. Pellets prepared on a manual press composed of 50% CAN + 50% P3HB were selected for preliminary coating solution tests. After preparing the individual P3HB coating solutions, the pre-weighed pellets were soaked in beakers with these individual solutions, pulled out after 2–5 s, and dried on glass material for at least 30 min (under laboratory conditions). After each soaking and drying (coating layer formation), the pellets were weighed again. Six layers of the coating were applied in this way. The pellets were weighed again the next day to determine the final weight of the pellets after applying six coats of coating to ensure all solvents were evaporated. For clarity, we present the final evaluation of coating weights of the pellets after applying six layers of coating (see Table 4) and the overall evaluation of individual coating solutions.

Table 4. Average coating weights (in%) and standard deviation of the weight of the original pellets (n = 10) after applying six layers of coating.

Coating Solution	Coating Weights of the Pellets (%)
7% P3HB in chloroform with ethanol	82.4 ± 4.6
6% P3HB in dioxolane	19.9 ± 1.5
6% P3HB in chloroform with amylene	Failed to form an acceptable coating

A solution of 7% P3HB in chloroform with ethanol seemed to be more gel-like. It was harder to apply and evaporated more slowly but created a quality coating that did not crack. Still, the resulting coating is too heavy. Boyandin et al. [23] acquired after the application of 6 layers approx. 60% extra pellet weight. A solution of 6% P3HB in dioxolane was applied well and quickly (lower viscosity), evaporated faster than chloroform, created a quality coating that does not crack, and the significant advantage was that the resulting 6-times coating was only (19.9 ± 1.5)% of the weight of the original pellet. A solution of 6% P3HB in chloroform with amylene was unusable; the solution was applied well at first, but the coating cracked at the second layer and was highly defective (see Figure S1). Although six-layer pellets were finally weighed, the resulting weight was irrelevant.

Although these individual solutions were not tested on a large number of pellets, the differences were so significant that in terms of quality and weight of the formed 6-times coating, a solution of 6% P3HB in dioxolane was a clear choice. A solution of 7% P3HB in chloroform with ethanol also produced a quality coating. However, due to the enormous weight of the applied coating (due to the high viscosity of the solution), this solution is disadvantageous and not so well used for manual coating on a laboratory scale, even when coating in a coating drum. Using a solution of P3HB in dioxolane, which has a suitable viscosity, we achieved a high-quality multiple coating: only 20% of the extra coating weight on the pellet is a significant improvement. Therefore, the solution of P3HB in dioxolane was best for application; dioxolane evaporated the fastest, and the resulting 6-times coating had excellent quality and the lowest weight. Legislative changes (Green Deal) have also contributed to the choice of dioxolane as a coating solvent. The advantage is also lower health risks when working with dioxolane. The disadvantages are its flammability and high price compared with chloroform.

3.2. Slow-Release Fertilizers

Pellets of various formulations were successfully prepared on a hydraulic press for further testing in aqueous and soil environments. These pellets contained CAN fertilizer, P3HB, struvite, and biomass in different composition. Approximately 120 pellets for each formulation were prepared. The pellets were coated manually with six layers using a solution of P3HB in dioxolane. Schematic representation of coated pellets can be seen in Figure 2 and coated pellets of different experimental formulations in Figure S2. Details of the pellet preparation and coating process were already described in the experimental part.

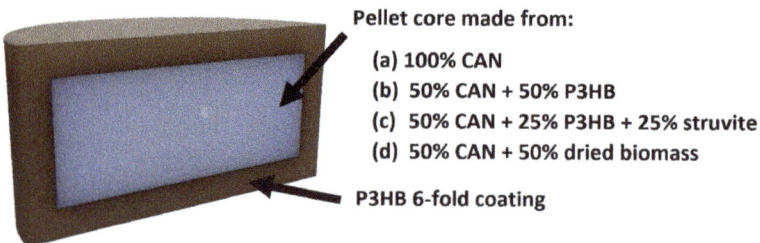

Figure 2. Schematic representation of coated pellets.

From each formulation, ten pellets were weighed before manual coating, and after pellets were coated with six layers of P3HB coating in dioxolane. The weighing of the coated pellets did not take place until the second day after manual coating so that the solvent could evaporate. Table 5 presents the calculated arithmetic mean ± standard deviation of the weight of the 6-fold coating applied to the pellets of different formulations (% of the original weight of the pellets). The resulting 6-fold coating of P3HB in dioxolane had finally even less weight than expected. In the case of pellets 50% P3HB with 50% CAN, the coating was approx. 17% by weight of the original pellet. For pellets that contained 50% CAN + 25% P3HB + 25% struvite or 50% CAN + 50% biomass, the resulting coating had approx. 14% by weight of the original pellet. Finally, for pellets containing 100% CAN fertilizer, the coating had only approximately 12% by weight of the original pellet.

Table 5. Calculated average weight (%, n = 10) and standard deviation of manually applied 6-fold coating for all pellet formulations prepared on a hydraulic press.

Formulation	6-Fold Additional Coating Applied (% of Original Pellet Weight, Average and Standard Deviation)
100% CAN	11.9 ± 1.6
50% CAN + 50% P3HB	16.9 ± 2.5
50% CAN + 25% P3HB + 25% struvite	13.8 ± 2.8
50% CAN + 50% biomass	14.4 ± 2.4

3.3. Pellets Encapsulated in Foils

A special foil was used for the manual encapsulation of 0.5 g pellets with a composition of 50% CAN + 50% P3HB instead of coating (details described in Section 2). Pellets were created on a hydraulic press in the laboratory, and an experimental biodegradable film marked NR/75/267 (Manufacturer Panara Nitra, Slovakia, P3HB content 13%, thermoplastic starch TPS 22%) was used for encapsulation. The film achieves 50% decomposition in the soil, measured on the basis of exhaled CO_2 after 420 days. The biodegradability test leading to the complete decomposition of the film has not yet been completed.

The first versions of the pellets with the starch-containing films were too permeable in the weld. Foil welds were first solved by folding the foil and closing with a weld on three sides. As it turned out, it was not possible to prepare a weld without a defect in the corners. Therefore, the encapsulation geometry was changed to a circular shape. Thus, an annular welding head was made to measure, which was electrically heated to a suitable temperature close to the melting point of the film (approx. 160 °C). Both welded foils were placed in the mating piece between the Teflon foils to prevent the molten foil from adhering to the welding head and the mating piece. A defect-free connection around the entire pellet was achieved by applying the appropriate pressure and time. Thus, 120 pellets with a composition of 50% fertilizer CAN + 50% P3HB were encapsulated in the film for further testing in aqueous and soil environments (see Figures S3 and S4).

3.4. The Nitrogen Release from Coated Fertilizer in the Aquatic Environment

Measuring the loss of fertilizer from the prepared defined carriers in the aquatic environment (distilled water, laboratory conditions, conductometry method described in the experimental part) provided us only an indicative prediction for the subsequent use of fertilizers in the soil in terms of their "slow-release" quality. Nevertheless, it could early on be deduced from these tests whether the pellets and their subsequent coating were designed correctly in terms of their chemical composition and whether they were well prepared and coated so that there would be a gradual release of nitrogen in the soil.

The compositions of the individual manually coated fertilizer formulations prepared on a hydraulic press that were tested in the aqueous medium can be seen in Table 6. The release of ammonium nitrate (AN) in the aqueous medium from our six-fold manually-coated pellets (containing CAN fertilizer) was investigated by the conductometric method. Figure 3 shows the results of the measurement completed on day 76.

Table 6. The composition of six-fold manually coated fertilizer formulations prepared on a hydraulic press and tested in the aqueous medium.

	Composition of 6-Fold Manually Coated Fertilizer Pellets
CAN/P/B-c	50% CAN + 50% biomass + coating P3HB in dioxolane
CAN/P-c	50% CAN + 50% P3HB + coating P3HB in dioxolane
CAN/P/S-c	50% CAN + 25% P3HB + 25% struvite + coating P3HB in dioxolane
CAN-c	100% CAN + coating P3HB in dioxolane
CAN	100% CAN (positive reference)
CAN/P-bf	50% CAN + 50% P3HB + biodegradable polymeric film (encapsulation)

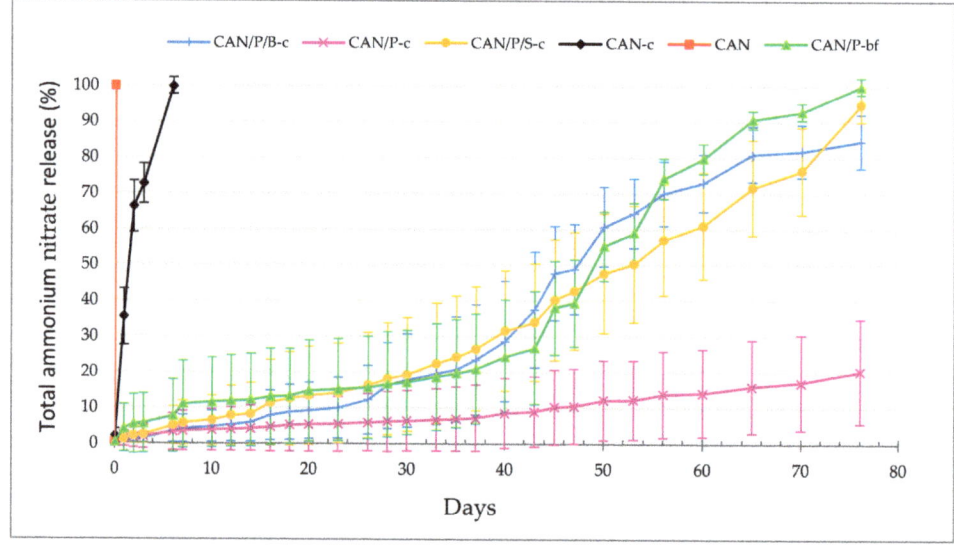

Figure 3. Release of ammonium nitrate (AN) from pellets of different compositions in the aquatic environment as a function of time. Total ammonium nitrate release in% is expressed as mean (n = 10), and the error bars represent the standard deviation. CAN-c: 100% CAN with P3HB coating; CAN/P-c: 50% CAN + 50% P3HB with P3HB coating; CAN/P/S-c: 50% CAN + 25% P3HB + 25% struvite with P3HB coating; CAN/P/B-c: 50% CAN + 50% biomass with P3HB coating; CAN/P-bf: 50% CAN + 50% P3HB encapsulated in biodegradable film; CAN: 100% CAN (positive reference).

As can be seen, all ammonium nitrate fertilizer was released immediately (in 1 h in the aquatic environment) from the reference (100% CAN without coating). The rapid release of fertilizer was also observed in coated pellets containing 100% CAN. Half of the AN was released from these pellets after approximately 1.5 days and all in 6 days. The adhesion of the 6-fold coating prepared on these pellets containing 100% CAN fertilizer was probably not as great as for pellets containing P3HB in addition to the fertilizer.

The release of AN from the coated pellets containing struvite, biomass, and pellets with foil instead of coating showed a similar course of fertilizer release. Half of the AN was released from pellets containing struvite on day 53, from pellets containing biomass approximately day 48, and from pellets with foil approximately day 49. However, at the end of the measurement (day 76), all ammonium nitrate was released from the foil pellets, while 95% of AN was released from the struvite-containing pellets and 85% of the AN from the pellets containing biomass. Significantly, the least AN was released from coated pellets containing 50% CAN and 50% P3HB. After 76 days, only 20% of the ammonium nitrate was released. The nitrogen ammonium release curve of the fertilizer has an almost linear course in this interval (lag phase and linear phase [4]), which indicates a perfectly flawless coating. This flawless coating could have been formed due to the high adhesion of the P3HB coating (dissolved in dioxolane) to the pellets of this composition. This indicates that P3HB inside the pellet is essential as an adhesion promotor. In this case, a well-made coating acts as a polymer layer almost impermeable to water.

The results of water tests show that these pellets have the potential for the gradual release of nitrogen even in the soil environment. It is also clear that for a quality coating that adhered well to our pellets, it was necessary to add P3HB as an additive. On the 100% CAN pellets without P3HB, no coating was formed that would withstand the action of water for a long time. In comparison, the coating on the pellets containing 50% P3HB showed high resistance even after 76 days in water.

3.5. Efect of Coated Fertilizers in Greenhouse Experiment

Based on the identified potential of the coated and encapsulated fertilizers under aqueous test conditions, the pot vegetation experiment aimed to verify the effect of our different fertilizer formulations on the development and growth of maize and the dynamics of N release in the soil. The experiment took place in Mitscherlich's vegetation pots in the vegetation hall at the Mendel University Brno (Figure S5).

3.5.1. The Nitrogen Release from the Fertilizers in the Soil

During plant growth, soil sampling was performed at regular 3-week intervals to determine the mineral nitrogen content (described in the experimental part). Based on the content of mineral N (NH_4^+ and NO_3^-) determined in individual soil layers of the pot (upper, middle, and bottom layer) over time, the effect of coating on the dynamics of nitrogen release from our tested fertilizers was assessed.

In the first term (t_1) of soil collection (3 weeks after sowing/fertilization), it is evident that from uncoated fertilizer (CAN), nitrogen is released very quickly and leaches in the lower layer of soil profile due to watering. The NH_4^+ nitrogen supply in this treatment is relatively low, which is a consequence of the rapid release of this form N from fertilizer and its rapid nitrification (Figure S6a,b).

Coated fertilizers (CAN-c, CAN/P-c, CAN/P/S-c, and CAN/P/B-c) showed the ability to gradually release N. This was evidenced by the significantly high contents of ammonium N in the upper layer of the soil. In contrast with uncoated fertilizer (CAN), NH_4^+ was later released from modified coated fertilizers. The gradual release of N from these fertilizers was also evidenced by the low content of nitrates in the lower soil layer (Figure S6b). Its lowest values were reached in treatment with encapsulated fertilizer CAN/P-bf. The hydrolysis of encapsulated fertilizer was slowed down, and thus, the nitrification of the released NH_4^+ did not occur so intensively). Thus, the CAN/P-bf fertilizer (encapsulated pellets) showed the slowest N release. This is evidenced by the content of ammonium N in the soil (all layers), which was statistically the lowest in this treatment option.

The gradual nutrient release from coated fertilizers is evident in the soil analysis results performed 6 weeks after sowing of maize (t_2). The significantly highest NH_4^+ content in the upper and middle soil layer was recorded for all treatments with coated (CAN-c, CAN/P-c, CAN/P/S-c, and CAN/P/B-c) and encapsulated fertilizer CAN/P-bf (Figure S7a). There was a quite change in the content of soil N, especially at CAN-c (100% CAN with P3HB coating). The determined NH_4^+ content was the lowest of the coated fertilizers. This indicates its nitrification and horizontal shift (between t_1 and t_2). In the treatment fertilized with uncoated fertilizer (CAN), the nitrogen released from the fertilizer has long been uptake by plants or nitrified and leached into the lower layers, as evidenced by the ammonium ion supply in the soil, which was at the level of the non-fertilized treatment (CON-nf).

The highest nitrate nitrogen content in the soil was also recorded in the upper and middle part of the soil profile in treatments fertilized with coated fertilizers, especially in treatments CAN/P/S-c, and CAN/P/B-c (Figure S7b). Its increased amount in the soil is due to the gradual release of nitrates from coated CAN and gradual nitrification of ammonium N. The content of nitrates in the lower soil layer was compared between the treatments. If we compare this amount of NO_3^- with the values found in the term t_1 for treatment CAN (79.6 mg/kg), the use of coated fertilizers significantly reduced the risk of its leaching.

In soil analyses performed 9 weeks after sowing (t_3), it is evident that the plants depleted mineral nitrogen from the soil. This is evidenced by the very low contents of the ammonium and nitrate forms of N in all soil layers (Figure S8a,b). However, for both forms of acceptable nitrogen, there are apparent differences between the fertilizers used at this time. The significantly highest supply of NH_4^+ and NO_3^- was found on treatments fertilized with coated fertilizers (especially CAN/P/S-c, CAN/P/B-c) and encapsulated fertilizer CAN/P-bf.

The dynamics of the release of the monitored forms of mineral nitrogen from fertilizers are presented in Figure 4a,b. It is clear that coated fertilizers tend to retain nitrogen in the soil (soil profile), delay its conversion into leachable N (nitrate), contribute to eliminating its loss by leaching, and thus increase the efficiency of fertilization using nitrogen by plants.

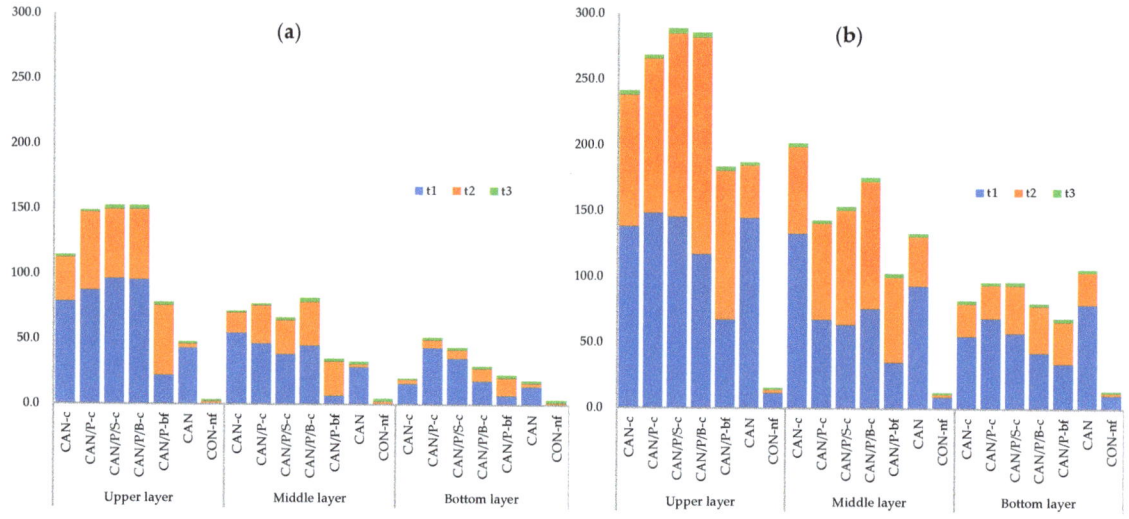

Figure 4. Development of ammonium (**a**) and nitrate (**b**) soil nitrogen content (mg/kg of soil) over time. CAN-c: 100% CAN with P3HB coating; CAN/P-c: 50% CAN + 50% P3HB with P3HB coating; CAN/P/S-c: 50% CAN + 25% P3HB + 25% struvite with P3HB coating; CAN/P/B-c: 50% CAN + 50% biomass with P3HB coating; CAN/P-bf: 50% CAN + 50% P3HB encapsulated in biodegradable film; CAN: 100% CAN (positive reference); CON-nf: without fertilizer (negative reference).

3.5.2. Effect of Coated and Encapsulated Fertilizers on Plant Biomass of Maize

In addition to the nitrogen content in the soil, the effect of the tested fertilizers on the production of plant matter and its quality was evaluated in a pot experiment. Plant biomass production was evaluated based on the assessment of the dry weight of maize plants in terms of t_1–t_3 and root system size (t_3). Biomass quality was determined by measuring chlorophyll content (N tester), development of vegetation index (NDVI), and selected photosynthetic parameters during vegetation (t_1–t_3).

Above-Ground Biomass Production and Root Size

The development of dry weight of maize aboveground biomass (AGB) production is presented in Figure 5. The dry weight of AGB was not significantly affected by fertilization in the initial stage (t_1). No significant differences between the fertilized treatments were found in other plant collection dates (t_2 and t_3) either. In this growth phase, the expected reduction in dry matter of AGB on the nitrogen unfertilized treatment (CON-nf) was confirmed. While the relatively highest dry matter yield of AGB was found on the CAN at t_2, at the end of the experiment (t_3) the highest dry matter production was obtained on the CAN/P/B-c coated fertilizer treatment.

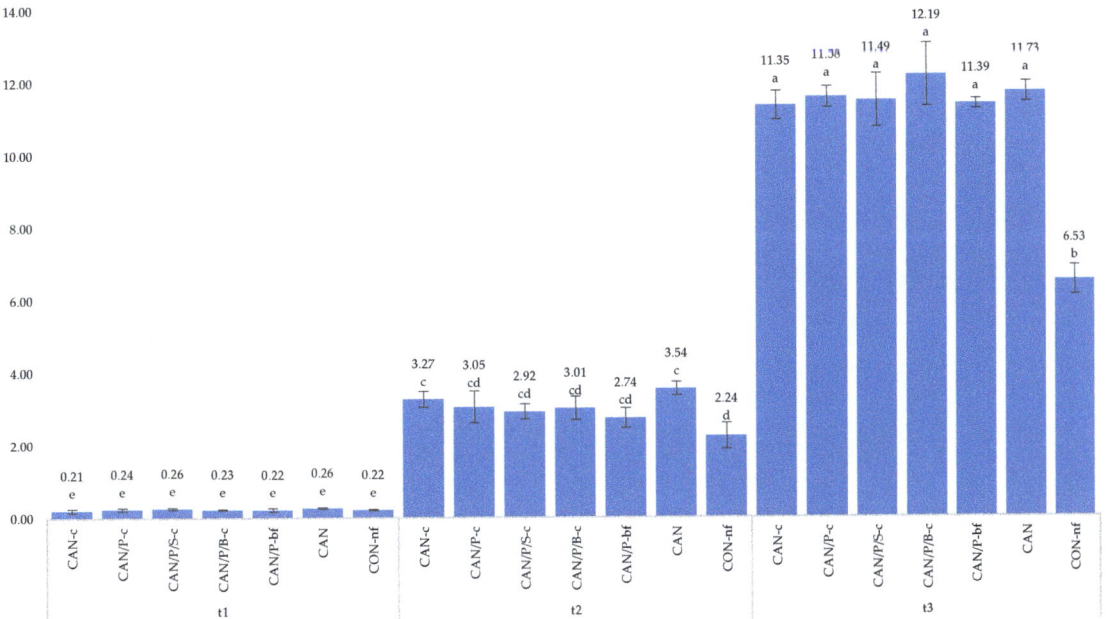

Figure 5. Dry matter weight of maize AGB (g/1 plant). The weights of AGB are expressed as mean (n = 4); the error bars represent the standard deviation. Different letters above error bars denote statistically significant differences among treatments using Tukey's post hoc tests. CAN-c: 100% CAN with P3HB coating; CAN/P-c: 50% CAN + 50% P3HB with P3HB coating; CAN/P/S-c: 50% CAN + 25% P3HB + 25% struvite with P3HB coating; CAN/P/B-c: 50% CAN + 50% biomass with P3HB coating; CAN/P-bf: 50% CAN + 50% P3HB encapsulated in biodegradable film; CAN: 100% CAN (positive reference); CON-nf: without fertilizer (negative reference).

The size (capacity) of the root system of maize plants was determined only at term t_3 using the so-called electric root capacity (C_R). The C_R values are presented in Figure 6. The size of plant roots strongly correlates with the weight of AGB dry matter in a given phase (r = 0.851; $p < 0.001$). The significantly highest value was reached in treatment with relatively highest AGB production (CAN/P/B-c). On this variant, C_R was 16.5% higher compared to maize grown on the variant fertilized with CAN.

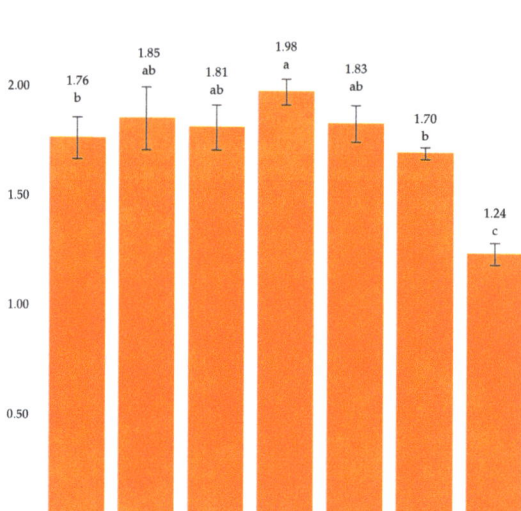

Figure 6. The maize root system size (nF) determined by root electrical capacitance (C_R). The values of C_R are expressed as mean (n = 4); the error bars represent the standard deviation. Different letters above error bars denote statistically significant differences among treatments using Tukey's post hoc tests. CAN-c: 100% CAN with P3HB coating; CAN/P-c: 50% CAN + 50% P3HB with P3HB coating; CAN/P/S-c: 50% CAN + 25% P3HB + 25% struvite with P3HB coating; CAN/P/B-c: 50% CAN + 50% biomass with P3HB coating; CAN/P-bf: 50% CAN + 50% P3HB encapsulated in biodegradable film; CAN: 100% CAN (positive reference); CON-nf: without fertilizer (negative reference).

Nitrogen Content in Plant, Chlorophyll Content, NDVI, and Quant Yield of PSII

The nitrogen content determined in AGB of maize plants decreased logically over time (t_1 to t_3). The effect of fertilizer application on its amount in tissues was observed mainly in treatment CAN-c (see Table 7). The nitrogen content in plants was the significantly highest in t_1 and t_2 after fertilization with this treatment. The nitrogen contents in the AGB of maize plants were equal in the fertilized variants at the end of the experiment (t_3). These contents of N in plant correlate with the state of mineral N amount in the soil, especially the NO_3^- form (t_1: r = 0.729, $p < 0.001$; t_2: r = 0.753, $p < 0.001$; t_3: r = 0.845, $p < 0.001$).

In addition to the N content in the AGB of maize plants, the normalized difference vegetation index (NDVI) and chlorophyll content, expressed as N-tester value, were determined (Table 7). The amount of chlorophyll in the plants was also determined using a Yara N-tester. N tester values significantly correlated with nitrogen contents in AGB (r = 0.740, $p < 0.001$). Since the chlorophyll content in the plant is directly dependent on the amount of nitrogen in the tissues, this fact can be explained by the gradual release of nutrients from coated fertilizers in time and its relative deficiency in the later stages of growth (t_3) in plants fertilized with conventional, non-coated fertilizers. While in the terms t_1 and t_2, the plants fertilized with uncoated CAN showed the significant highest N tester value, in the term t_3 the significant highest content of chlorophyll was found in treatments CAN/P-c, CAN/P/S-c and CAN/P-bf.

Table 7. Nitrogen content in the above-ground mass of maize plants (% DM), N tester value, NDVI, and quantum yield of photosystem II (Φ_{PSII}) in maize. The values represent the mean (n = 4) ± standard deviation. Different letters denote statistically significant differences among treatments using Tukey's post hoc tests. CAN-c: 100% CAN with P3HB coating; CAN/P-c: 50% CAN + 50% P3HB with P3HB coating; CAN/P/S-c: 50% CAN + 25% P3HB + 25% struvite with P3HB coating; CAN/P/B-c: 50% CAN + 50% biomass with P3HB coating; CAN/P-bf: 50% CAN + 50% P3HB encapsulated in biodegradable film; CAN: 100% CAN (positive reference); CON-nf: without fertilizer (negative reference).

Term of Measured	Treatments	N Content in AGB (% of DM ± SD)	N-Tester Value	NDVI	Φ_{PSII}
t1	CAN-c	5.99 a ± 0.25	519 def ± 8	0.75 cdefgh ± 0.01	0.835 abc ± 0.006
	CAN/P-c	5.81 ab ± 0.04	529 def ± 12	0.77 abcde ± 0.01	0.835 abc ± 0.006
	CAN/P/S-c	5.70 ab ± 0.20	518 def ± 8	0.75 defgh ± 0.02	0.838 abc ± 0.005
	CAN/P/B-c	5.58 bc ± 0.08	506 ef ± 7	0.76 abcdefg ± 0.02	0.825 abcd ± 0.006
	CAN/P-bf	5.57 bc ± 0.07	506 ef ± 13	0.76 abcdefg ± 0.02	0.825 abcd ± 0.006
	CAN	5.64 ab ± 0.06	565 abc ± 7	0.76 abcdefg ± 0.02	0.825 abcd ± 0.006
	CON-nf	5.25 c ± 0.08	498 f ± 10	0.77 abcdef ± 0.01	0.833 abc ± 0.005
t2	CAN-c	3.54 d ± 0.25	574 ab ± 20	0.79 abcd ± 0.01	0.833 abc ± 0.005
	CAN/P-c	3.32 de ± 0.13	538 cd ± 9	0.80 abc ± 0.01	0.840 ab ± 0.000
	CAN/P/S-c	3.47 de ± 0.11	530 de ± 12	0.80 a ± 0.00	0.840 ab ± 0.008
	CAN/P/B-c	3.30 de ± 0.16	543 bcd ± 16	0.80 ab ± 0.01	0.835 abc ± 0.013
	CAN/P-bf	3.14 e ± 0.23	541 cd ± 9	0.79 abcd ± 0.01	0.843 a ± 0.005
	CAN	3.29 de ± 0.12	579 a ± 3	0.78 abcd ± 0.02	0.825 abcd ± 0.010
	CON-nf	1.48 f ± 0.20	397 g ± 13	0.75 bcdefg ± 0.02	0.830 abc ± 0.008
t3	CAN-c	1.57 f ± 0.04	366 hi ± 19	0.71 gh ± 0.02	0.823 abcd ± 0.010
	CAN/P-c	1.50 f ± 0.04	405 g ± 7	0.71 gh ± 0.01	0.823 abcd ± 0.010
	CAN/P/S-c	1.58 f ± 0.06	399 g ± 3	0.70 h ± 0.01	0.820 bcd ± 0.000
	CAN/P/B-c	1.55 f ± 0.06	358 hi ± 23	0.73 efgh ± 0.02	0.818 cd ± 0.010
	CAN/P-bf	1.62 f ± 0.06	388 gh ± 8	0.72 gh ± 0.01	0.823 abcd ± 0.013
	CAN	1.52 f ± 0.06	350 i ± 7	0.72 fgh ± 0.00	0.808 de ± 0.010
	CON-nf	0.80 g ± 0.04	249 j ± 8	0.62 i ± 0.05	0.790 e ± 0.008

A significant dependence was found between the measured values of N tester and NDVI (r = 0.822, $p < 0.001$). However, fertilization did not have a significant effect on the values of the vegetation index, as shown in Table 7.

Quantum yield of photosystem II (Φ_{PSII}) is a measure of photosystem II (PSII) efficiency and corresponds to the F_V/F_M ratio, where F_V is the maximum variable chlorophyll fluorescence yield in the light-adapted state and F_M is the maximum chlorophyll fluorescence yield in the light-adapted state. The quantum yield thus provides an accurate estimate of photosynthetic activity. The value of Φ_{PSII} was not significantly affected by coated fertilizer fertilization (Table 7). The actual capacity of the PSII for photochemical processes by availability of reaction centers of the photosystem II significantly corelated with NDVI values (r = 0.713, $p < 0.001$). Nevertheless, its values at the end of the vegetation (t3) were relatively highest in plants fertilized with coated and encapsulated fertilizers (0.821), compared to the Φ_{PSI} value on the CAN (0.808) and unfertilized treatment (0.790).

4. Discussion

Several controlled-release fertilizer formulations have been successfully prepared: coated pellets containing CAN fertilizer and fillers that align with the circular economy's intentions (fully biodegradable P3HB, struvite, and biomass). The behavior and release dynamics of the fertilizer in aqueous environments and vegetation pot experiments were tested.

Boyandin et al. [23] investigated the release of ammonium nitrate in an aqueous medium in pellets with a 6-fold coating of P3HB in chloroform. Its pellets contained only 25% of ammonium nitrate, and the rest was made up of P3HB or other additives (wood flour). Testing in water in research lasted seven days. (18.3 ± 7.9)% of fertilizer was released after seven days from its coated pellets containing ammonium nitrate and P3HB, and (13.4 ± 7.9)% of fertilizer was released from its coated pellets also containing wood flour. Our coated pellets containing fertilizer with 50% P3HB, or struvite, or biomass had only 3.5–5.7% released ammonium nitrate after seven days. Pellets in foils released 11.5% of ammonium nitrate. The differences in a more extended experiment would be even more pronounced. In addition, we also achieved a small coating thickness on the pellets (approx. 12–17% of the weight of the original pellet for various formulations) due to using "green" dioxolane as a P3HB solvent [22].

The release of fertilizer into water is also described, e.g., by Rashidzadeh and Olad [47]. Their slow-released NPK fertilizer, encapsulated by superabsorbent nanocomposite and prepared via the in situ free radical polymerization of sodium alginate, acrylic acid, acrylamide, and montmorillonite in the presence of fertilizer compounds, also possessed excellent slow-release property. The release of fertilizer was 14.66% on the first day, 28.54% after one week, and 57.66% after one month.

Our coated pellets containing 50% fertilizer with 50% P3HB showed excellent resistance even after 76 days in water. After 76 days, only 20% of the ammonium nitrate was released. According to a review of controlled-release fertilizers by Lawrencia et al. [4], such fertilizer release in the water environment corresponds more to fertilizer with a synthetic polymer-based coating than fertilizer with a natural polymer-based coating. We proved the feasibility of P3HB as a filler and coating material. By adjusting filler content and coating thickness, we could design the period of the agrochemical release from the formulation. Effective P3HB coating could also be used for seed protection [48].

The pot experiment results proved coated CAN fertilizers as a possible option to improve nutrient use efficiency, reduce nitrogen losses, and minimize environmental pollution while providing nitrogen to plants more gradually during vegetation, which is also described by several authors [9,49–52]. The release of nitrogen from coated CAN fertilizers affected dynamic changes in the soil mineral N content during the vegetation of maize. Contents of N_{min} and its ionic forms (NO_3^-, NH_4^+) were determined in the soil in three terms (t_1–t_3). Although enough of the available nitrogen can be essential for direct plant consumption, the excessive content may inevitably increase its loss in soil [53]. One of the important aspects of coated fertilizers is the longevity of nutrient release at optimal levels for plant uptake. The application of coated CAN fertilizers showed a positive effect on the N_{min} (NO_3^-, NH_4^+) release pattern, as seen in Figures S6–S8. The statistically lowest values of nitrogen content in the soil in the t_1 term were observed after CAN/Pbf treatment, possibly because of the encapsulation, which slowed down the hydrolysis of this fertilizer and resulted in more gradual nitrogen release. This fact possibly resulted in slower nitrification and thereby reduced loss from leaching. Positive effects were found of fertilizers coated with different types of polymers such as polyolefin [54,55], multiorganic polymer, diamide of oxalic acid [55,56], and sulfur [57] on nitrate leaching. The results obtained from t_2 and t_3 clearly demonstrate the effects of coated fertilizers: each treatment provided a higher amount of nitrogen (both ammonia and nitrate) in comparison with common uncoated CAN. Nitrogen after this treatment was taken up by the plants and leached in the middle and lower layers of the soil, while the coated treatments continued to provide both forms of nitrogen more gradually. One of the highest supplies of nitrogen in both later terms was provided by coated treatments CAN/P/B-c and CAN/P/S-c; a similar result was also provided by encapsulated treatment CAN/Pbf. Zheng et al. [58] similarly describe the enhanced content of soil mineral nitrogen in later vegetation stages after applying sulfur-coated fertilizers. Figure 4 describes the development of nitrogen in the soil layers over time; a similar result with gradually increasing content of N in the top layer while maintaining the high amount of nitrogen in the middle layer is also presented by Xiao et al. [59]. Their controlled-release N fertilizer coated with paper-plastic composite material (composed of paper and polyethylene) reduced nitrogen leaching and ammonia evaporation from soil.

The aboveground biomass of maize after fertilization with coated CAN treatments was, as expected, higher in comparison with unfertilized treatment and comparable with the treatment without coating (CAN). These results proved that coated fertilizers present a potential alternative to common fertilizers, as their application has no negative effect on yield while minimizing the environmental losses of nitrogen. The same conclusions are presented by Škarpa et al. [60]. Trenkel [13] even describes the possibility of lowering the application doses of coated fertilizers while achieving similar yields. Some authors also describe increases in crop yields after coated fertilizer application [61–65]. After the use of CRF fertilizers based on polyhydroxyalkanoate, a significant increase in total fresh plant biomass was found compared with the quick-release NPK fertilizer [66]. In our experiment, the increase in aboveground biomass was observed only in t_3 after the treatment CAN/P/B-c in comparison with uncoated CAN. A strong correlation between aboveground biomass and root size was observed in our experiment. The highest value of root size was measured on the treatment CAN/P/B-c with the highest biomass production. The correlation between the nitrogen content (especially nitrate form) in the soil and N content in the aboveground biomass of plants was observed, the same as the correlation between nitrogen content in AGB and N-tester values. Similar findings were presented by Koning et al. [67]. The content of chlorophyll plants is dependent on the amount of nitrogen in the tissues [68]; these results can be explained by the gradual release of nitrogen from coated treatment in time and its relative deficiency in the later stages of growth (t_3) in plants fertilized with conventional, non-coated fertilizers. Although the effect of fertilization was not significant in terms of N-tester values and NDVI index, the coated treatment CAN/P/B-c and encapsulated treatment CAN/Pbf provided one of the highest values, especially in later terms of vegetation. A single pre-planting application of these controlled-release fertilizers can fill a crop's nutritional requirements throughout its growing season.

Our research on controlled-release fertilizers with biodegradable coating involved coated formulations with urea and P3HB filler. We were able to prepare these pellets on a quarter-operational pelletizing device and coat them in larger quantities in a coating drum. This technology could be transferable to the industry. A careful study of the biodegradation of these fertilizer formulations in soil and research into the long-term effect on soil composition under field conditions will also be needed.

Supplementary Materials: The following supporting information can be downloaded at: https://www.mdpi.com/article/10.3390/polym14204323/s1. Figure S1: Six coating layers applied manually to pellets composed of 50% P3HB with 50% CAN: coated with a solution of (a) 7% P3HB in chloroform with ethanol, (b) 6% P3HB in dioxolane, (c) 6% P3HB in amylene; Figure S2: Manually-coated pellets of different experimental formulations (prepared on a hydraulic press); Figure S3: Encapsulation of pellets containing 50% CAN and 50% P3HB into a biodegradable film; Figure S4: The final appearance of the encapsulated pellets; Figure S5: Grown maize in Mitscherlich vegetation pots in the rain shelter; Figure S6: The ammonium (a) and nitrate (b) soil nitrogen content (mg/kg of soil) in the first term (t_1) of soil collection (3 weeks after sowing/fertilization). The nitrogen contents of the different soil layers (up—upper, mi—middle, bo—bottom) are expressed as mean (n = 4); the error bars represent the standard deviation. The mean values marked with an asterisk are significantly different ($p \leq 0.05$) from the treatment without coated (CAN) by the Tukey test (each of the soil layers was statistically evaluated separately). CAN-c: 100% CAN with P3HB coating; CAN/P-c: 50% CAN + 50% P3HB with P3HB coating; CAN/P/S-c: 50% CAN + 25% P3HB + 25% struvite with P3HB coating; CAN/P/B-c: 50% CAN + 50% biomass with P3HB coating; CAN/P-bf: 50% CAN + 50% P3HB encapsulated in biodegradable film; CAN: 100% CAN (positive reference); CON-nf: without fertilizer (negative reference); Figure S7: The ammonium a) and nitrate b) soil nitrogen content (mg/kg of soil) in the second term (t_2) of soil collection (6 weeks after sowing/fertilization). The nitrogen contents of the different soil layers (up—upper, mi—middle, bo—bottom) are expressed as mean (n = 4); the error bars represent the standard deviation. The mean values marked with an asterisk are significantly different ($p \leq 0.05$) from the treatment without coated (CAN) by the Tukey test (each of the soil layers was statistically evaluated separately). CAN-c: 100% CAN with P3HB coating; CAN/P-c: 50% CAN + 50% P3HB with P3HB coating; CAN/P/S-c: 50% CAN + 25% P3HB + 25% struvite with

P3HB coating; CAN/P/B-c: 50% CAN + 50% biomass with P3HB coating; CAN/P-bf: 50% CAN + 50% P3HB encapsulated in biodegradable film; CAN: 100% CAN (positive reference); CON-nf: without fertilizer (negative reference); Figure S8: The ammonium (a) and nitrate (b) soil nitrogen content (mg/kg of soil) in the third term (t_3) of soil collection (9 weeks after sowing/fertilization). The nitrogen contents of the different soil layers (up—upper, mi—middle, bo—bottom) are expressed as mean (n = 4), the error bars represent the standard deviation. The mean values marked with an asterisk are significantly different ($p \leq 0.05$) from the treatment without coated (CAN) by the Tukey test (each of the soil layers was statistically evaluated separately). CAN-c: 100% CAN with P3HB coating; CAN/P-c: 50% CAN + 50% P3HB with P3HB coating; CAN/P/S-c: 50% CAN + 25% P3HB + 25% struvite with P3HB coating; CAN/P/B-c: 50% CAN + 50% biomass with P3HB coating; CAN/P-bf: 50% CAN + 50% P3HB encapsulated in biodegradable film; CAN: 100% CAN (positive reference); CON-nf: without fertilizer (negative reference).

Author Contributions: Conceptualization, R.P. and S.K.; methodology, R.P., S.K., S.F., P.Š. and V.J.; validation, P.M. and M.M.; formal analysis, S.K., R.P., Z.G., S.F., M.S., V.J., P.M., P.Š., T.K., J.A. and M.M.; investigation, S.K., M.S., Z.G., V.J., S.F., R.P. and P.M.; resources, R.P.; data curation, S.K., V.J., S.F., P.Š. and M.M.; writing—original draft preparation, S.K.; writing—review and editing, S.K., R.P., P.Š., T.K. and J.A.; supervision, R.P., S.K. and P.Š.; project administration, R.P., Z.G. and P.Š.; funding acquisition, R.P. All authors have read and agreed to the published version of the manuscript.

Funding: This research was funded by Ministry of Industry and Trade of the Czech Republic, grant number FV40095 and by the grant Internal Grant Agency AF-IGA2020-TP007: Use of superabsorbent polymers with controlled nutrient release in field-crop cultivation systems in arid areas.

Institutional Review Board Statement: Not applicable.

Informed Consent Statement: Not applicable.

Data Availability Statement: Not applicable.

Acknowledgments: We would like to thank Nafigate Corporation for the support in the form of supplies of P3HB material, struvite and biomass.

Conflicts of Interest: The authors declare no conflict of interest.

References

1. Smil, V. Nitrogen Cycle and World Food Production. *World Agric.* **2011**, *2*, 9–13.
2. United Nations, Department of Economic and Social Affairs, Population Division. World Population Prospects 2019: Highlights (ST/ESA/SER.A/423). 2019. Available online: https://population.un.org/wpp/publications/files/wpp2019_highlights.pdf (accessed on 1 July 2022).
3. Adam, D. How Far Will Global Population Rise. *Nature* **2021**, *597*, 462–465. [CrossRef]
4. Lawrencia, D.; Wong, S.K.; Low, D.Y.S.; Goh, B.H.; Goh, J.K.; Ruktanonchai, U.R.; Soottitantawat, A.; Lee, L.H.; Tang, S.Y. Controlled Release Fertilizers: A Review on Coating Materials and Mechanism of Release. *Plants* **2021**, *10*, 238. [CrossRef] [PubMed]
5. Galloway, J.N.; Townsend, A.R.; Erisman, J.W.; Bekunda, M.; Cai, Z.; Freney, J.R.; Martinelli, L.A.; Seitzinger, S.P.; Sutton, M.A. Transformation of the nitrogen cycle: Recent trends, questions, and potential solutions. *Science* **2008**, *320*, 889–892. [CrossRef]
6. Zhang, X.; Davidson, E.; Mauzerall, D.; Searchinger, T.D.; Dumas, P.; Shen, Y. Managing nitrogen for sustainable development. *Nature* **2015**, *528*, 51–59. [CrossRef] [PubMed]
7. Bisht, N.; Chauhan, P.S. Excessive and Disproportionate Use of Chemicals Cause Soil Contamination and Nutritional Stress. In *Soil Contamination—Threats and Sustainable Solutions*; Larramendy, M.L., Soloneski, S., Eds.; IntechOpen: London, UK, 2020. [CrossRef]
8. Fan, L.-T.; Singh, S.K. Introduction. In *Controlled Release: A Quantitative Treatment*, 1st ed.; Fan, L.-T., Singh, S.K., Eds.; Springer Science & Business Media: Berlin/Heidelberg, Germany, 2012; Volume 13, pp. 1–8.
9. Gil-Ortiz, R.; Naranjo, M.Á.; Ruiz-Navarro, A.; Atares, S.; García, C.; Zotarelli, L.; San Bautista, A.; Vicente, O. Enhanced Agronomic Efficiency Using a New Controlled-Released, Polymeric-Coated Nitrogen Fertilizer in Rice. *Plants* **2020**, *9*, 1183. [CrossRef] [PubMed]
10. Wang, Y.; Liu, M.; Ni, B.; Xie, L. κ-Carrageenan–sodium alginate beads and superabsorbent coated nitrogen fertilizer with slow-release, water-retention, and anticompaction properties. *Ind. Chem. Eng. Res.* **2012**, *51*, 1413–1422. [CrossRef]
11. Cole, J.C.; Smith, M.W.; Penn, C.J.; Cheary, B.S.; Conaghan, K.J. Nitrogen, phosphorus, calcium, and magnesium applied individually or as a slow release or controlled release fertilizer increase growth and yield and affect macronutrient and micronutrient concentration and content of field-grown tomato plants. *Sci. Hortic.* **2016**, *211*, 420–430. [CrossRef]

12. Cong, Z.; Yazhen, S.; Changwen, D.; Jianmin, Z.; Huoyan, W.; Xiaoqin, C. Evaluation of waterborne coating for controlled-release fertilizer using Wurster fluidized bed. *Ind. Eng. Chem. Res.* **2010**, *49*, 9644–9647. [CrossRef]
13. Trenkel, M.E. *Slow- and Controlled-Release and Stabilized Fertilizers: An Option for Enhancing Nutrient Efficiency in Agriculture*, 2nd ed.; IFA: Paris, France, 2010; p. 160.
14. Shaviv, A. Controlled release fertilizers. In Proceedings of the IFA International Workshop on Enhanced-Efficiency Fertilizers, Frankfurt, Germany, 28–30 June 2005.
15. AAPFCO. *Official Publication No. 48*; Association of American Plant Food Control Officials, Inc.: West Lafayette, IA, USA, 1995.
16. Lubkowski, K.; Smorowska, A.; Grzmil, B.; Kozłowska, A. Controlled-release fertilizer prepared using a biodegradable aliphatic copolyester of poly (butylene succinate) and dimerized fatty acid. *J. Agric. Food. Chem.* **2015**, *63*, 2597–2605. [CrossRef] [PubMed]
17. Fertilizers Europe. 2019/2020 Overview. Available online: https://www.fertilizerseurope.com/wp-content/uploads/2020/07/AR-2019_20_32-pager-screen.pdf (accessed on 16 July 2022).
18. Obruca, S.; Benesova, P.; Oborna, J.; Marova, I. Application of protease-hydrolyzed whey as a complex nitrogen source to increase poly(3-hydroxybutyrate) production from oils by *Cupriavidus necator*. *Biotechnol. Lett.* **2014**, *36*, 775–781. [CrossRef]
19. Silva, L.F.; Taciro, M.K.; Raicher, G.; Piccoli, R.A.M.; Mendonça, T.T.; Lopes, M.S.G.; Gomez, J.G.C. Perspectives on the production of polyhydroxyalkanoates in biorefineries associated with the production of sugar and ethanol. *Int. J. Biol. Macromol.* **2014**, *71*, 2–7. [CrossRef] [PubMed]
20. Dietrich, K.; Dumont, M.-J.; Rio, L.F.D.; Orsat, V. Producing PHAs in the bioeconomy—Towards a sustainable bioplastic. *Sustain. Prod. Consum.* **2017**, *9*, 58–70. [CrossRef]
21. Surendran, A.; Lakshmanan, M.; Chee, J.Y.; Sulaiman, A.M.; Thuoc, D.V.; Sudesh, K. Can Polyhydroxyalkanoates Be Produced Efficiently From Waste Plant and Animal Oils? *Front. Bioeng. Biotechnol.* **2020**, *8*, 169. [CrossRef]
22. Nuttipon, Y.; Suchada, C.N. Toward non-toxic and simple recovery process of poly(3-hydroxybutyrate) using the green solvent 1,3-dioxolane. *Process Biochem.* **2018**, *69*, 197–207. [CrossRef]
23. Boyandin, A.N.; Kazantseva, E.A.; Varygina, D.E.; Volova, T.G. Constructing Slow-Release Formulations of Ammonium Nitrate Fertilizer Based on Degradable Poly(3-hydroxybutyrate). *J. Agric. Food Chem.* **2017**, *65*, 6745–6752. [CrossRef] [PubMed]
24. Lovochemie, a.s. Available online: https://www.lovochemie.cz/en (accessed on 26 July 2022).
25. TianAn Biopolymer. Available online: www.tianan-enmat.com (accessed on 26 July 2022).
26. Nafigate Corporation. Available online: https://www.nafigate.com/ (accessed on 26 July 2022).
27. Merck. Available online: https://www.sigmaaldrich.com/CZ/en (accessed on 26 July 2022).
28. SPECAC. Available online: https://specac.com/ (accessed on 26 July 2022).
29. Lach:ner. Available online: https://www.lach-ner.cz/en (accessed on 26 July 2022).
30. PENTA Chemicals Unlimited. Available online: https://www.pentachemicals.eu/en/ (accessed on 26 July 2022).
31. AvantorTM Delivered by VWRTM. Available online: https://www.vwr.com/ (accessed on 26 July 2022).
32. Panara. Available online: https://panaraplast.com/ (accessed on 26 July 2022).
33. STU. Available online: https://www.stuba.sk/english.html?page_id=132 (accessed on 26 July 2022).
34. PubChem: Ammonium Nitrate. Available online: https://pubchem.ncbi.nlm.nih.gov/compound/Ammonium-nitrate (accessed on 26 July 2022).
35. Miller, R.L.; Bradford, W.L.; Peters, N.E. *Specific Conductance: Theoretical Considerations and Application to Analytical Quality Control*; US Government Printing Office: Washington, DC, USA, 1988.
36. Metex Corporation Limited. Available online: https://www.metexcorporation.com/hand-held-instruments.html (accessed on 26 July 2022).
37. OriginLab. Available online: https://www.originlab.com/ (accessed on 26 July 2022).
38. Oseva. Available online: https://oseva.com/about-oseva/ (accessed on 26 July 2022).
39. Gee, G.W.; Bauder, J.W. Particle-size analysis. In *Methods of Soil Analysis Part 1—Physical and Mineralogical Methods*; Klute, A., Ed.; ASA and SSSA: Madison, WI, USA, 1986; pp. 383–411.
40. Schumacher, B.A. *Methods for the Determination of Total Organic Carbon (TOC) in Soils and Sediments*; United States Environmental Protection Agency, Environmental Sciences Division National, Exposure Research Laboratory: Las Vegas, NV, USA, 2002.
41. Zbíral, J.; Malý, S.; Váňa, M. (Eds.) *Soil Analysis III*, 3rd ed.; Central Institute for Supervising and Testing in Agriculture: Brno, Czech Republic, 2011; pp. 18–52. (In Czech)
42. Zbíral, J. *Analysis of Soils I. Unified Techniques*, 2nd ed.; Central Institute for Supervising and Testing in Agriculture: Brno, Czech Republic, 2002; p. 197. (In Czech)
43. Netto, A.L.; Campostrini, E.; Goncalves de Oliverira, J.; Bressan-Smith, R.E. Photosynthetic pigments, nitrogen, chlorophyll a fluorescence and SPAD-502 readings in coffee leaves. *Sci. Hortic.* **2005**, *104*, 199–209. [CrossRef]
44. Škarpa, P.; Klofáč, D.; Krčma, F.; Šimečková, J.; Kozáková, Z. Effect of Plasma Activated Water Foliar Application on Selected Growth Parameters of Maize (*Zea mays* L.). *Water* **2020**, *12*, 3545. [CrossRef]
45. Zbíral, J. *Plant Analysis: Integrated Work Procedures*; Central Institute for Supervising and Testing in Agriculture: Brno, Czech Republic, 2005; p. 192. (In Czech)
46. StatSoft, Inc. STATISTICA (Data Analysis Software System), Version 12. 2013. Available online: http://www.statsoft.com/ (accessed on 13 May 2021).

47. Rashidzadeh, A.; Olad, A. Slow-released NPK fertilizer encapsulated by NaAlg-g-poly(AA-co-AAm)/MMT superabsorbent nanocomposite. *Carbohydr. Polym.* **2014**, *114*, 269–278. [CrossRef]
48. Grubbs, J.B., III; Locklin, J.J. PLA/PHA Biodegradable Coatings for Seeds and Fertilizers. U.S. Patent Application Number 16/880083, 26 November 2020.
49. Liu, C.; Chen, F.; Li, Z.; Cocq, K.L.; Liu, Y.; Wu, L. Impact of nitrogen practices on yield, grain quality, and nitrogen-use efficiencyof crops and soil fertility in three paddy-upland cropping systems. *J. Sci. Food Agric.* **2021**, *101*, 2218–2226. [CrossRef]
50. Zhu, S.; Liu, L.; Yang, Y.; Shi, R. Application of controlled release urea improved grain yield and nitrogen use efficiency: Ameta-analysis. *PLoS ONE* **2020**, *15*, e0241481. [CrossRef] [PubMed]
51. Zhang, K.; Wang, Z.; Yu, Q.; Liu, B.; Duan, M.; Wang, L. Effect of controlled-release urea fertilizers for oilseed rape (*Brassica napus* L.) on soil carbon storage and CO_2 emission. *Environ. Sci. Pollut. Res.* **2020**, *27*, 31983–31994. [CrossRef]
52. Liao, J.; Liu, X.; Song, H.; Chen, X.; Zhang, Z. Effects of biochar-based controlled release nitrogen fertilizer on nitrogen-use efficiency of oilseed rape (*Brassica napus* L.). *Sci. Rep.* **2020**, *10*, 11063. [CrossRef] [PubMed]
53. Zhu, Z.L. Loss of fertilizer N from plants-soil system and the strategies and techniques for its reduction. *Soil Environ. Sci.* **2000**, *9*, 1–6.
54. Zvomuya, F.; Rosen, C.J.; Russelle, M.P.; Gupta, S.C. Nitrate leaching and nitrogen recovery following application of polyolefin-coated urea to potato. *J. Environ. Qual.* **2003**, *32*, 480–489. [CrossRef]
55. Mikkelsen, R.L.; Williams, H.M.; Behel, A.D., Jr. Nitrogen leaching and plant uptake from controlled-release fertilizers. *Fertil. Res.* **1994**, *37*, 43–50. [CrossRef]
56. Cabrera, R.I. Comparative evaluation of nitrogen release patterns from controlled-release fertilizers by nitrogen leaching analysis. *Hort. Sci.* **1997**, *32*, 669–673. [CrossRef]
57. Zheng, W.; Wan, Y.; Li, Y.; Liu, Z.; Chen, J.; Zhou, H.; Gao, Y.; Chen, B.; Zhang, M. Developing water and nitrogen budgets of a wheat-maize rotation system using auto-weighing lysimeter: Effects of blended application of controlled-release an un-coated urea. *Environ. Pollut.* **2020**, *263*, 114383. [CrossRef] [PubMed]
58. Zheng, W.; Zhang, M.; Liu, Z.; Zhou, H.; Lu, H.; Zhang, W.; Yang, Y.; Li, C.; Chen, B. Combining controlled-release urea and normal urea to improve the nitrogen use efficiency and yield under wheat-maize double cropping system. *Field Crops Res.* **2016**, *197*, 52–62. [CrossRef]
59. Xiao, Y.; Peng, F.; Zhang, Y.; Wang, J.; Zhuge, Y.; Zhang, S.; Gao, H. Effect of bag-controlled release fertilizer on nitrogen loss, greenhouse gas emissions, and nitrogen applied amount in peach production. *J. Clean. Prod.* **2019**, *234*, 258–274. [CrossRef]
60. Škarpa, P.; Mikušová, D.; Antošovský, J.; Kučera, M.; Ryant, P. Oil-Based Polymer Coatings on CAN Fertilizer in Oilseed Rape (*Brassica napus* L.) Nutrition. *Plants* **2021**, *10*, 1605. [CrossRef] [PubMed]
61. Tian, C.; Zhou, X.; Liu, O.; Peng, J.; Wang, W.; Zhang, Z.; Yang, Y.; Song, H.; Guan, C. Effects of a controlled-release fertilizer on yield, nutrient uptake, and fertilizer usage efficiency in early ripening rapeseed (*Brassica napus* L.). *J. Zhejiang Univ. Sci.* **2016**, *17*, 775–786. [CrossRef]
62. Wei, H.; Chen, Z.; Xing, Z.; Zhou, L.; Liu, Q.; Zhang, Z.; Jiang, Y.; Hu, Y.; Zhu, J.; Cui, P.; et al. Effects of slow or controlled release fertilizer types and fertilization modes on yield and quality of rice. *J. Integr. Agric.* **2018**, *17*, 2222–2234. [CrossRef]
63. Ma, Q.; Wang, M.; Zheng, G.; Yao, Y.; Tao, R.; Zhu, M.; Ding, J.; Li, C.; Guo, W.; Zhu, X. Twice-split application of controlled-release nitrogen fertilizer met the nitrogen demand of winter wheat. *Field Crops Res.* **2021**, *267*, 108163. [CrossRef]
64. Ye, Y.; Liang, X.; Chen, Y.; Liu, J.; Gu, J.; Guo, R.; Li, L. Alternate wetting and drying irrigation and controlled-release nitrogen fertilizer in late-season rice. Effects on dry matter accumulation, yield, water and nitrogen use. *Field Crops Res.* **2013**, *144*, 212–224. [CrossRef]
65. Lu, Y.; Sun, Y.; Liao, Y.; Nie, J.; Yie, J.; Yang, Z.; Zhoiu, X. Effects of the application of controlled release nitrogen fertilizer on rapeseed yield, agronomic characters and soil fertility. *Agric. Sci. Technol.* **2015**, *16*, 1226.
66. Murugan, P.; Ong, S.Y.; Hashim, R.; Kosugi, A.; Arai, T.; Sudesh, K. Development and evaluation of controlled release fertilizer using P(3HB-co-3HHx) on oil palm plants (nursery stage) and soil microbes. *Biocatal. Agric. Biotechnol.* **2020**, *28*, 101710. [CrossRef]
67. Koning, L.A.; Veste, M.; Freese, D.; Lebzien, S. Effects of nitrogen and phosphate fertilization on leaf nutrient content, photosynthesis, and growth of the novel bioenergy crop *Fallopia sachalinensis* cv. 'Igniscum Candy'. *J. Appl. Bot. Food Qual.* **2015**, *88*, 22–28.
68. Gianquinto, G.; Goffart, J.P.; Olivier, M.; Guarda, G.; Colauzzi, M.; Dalla Costa, L.; Delle Vedove, G.; Vos, J.; Mackerron, D.K.L. The Use of Hand-held Chlorophyll Meters As a Tool to Assess the Nitrogen Status and to Guide Nitrogen Fertilization of Potato Crop. *Potato Res.* **2004**, *47*, 35–80. [CrossRef]

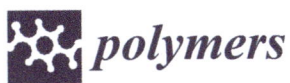

Article

Controlled Release Mechanism of Vancomycin from Double-Layer Poly-L-Lactic Acid-Coated Implants for Prevention of Bacterial Infection

Papon Thamvasupong [1] and Kwanchanok Viravaidya-Pasuwat [1,2,*]

[1] Department of Chemical Engineering, Faculty of Engineering, King Mongkut's University of Technology Thonburi, 126 Pracha-Utid Rd., Bangkok 10140, Thailand
[2] Biological Engineering Program, Faculty of Engineering, King Mongkut's University of Technology Thonburi, 126 Pracha-Utid Rd., Bangkok 10140, Thailand
* Correspondence: kwanchanok.vir@kmutt.ac.th; Tel.: +66-2-470-9222

Citation: Thamvasupong, P.; Viravaidya-Pasuwat, K. Controlled Release Mechanism of Vancomycin from Double-Layer Poly-L-Lactic Acid-Coated Implants for Prevention of Bacterial Infection. *Polymers* **2022**, *14*, 3493. https://doi.org/10.3390/polym14173493

Academic Editors: Arash Moeini, Pierfrancesco Cerruti and Gabriella Santagata

Received: 26 July 2022
Accepted: 23 August 2022
Published: 26 August 2022

Publisher's Note: MDPI stays neutral with regard to jurisdictional claims in published maps and institutional affiliations.

Copyright: © 2022 by the authors. Licensee MDPI, Basel, Switzerland. This article is an open access article distributed under the terms and conditions of the Creative Commons Attribution (CC BY) license (https://creativecommons.org/licenses/by/4.0/).

Abstract: Implantation failure due to bacterial infection incurs significant medical expenditure annually, and treatment tends to be complicated. This study proposes a method to prevent bacterial infection in implants using an antibiotic delivery system consisting of vancomycin loaded into poly-L-lactic acid (PLLA) matrices. A thin layer of this antibiotic-containing polymer was formed on stainless steel surfaces using a simple dip-coating method. SEM images of the polymeric layer revealed a honeycomb structure of the PLLA network with the entrapment of vancomycin molecules inside. In the in vitro release study, a rapid burst release was observed, followed by a sustained release of vancomycin for approximately 3 days. To extend the release time, a drug-free topcoat of PLLA was introduced to provide a diffusion resistance layer. As expected, the formulation with the drug-free topcoat exhibited a significant extension of the release time to approximately three weeks. Furthermore, the bonding strength between the double-layer polymer and the stainless steel substrate, which was an important property reflecting the quality of the coating, significantly increased compared to that of the single layer to the level that met the requirement for medical coating applications. The release profile of vancomycin from the double-layer PLLA film was best fitted with the Korsmeyer–Peppas model, indicating a combination of Fickian diffusion-controlled release and a polymer relaxation mechanism. More importantly, the double-layer vancomycin-PLLA coating exhibited antibacterial activity against *S. aureus*, as confirmed by the agar diffusion assay, the bacterial survival assay, and the inhibition of bacterial surface colonization without being toxic to normal cells (L929). Our results showed that the proposed antibiotic delivery system using the double-layer PLLA coating is a promising solution to prevent bacterial infection that may occur after orthopedic implantation.

Keywords: PLLA coating; orthopedic coating; vancomycin; antibacterial coating; controlled release; double-layer coating; adhesive strength

1. Introduction

Implant-associated osteomyelitis is a common problem in orthopedic surgery and is caused by bacterial infection, mostly from the Gram-positive bacterium *Staphylococcus aureus* (*S. aureus*), leading to bone destruction and necrosis [1]. The standard treatment for infections caused by implants consists of the removal of the infected implant, debridement of damaged tissue, and long-term antibiotic administration [2]. However, systemic antibiotic therapy may not be effective, as the bacteria can adhere to the surface of the implant and form a biofilm, preventing antibiotics from penetrating the infected area and causing bacterial resistance toward antibiotic treatment [3]. Local antibiotic delivery using poly (methyl methacrylate) (PMMA) or bone cement was then suggested. Although PMMA bone cements containing antibiotics are commercially available, many surgeons still manually mix antibiotics into PMMA during a revision procedure [4]. Only a limited amount of antibiotics (approximately 10–15 wt%) is recommended to be loaded into PMMA cement, as it can significantly compromise the mechanical strength of PMMA cement [5]. Another major shortcoming of this method is its drug elution kinetics. An initial burst release of antibiotic drugs from PMMA cement was observed in the first 4 h, followed by a steep decrease until day 9 [6]. A burst release in drug delivery is undesirable, as it can lead to a high antibiotic concentration at the target site, which can cause tissue toxicity [7]. In addition, since PMMA is nonbiodegradable, a second surgery is required to remove the bone cement, which could impair healing and increase treatment costs as well as patient risk. Therefore, the development of prostheses with local antibiotic delivery in a controlled manner could aid in successful infection management.

Recently, antibiotic delivery systems for alloy medical devices have been developed to prevent bacterial infection. One approach is to permanently crosslink an antibiotic drug onto a metal surface of an implant. A study by Martin Rottman et al., in which vancomycin was tethered on titanium discs, showed that the tethered discs displayed satisfactory outcomes to inhibit bacterial growth [8]. However, since the drug molecules are permanently fixed to the metal surface, they are not able to diffuse to the adjacent tissues or to body fluids through the biofilm [9]. In addition, the antimicrobial activity of the active agents can be lost due to irreversible binding to bacteria, making the antibiotic surface spent [10].

The concept of antibiotic release coatings was proposed to reduce bacterial adhesion and avoid biofilm formation on implant surfaces. In this technology, antimicrobial compounds are mixed with polymeric materials (biodegradable or nonbiodegradable) and deposited on the surface of an implant by impregnation, physical adsorption, conjugation or complexation [11]. The clear advantage of this method is that the polymer used for coating can be tailored to control the release of a drug molecule at a desired rate and duration [12]. One of the most commonly used polymers in medical applications as an antibiotic carrier is polylactic acid-based polymer (PLA) because of its biodegradable properties and its approval by the FDA for use in medical devices [13]. PLA exists in three forms: poly(L-lactic acid) (PLLA), poly(D-lactic acid) (PDLA), and their racemic mixture, poly(DL-lactic acid) (PDLLA). While both are semi-crystalline, PLLA exhibits higher crystallinity than PDLA, leading to better chemical stability, more structural integrity and a slower degradation rate, which is suitable for use in drug delivery applications [14,15]. More importantly, PLLA degrades to L-lactic acid, which is the only form of lactic acid already produced by humans and mammals. On the other hand, D-lactic acid, the by-product of PDLA degradation, can cause health problems [16]. These properties make PLLA attractive for use with orthopedic implants due to its high tensile strength, which can withstand the mechanical loading applied in implant applications.

Although PLA-based coatings have already been studied by various groups to deliver a variety of antibiotic active agents [17,18], the isoforms of these coatings have not been specified. Different PLA stereoisomers possess different characteristics, specifically the degree of crystallinity, which directly affects their degradation rates, leading to entirely different drug release behaviors from the polymer matrix [19]. Another important characteristic of drug delivery coatings that has rarely been studied is the adhesion strength between the coatings and the substrates. Not only does the adhesion strength reveal the ability of the coating to withstand the process of implantation, it also controls the performance and outcome of any coated orthopedic implants [20]. The separation of the coating from the substrate, either by cracking or delamination, can lead to decreased levels of drug release and poor performance of the drug delivery system.

In this study, a controlled release system using double-layer PLLA for antibiotic delivery from a metal implant was developed. Vancomycin, a commonly used antibiotic to treat osteomyelitis [21], was selected as an antibiotic drug model in this study. A dip coating method was used to fabricate the drug-containing PLLA film with a drug-free topcoat on stainless steel with prolonged drug release. A mathematical model was developed to predict and understand the mechanism of vancomycin release from the double-layer PLLA system. In addition, the bonding strength between the polymer layer and the stainless steel substrate was evaluated. The antibiotic activities against *S. aureus* and the biocompatibility of the vancomycin-PLLA delivery system were also investigated.

2. Materials and Methods

2.1. Coating of Vancomycin-PLLA on Stainless Steel Plates

A 316 L stainless steel sheet purchased from a local supplier was cut to a size of 3 cm × 3 cm using an electrical discharge machining (EDM) wire cut. Abrasive scrubbing was employed to ensure a smooth surface, followed by immersion in 65% nitric acid at 60 ± 5 °C for 30 min.

Vancomycin hydrochloride powder (Siam Bheasach, Bangkok, Thailand) and PLLA (5 dL/g pellets, M.W. ~325,000–460,000), supplied by Polysciences, Warrington, PA, USA, were mixed at a ratio of 1:5 by weight using dichloromethane (Sigma-Aldrich, Singapore) as a solvent. The concentration of vancomycin was maintained at 10 mg/mL for all experiments. The completely mixed solution was used for coating immediately. The pretreated stainless steel plates were dipped vertically in the polymer solution using forceps. The thickness of the coating was controlled by an immersion duration of 10 s with 15 s of withdrawal time. The coated plates were then dried in a vacuum oven at 37 °C for 24 h. To determine the loading of vancomycin in the PLLA coating, stainless steel substrates were weighed before and after the coating procedure. The amount of vancomycin in the coating was calculated according to the formulation of the PLLA solution. To fabricate a drug-free topcoat, the vancomycin-polymer-coated plates were dipped into the PLLA solution without vancomycin as described previously. Afterwards, the plates were dried in a vacuum at 37 °C for an additional 24 h. A schematic diagram of the preparation procedure for the double-layer PLLA coating is presented in Supplementary Materials Figure S1.

2.2. Characterization of the Vancomycin-Loaded PLLA Coating

2.2.1. Morphology and Thickness

Both the surface morphologies and the cross-section of the coating were observed using a scanning electron microscope (SEM, JEOL, JSM-6610LV, USA). Image analysis software (ImageJ, NIH) was used to estimate the average pore diameter and thickness. The degree of uniformity was indicated by a coefficient of variance (%CV), as shown in Equation (1):

$$\%CV = \frac{SD_d}{d_{ave}} \times 100 \qquad (1)$$

where d_{ave} is the average thickness of the coating and SD_d is the standard deviation of the thicknesses of the coating.

The dispersion of vancomycin in the PLLA matrix was investigated using elemental mapping of chlorine atoms, as chlorine was only present in vancomycin.

2.2.2. Adhesive Strength between the Vancomycin-PLLA Layer and Stainless Steel Substrate

The adhesive strength of the coating was measured by a universal tensile testing machine (SHIMADZU, AG-X, Japan). The test method was modified from ASTM D5179-16, a standard test method for measuring the adhesion of organic coatings in the laboratory by the direct tensile method [22]. The one-sided-coated stainless steel plate was attached to the holder using Epoxy-2216 Gray. The samples were heated to 60 °C for 10 min and left at room temperature for at least 24 h to accelerate the curing process. The test was carried out at a pulling rate of 0.5 mm/min under dry conditions at room temperature and terminated at the point of coating separation.

2.3. In Vitro Release of Vancomycin

The coated plates were immersed in simulated body fluid (SBF) to simulate the human body's condition and placed in a 37 °C incubator. The composition and preparation of SBF are described in Supplementary Materials Table S1. SBF solution was removed every hour and replenished with an equal volume of fresh SBF solution. The amount of vancomycin released was analyzed and fitted to various kinetic models.

The concentration of vancomycin was determined by the colorimetric method described by Fooks, J. R. et al. [23]. The sample containing vancomycin was treated with 5% sodium carbonate solution and 25% Folin–Ciocalteu reagent at a ratio of 2:2:1 by volume. Next, sonication was used to homogenize the solution. Color development occurred after the addition of the Folin–Ciocalteu reagent. Because the color intensity changed over time, the absorbance of the sample at a wavelength of 725 nm had to be measured two hours after the reaction for signal stability. The calibration curve between vancomycin concentration and its absorbance is shown in Supplementary Materials Figure S2.

2.4. Antibacterial Activity

The methods to test the antibacterial activity of the polymeric controlled release system using a survival assay and an agar diffusion assay were modified from Ordikhani et al. [24]. The test samples were 3 m × 3 cm stainless steel plates coated with vancomycin-PLLA, drug-free PLLA as a negative control, and a 10 mg/mL vancomycin solution-immersed paper filter as a positive control. The surfaces of these samples were sterilized by 70% ethanol, according to the protocol by Graziano et al. [25] Briefly, 70% ethanol was applied directly on all the sample surfaces, with a period of contact of at least 10 s, before being wiped clean with a sterile cloth. This process was repeated 3 times to ensure that there was no contamination. For the survival assay, S. aureus (ATCC25923) was inoculated in nutrient broth at 37 °C overnight. Five hundred microliters of S. aureus bacterial suspension at 5.0×10^5 CFU/mL was dropped on the coated plate samples and incubated at 37 °C. After 12 h of incubation, 100 µL of the exposed culture was serially diluted, plated on nutrient agar, and incubated overnight at 37 °C to determine the residual bacteria. The CFUs of the surviving bacteria were counted, and the bacterial reduction percentage was calculated, according to Equation (2).

$$\%\text{Reduction} = \frac{B - A}{B} \times 100 \qquad (2)$$

where B is the initial concentration of bacteria and A is the concentration of bacteria after 12 h of incubation.

In the agar diffusion assay, S. aureus bacteria at a density of 3.8×10^{10} CFU/mL were applied uniformly to an agar plate and further incubated for two hours. The coated plate samples, as described previously, were placed on the surface of each plate. After 24 h at 37 °C incubation, the average inhibition area was calculated with three replications. The equivalent diameter (D_{eq}) of inhibition was calculated as in Equation (3).

$$D_{eq} = \sqrt{\frac{\pi \cdot Area}{4}} \qquad (3)$$

The relative percentage of inhibition was calculated using Equation (4), where D_{eq+} is the equivalent diameter of inhibition of the positive control. In this study, the control was a 10 mg/mL vancomycin solution-immersed paper filter.

$$\text{Relative Percentage of Inhibition} = \frac{D_{eq,sample}}{D_{eq,+}} \times 100 \qquad (4)$$

Another antibacterial activity was crystal violet staining to study bacterial attachment to substrate surfaces. S. aureus was grown overnight in trypticase soy broth (TSB) at 37 °C and 200 rpm. Bacterial cells were collected by centrifugation at $10,000 \times g$ for three minutes and resuspended in normal saline solution (0.85% NaCl) to a final cell concentration of 10^8 CFU/mL (OD_{600} = 0.08–0.12). The cell suspensions were 100-fold diluted in TSB to a final cell concentration of 10^6 CFU/mL. Next, the broth was poured into Petri dishes containing samples and incubated at 37 °C for 24 h without shaking. The assay was performed in parallel with negative controls, which were an uncoated stainless steel plate and a PLLA-coated stainless steel plate without vancomycin. After incubation, the bacterial culture was removed by pipetting, and the samples were rinsed with PBS to remove planktonic cells. The contact surfaces were stained with 0.1% aqueous crystal violet for 30 min at room temperature. Bacterial colonization was determined by the presence of the violet color on the contact surface.

2.5. Cytotoxicity Study

The biocompatibility of the vancomycin-loaded PLLA coating was evaluated following the protocol of ISO-10993-5 [26] using the MTT assay. In this study, 3×3 cm^2 plain stainless steel plates, PLLA with and without vancomycin coated on stainless steel plates, were sterilized using 70% ethanol, as previously described, before being incubated in Dulbecco's modified Eagle's medium (DMEM) with low glucose (Sigma Aldrich, St. Louis, MO, USA) for 24 h at 37 °C. L929 mouse fibroblast cells (Mouse C3H/An) were seeded on a 96-well plate at a density of 10,000 cells/well and allowed to grow until confluence overnight. Afterwards, 100 µL of the media previously incubated with the test samples was added to monolayers of confluent L929 cells and further incubated for another 24 h. The cells cultured in growth medium served as a positive control, while the cells treated with 1% phenol served as a negative control. After 24 h of incubation, the spent medium was discarded and replaced with MTT solution according to the manufacturer's instructions (Thermo Fisher Scientific, Waltham, MA, USA). The absorbance of dissolved formazan salt was measured at 570 nm (Infinite® 200 Tecan, Grödig, Austria). The cell viability is presented as a percentage of the control cells, as shown in Equation (5).

$$Cell\ Viability\ (\%) = \frac{Absorbance\ at\ 570\ nm\ of\ the\ treated\ sample}{Absorbance\ at\ 570\ nm\ of\ the\ control\ sample} \times 100 \qquad (5)$$

2.6. Statistical Analysis

Data are presented as the mean ± standard deviation (SD). All results were analyzed by an unpaired *t* test. A statistically significant difference was defined at a 95% confidence level ($p < 0.05$).

3. Results and Discussion

3.1. Characterization of the Vancomycin-Loaded PLLA Coating

Fabrication of polymeric coatings can be achieved using various techniques, including dip coating, spin coating, spray coating, and solvent casting. Among these methods, dip coating is the most commonly used in laboratories and industries because of its simplicity, cost-effectiveness, reliability and reproducibility [27]. This method is especially useful for researchers to quickly and inexpensively optimize processing parameters for their studies. The thickness of the films fabricated by dip coating can be controlled with the duration of immersion, speed of withdrawal, and rheological properties of the coating solution [28].

The PLLA coating in this study appeared opaque and was uniformly distributed on the surface of the stainless steel. According to Figure 1, the SEM images of the PLLA coating on stainless steel without vancomycin (Figure 1A,B) show a honeycomb network of the polymer with pore sizes of 14.85 ± 3.07 µm. The formation of the honeycomb structure, previously explained by Escalé, L. et al. [29], began with the rapid evaporation of the organic solvent causing a quick temperature drop at the surface under the dew point, leading to fast condensation of water vapor into water droplets on the cold surface. The water droplets were, then, self-organized into an ordered hexagonal lattice. In a solution with a thermal gradient, the convection flow promoted the regular stacking of the water droplets. After the evaporation of both organic solvent and water droplets was completed, a polymer film was formed with an organized pattern of small pores, which were previously occupied by water droplets. Factors influencing the formation and morphology of a honeycomb structure include the relative humidity, volatility of solvents, air velocity, and polymer concentration [30]. For example, the pore size can be decreased with an increase in the polymer concentration due to less space for water droplets to occupy, while a low air velocity results in a larger pore diameter.

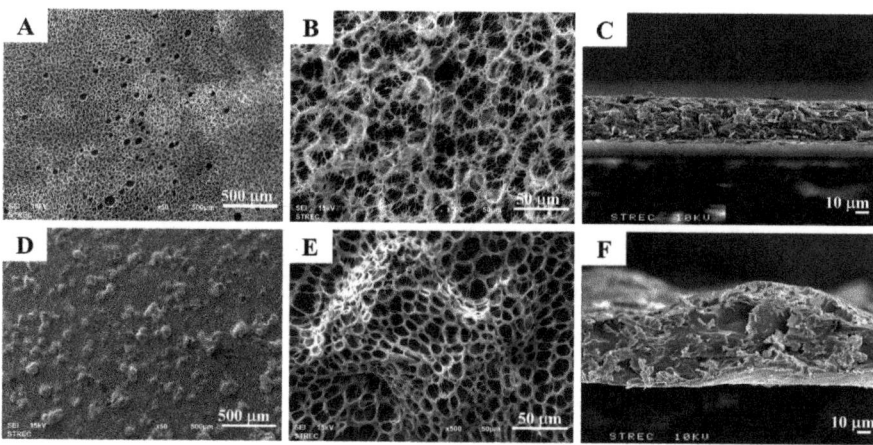

Figure 1. SEM images of the PLLA-coated layer: top view at (**A**) 50× and (**B**) 500×, and (**C**) cross-sectional view at 500× (the scale bars represent 500 μ, 50, and 10 μm, respectively). SEM images for the PLLA-coated layer containing vancomycin: top view at (**D**) 50× and (**E**) 500×, and (**F**) cross-sectional view at 500× (the scale bars represent 500, 50, and 10 μm, respectively).

The cross-sectional view revealed that the polymer had uniform thickness (Figure 1C), as confirmed by %CV in Table 1. When vancomycin was incorporated into the system, uniformly distributed drug particles were observed on the polymer surface (Figure 1D). When focusing on each drug particle, it was found that the particles did not incorporate themselves into the polymer network but rather were trapped inside the polymer layer (Figure 1E). This was confirmed by a cross-sectional view in which the vancomycin molecules filled in the space within the PLLA network structure, resulting in a denser matrix (Figure 1F). The porous network of PLLA remained the same, with an average pore size of 12.43 ± 4.08 μm. Judging from a cross-sectional view (Figure 1C,F), vancomycin clearly did not affect the average thickness of the coating, but the uniformity of the coating's thickness decreased, which was indicated by an increase in the coefficient of variance (Table 1). As the porous network of PLLA remained unchanged, it indicated that vancomycin did not interfere with the microscopic structure of the polymer. This was possibly because of the immiscibility of the hydrophobic PLLA and hydrophilic vancomycin molecules [31]. One of the important factors for long-term drug release is the hydrophobicity of the matrix, which provides a significantly slow degradation rate [32]. The loading of vancomycin into the PLLA coating was measured to be 0.57 ± 0.07 mg/cm^2. The amount of vancomycin loaded in the PLLA coating in this study was consistent with other studies using different polymer matrices, which demonstrated successful outcomes in treating bacterial infection [33,34].

The SEM images of the PLLA-vancomycin layer with a drug-free topcoat (Figure 2A) show a morphology similar to that of the coating of the PLLA network (Figure 1B). The drug particles were not easily seen from the top view due to the topcoat. In the cross-sectional view, a clear interface between the base layer and the topcoat could not be seen, possibly because the coating redissolved into the solvent during the second coating process, resulting in no clear interface between the two layers (Figure 2B). The thickness of the coating was found to be 54.09 ± 3.46 µm, which consisted of 41.35 ± 2.53 µm of the base vancomycin-PLLA layer and 12.74 ± 0.93 µm of the drug-free PLLA topcoat. According to the %CV, adding the drug-free topcoat remarkably improved the thickness uniformity from that of the PLLA containing vancomycin coating.

Table 1. Thickness of the PLLA films and their coefficients of variance.

Formulation	Average Thickness (µm)	Coefficient of Variance (%)
PLLA coating	44.57 ± 2.51	5.63
Vancomycin-PLLA coating	44.43 ± 10.13	22.79
Vancomycin-PLLA coating with a drug-free topcoat	54.09 ± 3.46	6.39

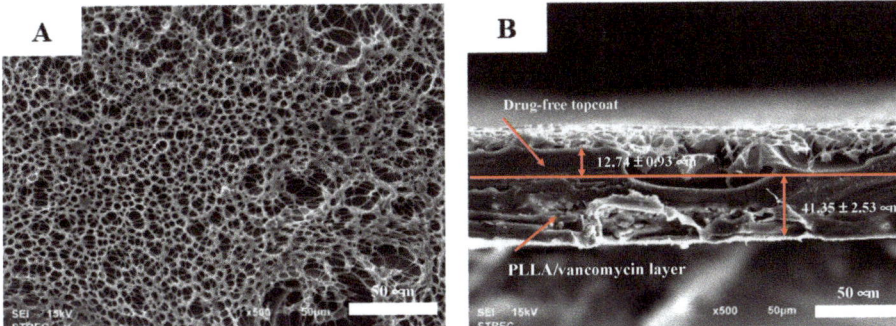

Figure 2. SEM images of the PLLA layer containing vancomycin with the drug-free topcoat: (**A**) top view at 500×, and (**B**) cross-sectional view at 500× (scale bars represent 50 µm).

The adhesive bond strength between the coating layer and the substrate is a major consideration in practice. The detachment of the coating from the substrate incurs unpleasant effects on the implants and the surrounding tissue [35,36]. The adhesion strengths between the coating and the stainless steel substrate determined by the direct tensile method are reported in Table 2. The bonding strength between the PLLA coating and a stainless steel substrate was 1.80 ± 0.49 MPa. The adhesion strength between the PLLA coating loaded with vancomycin and a stainless steel substrate was 1.87 ± 0.45 MPa, which was not significantly different from that of the PLLA coating without vancomycin. The bond strength between the PLLA layer and the stainless steel substrate in this study was of the same level as the bond strength between PLLA and metallic magnesium [37,38]. However, the double-layer PLLA coating with a topcoat exhibited a considerably higher bonding strength, which exceeded the epoxy strength (approximately 5 MPa). This could be explained by the longer processing time of the coating. Similar to the study by Morris B., an increase in the processing time reduced the stress within the coating layer, leading to a thicker film and stronger bond strength [39,40]. Although there is currently no ISO requirement for orthopedic implant coatings, the requirement for dental implant coatings can be used as a reference. According to ISO 10477, the adhesive bond strength at the interface between biocomposites and substrate should be greater than 5 MPa [41]. Only the adhesive bond strength of the vancomycin-PLLA coating with a topcoat met the ISO requirement. The topcoat layer clearly produced a stronger and more durable bond with the metal substrate, leading to the long-term reliability of the coating in implant applications.

Table 2. Bonding strength between the polymeric coatings and stainless steel substrates using a direct tensile method.

Sample	Bonding Strength (MPa)
PLLA coating and stainless steel	1.80 ± 0.49
Vancomycin-PLLA coating and stainless steel	1.87 ± 0.45
Vancomycin-PLLA coating with a topcoat and stainless steel	>5 MPa

3.2. Release Profile and Release Mechanism of Vancomycin from the PLLA Coating

The in vitro cumulative release of vancomycin from PLLA-coated stainless steel was investigated for up to 72 h (Figure 3A). The release profile of vancomycin shows a sudden burst release of approximately 95% of the total release in the first 12 h. After 72 h, there was no noticeable amount of vancomycin released into the receiving medium. The cumulative amount of vancomycin released from the system was calculated to be approximately 55% of the amount of vancomycin loaded. To investigate the kinetics of the drug release, several theoretical and empirical models were used to fit the experimental release data up to 24 h, as the amount of vancomycin release after 24 h was considered negligible. Table 3 shows the equations, rate constants (k) and exponents (n) of several selected kinetic models with their coefficients of determination (R^2) to evaluate the goodness of fit of different models. The release data of vancomycin from the PLLA coating fit best with the Korsmeyer–Peppas kinetic model, with the highest coefficient of determination (R^2) close to 0.99. Using the Korsmeyer–Peppas model, the initial release rate of vancomycin during the burst release period in the first six hours was calculated to be 1.6 mg/h with a calculated release exponent (n) of 0.21.

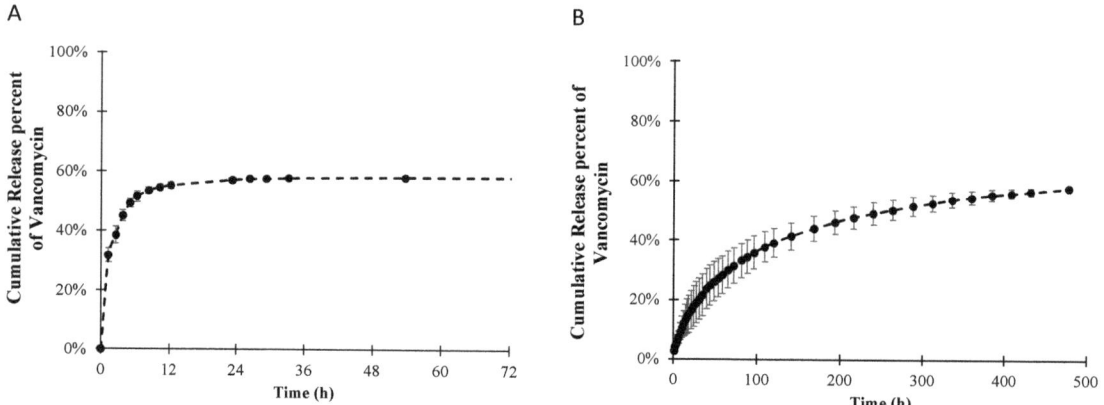

Figure 3. Release profiles of vancomycin from (**A**) a single PLLA layer and (**B**) double layer of PLLA with a drug-free topcoat.

A drug-free PLLA topcoat was added to the existing vancomycin-PLLA coating to extend the drug release, as shown in Figure 3B. No initial burst release was observed from the double-layer coating. In the first 2 days, the initial release rate was reduced from 1600 µg/h in the single-layer vancomycin-PLLA coating to 50 µg/h with another layer of drug-free PLLA topcoat (double layer), which was an approximately 95% reduction. Furthermore, the release duration was remarkably extended to longer than 20 days. Postoperative bacterial infection leading to acute osteomyelitis typically occurs within 2 weeks [42]. The administration of antibiotics was recommended for the first three weeks after the surgery [43]. The current vancomycin release duration would be able to treat an early bacterial infection, which could subsequently prevent osteomyelitis.

Elemental mapping of chlorine was performed to track the location of the vancomycin molecules and allow for the observation of vancomycin dispersion in the polymer matrix before and after the release experiment. Figure 4A displays the uniform dispersion of vancomycin in the polymer system before the in vitro release study. The cross-sectional view confirmed that the vancomycin molecules were entrapped under the layer of the PLLA matrix (Figure 4B). After immersion in the release medium for three weeks, chlorine element mapping illustrated a significant change in the vancomycin density in the polymer matrix (Figure 4C). Significantly less vancomycin was observed in the PLLA layer after three weeks, and the vancomycin molecules were mostly located near the surface of the coating.

Table 3. Equations and parameters of the theoretical and empirical kinetics models for the vancomycin-PLLA coatings: single layer and double layer.

Model	Equation	Vancomycin-PLLA Coating			Vancomycin-PLLA Coating with a Drug-Free Topcoat		
		K	n	R^2	k	n	R^2
Zero order	$\frac{M_t}{M_\infty} = k_0 t$	0.0702	-	0.7284	0.0032	-	0.3945
First-order	$\frac{M_t}{M_\infty} = 1 - e^{-k_1 t}$	0.2253	-	0.9191	0.0096	-	0.9744
Higuchi	$\frac{M_t}{M_\infty} = k_H \sqrt{t}$	0.2882	-	0.9401	0.0557	-	0.9689
Korsmeyer–Peppas	$\frac{M_t}{M_\infty} = k t^n$	0.315	0.206	0.9907	0.049	0.54	0.9995

Note: where M_t is the cumulative mass of an active agent in the receiving medium over time t; M_∞ is the total mass of the active agent in the receiving medium at time ∞; k_0, k_1, k_H and k are the release constants for the zero-order, first-order, Higuchi, and Korsmeyer–Peppas models, respectively; and n is the release exponent.

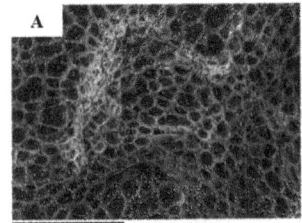
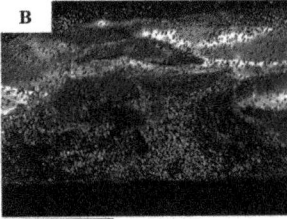
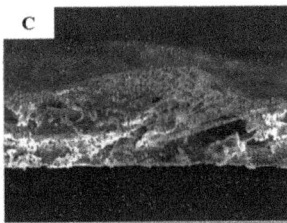

Figure 4. Chlorine elemental mapping of the PLLA layer containing vancomycin before the in vitro release experiment: (**A**) top view and (**B**) cross-sectional view, and (**C**) the cross-sectional view of PLLA containing vancomycin after 3 weeks in SBF (the scale bars represent 100 µm).

To further investigate the release mechanism of vancomycin from the PLLA layer with the drug-free topcoat, the experimental data from Figure 3B were fitted with the same release kinetic models. According to Table 3, based on the coefficient of determination (R^2), the experimental data would fit best with the Korsmeyer–Peppas model with an exponent of 0.54 (n = 0.54).

In the Korsmeyer–Peppas kinetic model, the exponent n is used to characterize the mechanism of drug release. For a thin film, $n < 0.5$ corresponds to a Fickian diffusion mechanism, $0.5 < n < 1$ to non-Fickian transport, and $n > 1$ to case II transport or zero order release [44]. Since the release exponent of the formulation without a topcoat was less than 0.5, the release mechanism of vancomycin from the PLLA coating was thought to be Fickian diffusion, in which the release of an active compound depended on the drug content in the system. At the earlier stage, the vancomycin content was high, leading to the presence of burst release. The release rate was significantly lower after the reduction in the vancomycin content in the system. As shown by element mapping of chlorine in Figure 4, it is believed that the movement of the vancomycin molecules was controlled by the concentration gradient driving force [40]. This phenomenon implied that vancomycin diffused through the polymer matrix toward the surface of the PLLA coating and was released into the receiving medium.

Typically, a burst release in a drug delivery system is undesirable and cannot be used for a long-term release scenario, not to mention the toxic side effects due to the high initial concentration of a drug, leading to a decrease in the efficiency of the drug delivery system. In this study, the initial burst release of vancomycin from the PLLA layer was observed in the first 12 h, resulting in the released concentration being far above its minimum inhibitory concentration (MIC), which was approximately 2 μg/mL [45]. Too high a concentration of an antibiotic drug in the first twelve hours may be toxic to human cells [46]. Therefore, the release rate of vancomycin must be reduced to achieve the sustained drug delivery goal. Because the release mechanism is diffusion controlled, the release rate can be reduced by adding mass transfer resistance. One method to increase the mass transfer resistance is to fabricate a drug-free barrier, increasing the diffusion distance. In a previous study, the release kinetics for two different formulations of sirolimus drug-eluting stents showed a remarkable reduction in the release rate by adding a drug-free topcoat layer [47]. Another method used to reduce the burst release effect is the crosslinking of the polymer layer using crosslinking agents, which results in the creation of additional diffusional resistance [48]. The crosslinking agents are normally cytotoxic and usually need further validation of safety. To avoid using toxic crosslinking agents, heat treatment or ultraviolet radiation are alternatives [49]. However, high temperatures and exposure to radiation might cause some antibiotics to become inactive or degraded [50,51]. On the other hand, the topcoat method was able to provide the same level of safety, but the release mechanism tended to be more complicated. The release kinetics could be based on more than one mechanism [52]; for example, the combination of release mechanisms, including Fickian diffusion control and polymer relaxation control, was observed [53].

In this study, a drug-free PLLA topcoat was applied to the existing vancomycin-loaded PLLA layer to provide an additional mass transfer resistance layer that was believed to be able to reduce the release of vancomycin during the burst period and extend the release of vancomycin in a controlled manner. By adding the drug-free topcoat, the diffusion distance increased, leading to more resistance to mass transfer. As a result, a lower amount of vancomycin was delivered, leading to a lower release rate but longer release duration as a consequence of the resistant layer provided by the topcoat. The n value (0.54) indicates that the vancomycin delivery system was possibly controlled by two mechanisms, including both Fickian diffusion-controlled and polymer relaxation-controlled mechanisms. To further understand the release mechanism, the Peppas and Sahlin equation [48], a modification of the Korsmeyer–Peppas equation, as shown in Equation (6), was applied.

$$\frac{M_t}{M_\infty} = k_1 t^m + k_2 t^{2m} \qquad (6)$$

where k_1 and k_2 are the Fickian contribution coefficient and relaxation contribution coefficient, respectively. The first term of the exponent m of Equation (6) represents the release due to the Fickian diffusion mechanism, while the second term of the exponent $2m$ represents the release due to the polymer relaxation mechanism. By fitting the release profile of vancomycin from the PLLA layer with drug-free topcoat, k_1 and k_2 were found to be 0.058 h$^{-0.5}$ and 0.001 h^{-1}, respectively, with a release exponent (m) of 0.5 and R^2 of 0.9954. These coefficients, k_1 and k_2, were then used to calculate the ratio of relaxational (R) over Fickian (F) contributions as a function of time, as shown in Equation (7) [48].

$$\frac{R}{F} = \frac{k_2}{k_1} t^m \qquad (7)$$

The Peppas–Sahlin model shows that the ratio of relaxation over the Fickian diffusion contributions increased over time, indicating that the effect of polymer relaxation became more pronounced on controlling the release of vancomycin as the time increased (Figure 5). However, the release of vancomycin from the system was still largely controlled by Fickian diffusion even after 16 days, as the ratio was still less than 0.5. According to the Peppas–Sahlin model, 98.30% of vancomycin release in the first hour was the result of Fickian diffusion, and the rest (1.70%) was from the polymer relaxation mechanism, while at later times, a lower amount of vancomycin (73.55%) was released by Fickian diffusion, and approximately 26.45% of vancomycin was released because of polymer relaxation. Fickian diffusion showed a smaller effect on vancomycin release at later times, possibly because the vancomycin content in the matrix was lower than that at earlier time points, resulting in a lower concentration gradient for diffusion. Additionally, previous studies revealed that PLLA tended to be slowly swollen and degraded after water molecules diffused into the matrix. The incorporation of drug particles in PLLA could also lead to an increase in the swelling index [54,55]. Therefore, our PLLA delivery system might have been swollen slightly after being incubated in aqueous solution for a few weeks. Compared with the formulation without the topcoat, the polymer was not exposed to the receiving medium long enough to be swollen. As a result, the mechanism of vancomycin release was expected to be only Fickian diffusion-controlled.

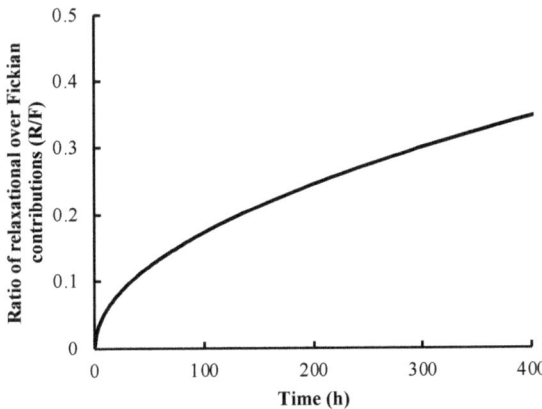

Figure 5. The ratio of relaxation over the Fickian contributions from the Peppas–Sahlin model for the vancomycin-PLLA coating with a drug-free topcoat.

3.3. Antibacterial Activity of Vancomycin-Loaded PLLA

According to the survival assay results in Table 4, the vancomycin-PLLA coating with and without a drug-free topcoat was able to reduce the *S. aureus* population by approximately 81.77% and 91.77%, respectively. The PLLA coating without vancomycin, which served as a negative control, did not show any antibacterial activities against *S. aureus*, while the positive control, which was vancomycin-immersed filter paper, could completely eliminate bacteria from the system. An agar diffusion assay was also performed to confirm the antibacterial activities of the proposed vancomycin delivery system (Table 4 and Figure 6). Similarly, the vancomycin-PLLA coating without the topcoat gave a larger inhibition zone of 19.71 cm^2 (or D_{eq} of 3.93 cm), while the inhibition area of the vancomycin-PLLA coating with the drug-free topcoat was estimated to be 12.97 cm^2 (or D_{eq} of 3.19 cm). As expected, there was no inhibition zone for the negative control, but the positive control provided the largest inhibition zone of 24.56 cm^2 (or D_{eq} of 4.39 cm). Compared to the control, the relative percentages of inhibition were calculated to be 72.67% and 89.52% for the vancomycin-PLLA coating with and without a topcoat, respectively. The presence of bacterial colonization, as indicated by crystal violet staining was observed on the stainless steel substrate (Figure 7A) and the PLLA coating without vancomycin (Figure 7B). The violet color was clearly visible on both samples, which represented the presence of the *S. aureus* culture. In contrast, the vancomycin-PLLA coating with and without a topcoat showed colorless samples, indicating no bacterial attachment on either surface (Figure 7C,D). Our results confirmed the antibacterial activity of the proposed formulations and that the concentration of vancomycin released from the system was sufficient to prevent bacterial surface attachment and growth on the material surfaces.

Table 4. Results of the survival assay and agar diffusion assay, which are reported as the average ± standard deviation.

	% Reduction in *S. aureus* Population	Inhibition Area (cm^2)
PLLA-vancomycin without drug-free topcoat	91.77 ± 0.42	19.71 ± 2.25
PLLA-vancomycin with drug-free topcoat	81.77 ± 9.57	12.97 ± 1.3
Positive control	100.00 ± 0.00	24.56 ± 4.96
Negative control	Increase in the number of CFU	No inhibition zone

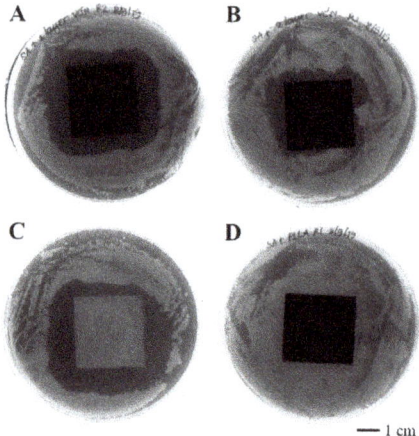

Figure 6. Inhibition zones of (A) vancomycin-loaded PLLA-coated plate without the drug-free topcoat, (B) vancomycin-loaded PLLA-coated plate with the drug-free topcoat, (C) 10 mg/mg vancomycin-loaded PVDF membrane as a positive control, and (D) PLLA-coated plate without vancomycin as a negative control.

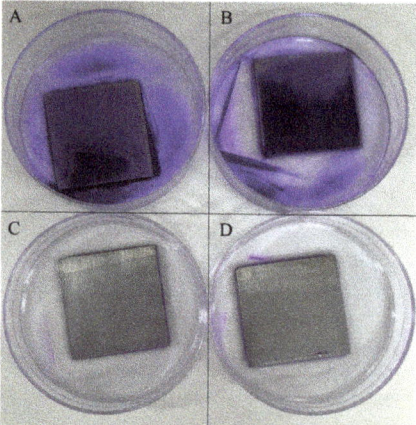

Figure 7. Crystal violet staining assay (A) stainless steel (B) PLLA-coated stainless steel (C) vancomycin loaded PLLA-coated stainless steel, and (D) vancomycin loaded PLLA-coated stainless steel with a drug-free topcoat.

3.4. Cytotoxicity of the Vancomycin-PLLA Delivery System

The cytotoxicity of the vancomycin-PLLA delivery system was evaluated using an MTT assay, as shown in Figure 8A. Mouse fibroblast cells (L929) were exposed to the extracted media from plain stainless steel, PLLA coating without vancomycin, vancomycin-PLLA coating on stainless steel substrates, and 1% phenol (a negative control) for 24 h. As shown in Figure 8, the cell viability of L929 cells incubated with the spent medium from the stainless steel substrate was the highest at 97.35 ± 2.17%, while a slightly lower cell viability of L929 of approximately 90% was observed when the cells were incubated with the spent medium from the stainless steel substrates coated with drug-free PLLA and vancomycin-PLLA. According to ISO 10993-5: Biological evaluation of medical devices—Part 5, which stated that any materials with cell viability greater than 70% are considered noncytotoxic, our results showed that the cell viability of the PLLA coating with and without vancomycin was well above 70%. The morphology of L929 cells also confirmed that the extracts of all test samples were nontoxic (Figure 8B). The cells exposed to the extracts had normal fibroblast-like morphology and were well spread, similar to that of the cells maintained in the culture medium. Therefore, both quantitative and qualitative cytotoxicity assessments indicated that the vancomycin-loaded PLLA delivery system was not cytotoxic and could be considered safe for use in future applications.

There has been considerable interest in coating orthopedic implants with antibiotic impregnated PMMA bone cement, which has already been approved by the FDA [56]. Coating of antibiotic-PMMA on implants is usually carried out using a molding technique, resulting in a film with a thickness between 1–3 mm [56]. The release characteristic of PMMA consists of an initial burst release, followed by a decrease in the drug release to a level below its therapeutic range [6]. The low efficiency of the antibiotic-PMMA coating for local drug delivery is the main barrier for its clinical use. Another complication arising from the use of PMMA-coated implants is implant-PMMA debonding, which is reported to occur in approximately 10–30% of cases [56]. In this proof-of-concept study, PLLA was shown to represent a possible alternative solution, as it has been demonstrated to be able to deliver an antibacterial agent in controlled and sustained manners and is effective against *S. aureus*, the main cause of osteomyelitis. However, more studies are needed to thoroughly evaluate the potential use of this delivery system in preclinical and clinical settings.

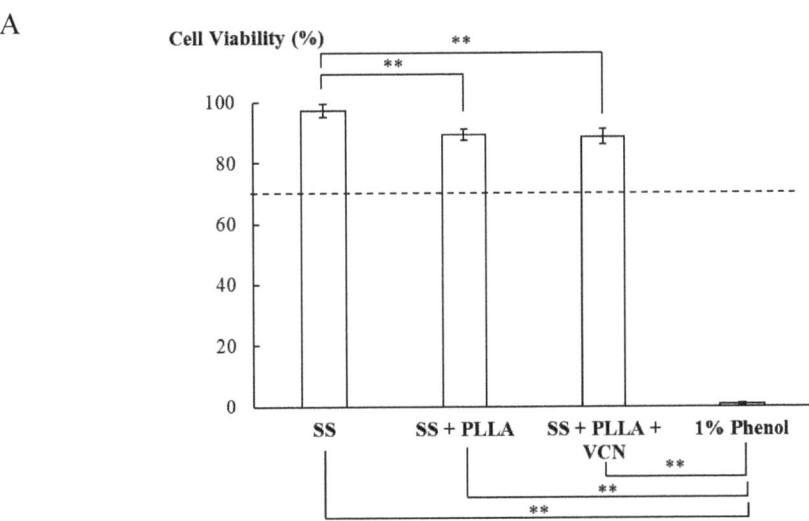

Figure 8. *Cont.*

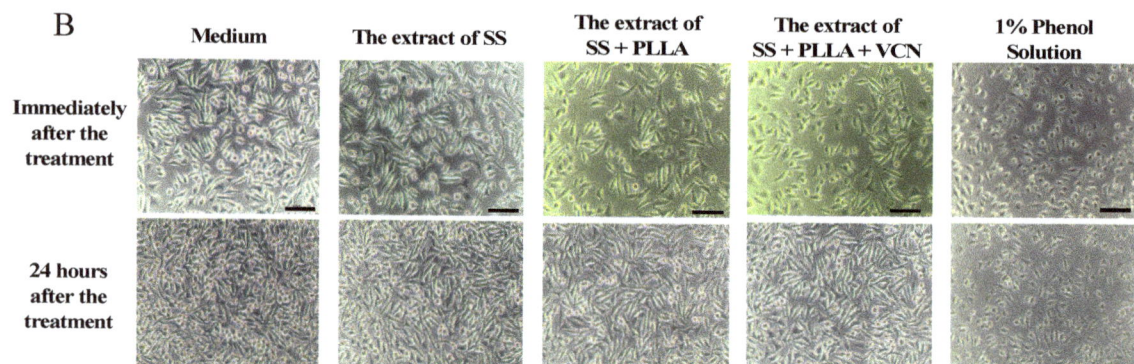

Figure 8. (**A**) Cell viability of different substrates including stainless steel (SS) as a negative control, PLLA coated on stainless steel (SS + PLLA), PLLA containing vancomycin coated on stainless steel (SS + PLLA + VCN), and 1% phenol as a positive control (** indicates $p < 0.001$). (**B**) Morphology of L929 cells after exposure to its culture medium (control), the extract of stainless steel (SS), the extract of PLLA coated on stainless steel (SS + PLLA), the extract of PLLA containing vancomycin coated on stainless steel (SS + PLLA + VCN) and the culture medium containing 1% phenol.

4. Conclusions

In this study, we successfully developed a vancomycin delivery system using PLLA coated on stainless steel. The PLLA film showed a honeycomb structure with a thickness of approximately 44 μm. After incorporation with vancomycin, the structure of the polymer film changed slightly, with vancomycin molecules trapped inside the polymer network, resulting in nonuniformity of the coating thickness. However, the thickness of the PLLA layer after vancomycin loading remained the same, leading to no change in the bonding strength between the coating and the stainless steel substrate. In the in vitro release experiment, the burst release of vancomycin was observed at an earlier time, and the sustained release was present afterwards. The release period of vancomycin was approximately three days for the single-layer coating. The release kinetics of vancomycin from the system fit best with the Korsmeyer–Peppas model, indicating that vancomycin release was dominantly controlled by Fickian diffusion. To solve the problem of undesirable burst release and to extend the release of vancomycin, a drug-free layer of PLLA was added to the existing vancomycin-loaded PLLA layer, acting as a diffusion resistance layer. With the drug-free topcoat, the bonding strength between the coating and the substrate significantly increased, indicating a stronger and more durable bond, which would be desirable in real applications. Moreover, the new system was able to extend the release period of vancomycin for more than 20 days. Although the mechanism of vancomycin release remained primarily controlled by Fickian diffusion, the contribution of the polymer relaxation to the release mechanism became more pronounced at later time points. The antibacterial assays confirmed that vancomycin embedded in a PLLA coating with and without a drug-free topcoat had the ability to inhibit the growth of bacteria without being cytotoxic to normal cells. Thus, the double-layer PLLA-vancomycin coating may represent a promising approach that can possibly prevent bacterial infection associated with implants.

Supplementary Materials: The following supporting information can be downloaded at: https://www.mdpi.com/article/10.3390/polym14173493/s1, Figure S1: Schematic diagram of the preparation procedure of the double-layer PLLA-vancomycin coating on a stainless steel substrate; Figure S2: Calibration curve of vancomycin in SBF for colorimetric quantitation; Table S1: Preparation and composition of simulated body fluid (SBF).

Author Contributions: Conceptualization and methodology, P.T. and K.V.-P.; data curation and investigation, P.T.; writing—original draft preparation, P.T.; writing—review and editing, K.V.-P.;

supervision, K.V.-P.; funding acquisition, K.V.-P. All authors have read and agreed to the published version of the manuscript.

Funding: This research was supported by the Petchra Pra Jom Klao Ph.D. Research Scholarship from King Mongkut's University of Technology Thonburi (Agreement No. 5/2015), FY2017 Thesis Grants for Doctoral Degree Students from the National Research Council of Thailand (Agreement No. NRCT(A)(DPARB) 8/2017, Appendix 7, No. 4 on the grantee list) and Thailand Science Research and Innovation (TSRI) Basic Research Fund: Fiscal year 2021 under project number FRB640008.

Institutional Review Board Statement: Not applicable.

Informed Consent Statement: Not applicable.

Data Availability Statement: Data presented in this study are available on request from the corresponding author.

Conflicts of Interest: The authors declare no conflict of interest.

References

1. Jennsen, L.K.; Koch, J.; Aalbaek, B.; Moodley, A.; Bjarnsholt, T.; Kragh, K.N.; Petersen, A.; Jensen, H.E. Early implant-associated osteomyelitis in a peri-implanted bacterial reservoir. *APMIS* **2017**, *125*, 38–45. [CrossRef]
2. Darouiche, R.O. Treatment of infections associated with surgical implants. *N. Engl. J. Med.* **2004**, *350*, 1422–1429. [CrossRef] [PubMed]
3. Gimza, B.D.; Cassat, J.E. Mechanism of antibiotic failure during Straphylococcus aureus osteomyelitis. *Front. Immunol.* **2021**, *12*, 638085. [CrossRef] [PubMed]
4. Jaeblon, T. Polymethylmethacrylate: Properties and contemporary uses in orthopaedics. *J. Am. Acad. Orthop. Surg.* **2010**, *18*, 297–305. [CrossRef] [PubMed]
5. Arora, M.; Chan, E.K.; Gupta, S.; Diwan, A.D. Polymethylmethacrylate bone cements and additives: A review of the literature. *World J. Orthop.* **2013**, *4*, 67–74. [CrossRef]
6. Gálvez-López, R.; Peña-Monje, A.; Antelo-Lorenzo, R.; Guardia-Olmedo, J.; Moliz, J.; Hernández-Quero, J.; Parra-Ruiz, J. Elution kinetics, antimicrobial activity, and mechanical properties of 11 different antibiotic loaded acrylic bone cement. *Diagn. Microbiol. Infect. Dis.* **2014**, *78*, 70–74. [CrossRef]
7. Duewelhenke, N.; Krut, O.; Eysel, P. Influence on Mitochondria and Cytotoxicity of Different Antibiotics Administered in High Concentrations on Primary Human Osteoblasts and Cell Lines. *Antimicrob. Agents Chemother.* **2007**, *51*, 54–63. [CrossRef]
8. Rottman, M.; Goldberg, J.; Hacking, S.A. Titanium-Tethered Vancomycin Prevents Resistance to Rifampicin in Staphylococcus aureus in vitro. *PLoS ONE* **2012**, *7*, e52883. [CrossRef]
9. Zhang, B.G.X.; Myers, D.E.; Wallace, G.G.; Brandt, M.; Choong, P.F.M. Bioactive Coatings for Orthopaedic Implants—Recent Trends in Development of Implant Coatings. *Int. J. Mol. Sci.* **2014**, *15*, 11878–11921. [CrossRef]
10. Antoci, V.J.; Adams, C.S.; Hickok, N.J.; Shapiro, I.M.; Parvizi, J. Antibiotics for Local Delivery Systems Cause Skeletal Cell Toxicity In Vitro. *Clin. Orthop. Relat. Res.* **2007**, *462*, 200–206. [CrossRef]
11. Shahid, A.; Aslam, B.; Muzammil, S.; Aslam, N.; Shahid, M.; Almatroudi, A.; Allemailem, K.S.; Saqalein, M.; Nisar, M.A.; Rasool, M.H.; et al. The prospects of antimicrobial coated medical implants. *J. Appl. Biomater. Funct. Mater.* **2021**, *19*, 22808000211040304. [CrossRef] [PubMed]
12. Qiu, H.; Si, Z.; Lou, Y.; Feng, P.; Wu, X.; Hou, W.; Zhu, Y.; Chan-Park, M.B.; Xu, L.; Huang, D. The mechanisms and the applications of antibacterial polymers in surface modification on medical devices. *Front. Bioeng. Biotechnol.* **2020**, *8*, 910. [CrossRef] [PubMed]
13. DeStefano, V.; Khan, S.; Tabada, A. Applications of PLA in modern medicine. *Eng. Regen.* **2020**, *1*, 76–87. [CrossRef]
14. Ribeiro, D.; Sencadas, V.; Costa, C.M.; Ribelles, J.L.G.; Lanceros-Mendez, S. Tailoring the morphology and crystallinity of poly(L-lactide acide) electrospun membranes. *Sci. Technol. Adv. Mater.* **2011**, *12*, 015001. [CrossRef] [PubMed]
15. Farah, S.; Anderson, D.G.; Langer, R. Physical and mechanical properties of PLA, and their functions in widespread applications—A comprehensive review. *Adv. Drug Deliv. Rev.* **2016**, *107*, 367–392. [CrossRef]
16. Pohanka, M. D-lactic acid as a metabolite: Toxicity, diagnosis, and detection. *BioMed Res. Int.* **2020**, *2020*, 3419034. [CrossRef]
17. He, C.; Chen, Q.; Yarmolenko, M.A.; Rogachev, A.A.; Piliptsou, D.G.; Jiang, X.; Rogachev, A.V. Structure and antibacterial activity of PLA-based biodegradable nanocomposite coatings by electron beam deposition from active gas phase. *Prog. Org. Coat.* **2018**, *123*, 282–291. [CrossRef]
18. Dwivedi, A.; Mazumder, A.; Nasongkla, N. Layer-by-layer nanocating of antibacterial noisome on orthopedic implants. *Int. J. Pharm.* **2018**, *547*, 235–243. [CrossRef]
19. Harting, R.; Johnston, K.; Petersen, S. Correlating in vitro degradation and drug release kinetics of biopolymer-based drug delivery systems. *Int. J. Biobased Plast.* **2019**, *1*, 8–21. [CrossRef]
20. Lei, W.-S.; Mittal, K.; Yu, Z. Adhesion measurement of coatings on biodevices/implants: A critical review. *Rev. Adhes. Adhes.* **2016**, *4*, 367–397. [CrossRef]

21. Cevher, E.; Orhan, Z.; Mulazimmouglu, L.; Sensoy, D.; Alper, M.; Yildiz, A.; Ozsoy, Y. Characterization of biodegradable chitosan microspheres containing vancomycin and treatment of experimental osteomyelitis caused by methicillin-resistant *Staphylococcus aureus* with prepared microspheres. *Int. J. Pharm.* **2006**, *317*, 127–135. [CrossRef] [PubMed]
22. Arizmendi-Morquecho, A.; Chavez-Valdez, A.; Navarro, C.H.; Moreo, K.J. Performance evaluation of chitosan/hydroxyapatite composite coating on ultrahigh molecular weight polyethylene. *Polym. Test.* **2013**, *32*, 32–37. [CrossRef]
23. Fooks, J.R.; McGilveray, I.J.; Strickland, R.D. Colorimetric assay and improved method for identification of vancomycin hydrochloride. *J. Pharm. Sci.* **1968**, *57*, 314–317. [CrossRef] [PubMed]
24. Ordikhani, F.; Tamjid, E.; Simchi, A. Characterization and antibacterial performance of electrodeposited chitosan–vancomycin composite coatings for prevention of implant-associated infections. *Mater. Sci. Eng. C* **2014**, *41*, 240–248. [CrossRef]
25. Graziano, M.U.; Graziano, K.U.; Pinto, F.M.G.; Bruna, C.Q.M.; Souza, R.Q.; Lascala, C.A. Effectiveness of disinfection with alcohol 70% (w/v) of contaminated surfaces not previously cleaned. *Rev. Latino-Am. Enferm.* **2013**, *21*, 618–623. [CrossRef]
26. Biological Evaluation of Medical Devices—Part 5: Tests for in Vitro Cytotoxicity (ISO Standard No. 10993-5:2009). Available online: https://www.iso.org/standard/36406.html (accessed on 6 July 2022).
27. Grosso, D. How to exploit the full potential of the dip-coating process to better control film formation. *J. Mater. Chem.* **2011**, *21*, 17033–17038. [CrossRef]
28. Figueira, R.B.; Silva, C.J.R.; Pereira, E.V. Influence of experimental parameters using the dip-coating method on the barrier performance of hybrid sol-gel coatings in strong alkaline environments. *Coatings* **2015**, *5*, 124–141. [CrossRef]
29. Escalé, P.; Rubatat, L.; Billon, L.; Save, M. Recent advances in honeycomb-structured porous polymer films prepared via breath figures. *Eur. Polym. J.* **2012**, *48*, 1001–1025. [CrossRef]
30. Heng, L.; Wang, B.; Li, M.; Zhang, Y.; Jiang, L. Advances in Fabrication Materials of Honeycomb Structure Films by the Breath-Figure Metho. *Materials* **2013**, *6*, 460–482. [CrossRef]
31. Loiola, L.M.D.; Más, B.A.; Duek, E.A.R.; Felisberti, M.I. Amphiphilic multiblock copolymers of PLLA, PEO and PPO blocks: Synthesis, properties and cell affinity. *Eur. Polym. J.* **2015**, *68*, 618–629. [CrossRef]
32. Ali, T.; Shoaib, M.H.; Yousuf, R.I.; Jabeen, S.; Muhammad, I.N.; Tariq, A. Use of hydrophilic and hydrophobic polymers for the development of controlled release tizanidine matrix tablets. *Braz. J. Pharm. Sci.* **2014**, *50*, 799–818. [CrossRef]
33. Zhang, L.; Yan, J.; Yin, Z.; Tang, C.; Guo, Y.; Li, D.; Wei, B.; Xu, Y.; Gu, Q.; Wang, L. Electrospun vancomycin-loaded coating on titanium implants for the prevention of implant-associated infections. *Int. J. Nanomed.* **2014**, *9*, 3027–3036. [CrossRef]
34. Boot, W.; Gawlitta, D.; Nikkels, P.G.J.; Pouran, B.; van Rijen, M.H.P.; Dhert, W.J.A.; Vogely, H.C. Hyaluronic acid-based hydrogel coating does not affect bone apposition at the implant surface in a rabbit model. *Clin. Orthop. Relat. Res.* **2017**, *475*, 1911–1919. [CrossRef] [PubMed]
35. Mohseni, E.; Zalnezhad, E.; Bushroa, A.R. Comparative investigation on the adhesion of hydroxyapatite coating on Ti-6Al-4 V implant: A review paper. *Int. J. Adhes. Adhes.* **2014**, *48*, 238–257. [CrossRef]
36. Dashevskiy, I.; Balueva, A.; Todebush, P.; Campbell, C.; Magana, J.; Clement, N. On Estimation of Adhesive Strength of Implants Bioactive Coating with Titanium by Density Functional Theory and Molecular Dynamics Simulations. *Mater. Res.* **2019**, *22*, e20190030. [CrossRef]
37. Yamada, S.; Maeda, H.; Obata, A.; Lohbauer, U.; Yamamoto, A.; Kasuga, T. Cytocompatibility of Siloxane-Containing Vaterite/Poly(l-lactic acid) Composite Coatings on Metallic Magnesium. *Materials* **2013**, *6*, 5857–5869. [CrossRef]
38. Kannan, M.B. Biodegradable polymeric coatings for surface modification of magnesium-based biomaterials. In *Surface Modification of Magnesium and Its Alloys for Biomedical Applications*; Narayanan, T.S.N.S., Park, I.-S., Lee, M.-H., Eds.; Woodhead Publishing: Sawston, UK, 2015; pp. 355–376.
39. Morris, B.A. Influence of Stress on Peel Strength of Acid Copolymers to Foil. *J. Reinf. Plast. Compos.* **2002**, *21*, 1243–1255. [CrossRef]
40. Fu, Y.; Kao, W.J. Drug release kinetics and transport mechanisms of non-degradable and degradable polymeric delivery systems. *Expert Opin. Drug Deliv.* **2010**, *7*, 429–444. [CrossRef]
41. ISO 10477:2020; Dentistry-Polymer-Based Crown and Bridge Materials. International Organization for Standardization: Geneva, Switzerland, 2020. Available online: https://www.iso.org/obp/ui/#iso:std:iso:10477:ed-4:v1:en (accessed on 6 July 2022).
42. Hatzenbuehler, J.; Pulling, T.J. Diagnosis and management of osteomyelitis. *Am. Fam. Physician* **2011**, *84*, 1027–1033.
43. Fang, C.; Wong, T.M.; To, K.K.; Wong, S.S.; Lau, T.W.; Leung, F. Infection after fracture osteosynthesis—Part II. *J. Orthop. Surg. (Hong Kong)* **2017**, *25*, 2309499017692714. [CrossRef]
44. Dash, S.; Murthy, N.P.; Nath, L.; Chwdhury, P. Kinetic modeling on drug release from kinetic controlled drug delivery systems. *Acta Pol. Pharm.* **2010**, *67*, 217–223.
45. Prakash, V.; Lewis, J.S.; Jorgensen, J.H. Vancomycin MICs for Methicillin-Resistant Staphylococcus aureus Isolates Differ Based upon the Susceptibility Test Method Used. *Antimicrob. Agents Chemother.* **2008**, *52*, 4528. [CrossRef] [PubMed]
46. Huang, X.; Brazel, C.S. On the importance and mechanisms of burst release in matrix-controlled drug delivery systems. *J. Control. Release* **2001**, *73*, 121–136. [CrossRef]
47. Venkatraman, S.; Boey, F. Release profiles in drug-eluting stents: Issues and uncertainties. *J. Control. Release* **2007**, *120*, 149–160. [CrossRef]
48. Thote, A.J.; Chappell, J.T.; Kumar, R.; Gupta, R.B. Reduction in the Initial-Burst Release by Surface Crosslinking of PLGA Microparticles Containing Hydrophilic or Hydrophobic Drugs. *Drug Dev. Ind. Pharm.* **2005**, *31*, 43–57. [CrossRef] [PubMed]

49. Kumbar, S.G.; Kulkarni, A.R.; Aminabhavi, T.M. Crosslinked chitosan microspheres for encapsulation of diclofenac sodium: Effect of crosslinking agent. *J. Microencapsul.* **2002**, *19*, 173–180. [CrossRef] [PubMed]
50. Traub, W.H.; Leonhard, B. Heat stability of the antimicrobial activity of sixty-two antibacterial agents. *J. Antimicrob. Chemother.* **1995**, *35*, 149–154. [CrossRef]
51. Zafalon, A.; Juvino, V.; Santos, D.; Lugão, A.B.; Rangari, V.; Temesgen, S.; Parra, D.F. Stability of the Neomycin Antibiotic in Irradiated Polymeric Biomaterials. *Eur. J. Biomed. Pharm. Sci.* **2018**, *5*, 49–57.
52. Livingston, M.; Tan, A. Coating Techniques and Release Kinetics of Drug-Eluting Stents. *J. Med. Devices* **2019**, *10*, 010801. [CrossRef]
53. Raval, A.; Parikh, J.; Engineer, C. Mechanism of controlled release kinetics from medical devices. *Braz. J. Chem. Eng.* **2010**, *27*, 211–225. [CrossRef]
54. Domenek, S.; Fernandes-Nassar, S.; Ducruet, V. Rheology, Mechanical Properties, and Barrier Properties of Poly(lactic acid). In *Synthesis, Structure and Properties of Poly (Lactic Acid)*; Maria, D.L., Androsch, R., Eds.; Springer International Publishing: Cham, Switzerland, 2018; pp. 303–341.
55. Proikakis, C.; Mamouzelos, N.; Tarantili, P.; Andreopoulos, A. Swelling and hydrolytic degradation of poly(d, l-lactic acid) in aqueous solutions. *Polym. Degrad. Stab.* **2006**, *91*, 614–619. [CrossRef]
56. Ismat, A.; Walter, N.; Baertl, S.; Mika, J.; Lang, S.; Kerschbaum, M.; Alt, V.; Rupp, M. Antibiotic cement coating in orthopedic surgery: A systematic review of reported clinical techniques. *J. Orthop. Traumatol.* **2021**, *22*, 56. [CrossRef]

Microfluidic-Assisted Formulation of ε-Polycaprolactone Nanoparticles and Evaluation of Their Properties and In Vitro Cell Uptake

Ewa Rybak [1,*], Piotr Kowalczyk [1], Sylwia Czarnocka-Śniadała [2], Michał Wojasiński [1], Jakub Trzciński [1,3,†] and Tomasz Ciach [1,2,†]

[1] Faculty of Chemical and Process Engineering, Warsaw University of Technology, Waryńskiego 1, 00-645 Warsaw, Poland; piotr.kowalczyk.dokt@pw.edu.pl (P.K.); michal.wojasinski@pw.edu.pl (M.W.); jakub.trzcinski@pw.edu.pl (J.T.); tomasz.ciach@pw.edu.pl (T.C.)
[2] Nanosanguis S.A., Rakowiecka 36, 02-532 Warsaw, Poland; s.czarnocka@nanogroup.eu
[3] Centre for Advanced Materials and Technologies CEZAMAT, Warsaw University of Technology, Poleczki 19, 02-822 Warsaw, Poland
* Correspondence: ewa.rybak.dokt@pw.edu.pl
† These authors contributed equally to this work.

Abstract: The nanoprecipitation method was used to formulate ε-polycaprolactone (PCL) into fluorescent nanoparticles. Two methods of mixing the phases were evaluated: introducing the organic phase into the aqueous phase dropwise and via a specially designed microfluidic device. As a result of the nanoprecipitation process, fluorescein-loaded nanoparticles (NPs) with a mean diameter of 127 ± 3 nm and polydispersity index (PDI) of 0.180 ± 0.009 were obtained. The profiles of dye release were determined in vitro using dialysis membrane tubing, and the results showed a controlled release of the dye from NPs. In addition, the cytotoxicity of the NPs was assessed using an MTT assay. The PCL NPs were shown to be safe and non-toxic to L929 and MG63 cells. The results of the present study have revealed that PCL NPs represent a promising system for developing new drug delivery systems.

Keywords: polymeric nanoparticles; nanoprecipitation; microfluidic; cytotoxicity

1. Introduction

In recent years, the evolution of polymeric particles as drug delivery carriers has promoted the development of nano- and micro-medicine [1]. Due to their size, properly designed nanoparticles (NPs) can freely move throughout the body via the smallest capillaries, are easily administered by oral, pulmonary, vascular, and parenteral routes, and do not require surgical resection after complete administration. Due to the unique properties of some NPs, the biodistribution and pharmacokinetics of the encapsulated drug molecules can be altered, leading to improved efficacy, reduced side effects, and improved patient compliance [2]. NPs can be made to target desired cells and achieve controlled drug release; they could also bring about significant changes in medicine [3]. However, after first promising results and plans, we now understand that nanoparticle drug delivery technology still demands development and understanding to improve the final rate of delivery to the targeted cells [4].

Nanoparticles can be formulated from inorganic or organic–polymeric materials. Inorganic materials like gold, silica, or iron oxide are widely developed due to the vast number of synthetic protocols available. However, polymeric NPs have gained much attention as they can be precisely designed to achieve prominent biodegradability and biocompatibility [5]. The degradation in vivo of some desired polymers results in toxicologically safe side products that are further removed via the normal metabolic pathways or reused as nutrients. Biodegradable polymers are advantageous over other materials for use in drug delivery systems, such as nanoparticles. NPs can be customized into various shapes and sizes, with tailored pore morphologies, mechanical properties, and degradation kinetics. By selecting the appropriate polymer type, molecular weight, and copolymer blend ratio, the degradation/erosion rate of the nanoparticles can be controlled to achieve the desired style and rate of release of the encapsulated payload [6]. Biodegradable polymers can be generally classified as natural polymers, such as chitosan, hyaluronan, etc., and synthetic polymers that include poly-lactic-*co*-glycolic acid (PLGA) [7,8], polylactic acid (PLA) [9,10], or polycaprolactone (PCL) [11,12]. The aforementioned polymers have been widely used for drug encapsulation studies. PCL is a semi-crystalline polyester that is hydrophobic, biodegradable, and biocompatible. The glass transition temperature (T_g) of $-60\ °C$ and low melting point (59–64 °C) of PCL allow for the easy fabrication of delivery systems at reasonably low temperatures [13]. Furthermore, PCL has excellent blend compatibility with other polymers, facilitating the tailoring of desired properties like degradation kinetics, hydrophilicity, and mucoadhesion [14,15]. PCL is an advantageous material for its high permeability to small drug molecules and, in comparison to polylactic and polyglycolic acid polymers, has an inessential tendency to generate an acidic environment during the degradation process, a problem that contributes to the generation of inflammatory reactions. The degradation of PCL is very slow compared to the other polyesters, making it more suitable for long-term delivery systems with the advantages of less frequent administrations, an increase in patient compliance, and the reduction of discomfort [16].

The methods of nanoparticle production have evolved in the last few decades. However, nanoparticles are primarily synthesized in a bench-top batch mode using basic experimental techniques and equipment, i.e., traditional beaker or stirred flask methods. These techniques involve various drawbacks, resulting in polydispersity in size distribution, particle structure, and particle properties [17]. Depending on the particular application and the formulation method, it is crucial to achieve the required characteristics of NPs [8]. There are various methods for NP formulation using biodegradable polymers, such as salting out, emulsification, solvent evaporation, monomer polymerization, or nanoprecipitation [18]. The nanoprecipitation method was first described by Fessi et al. [19], who reported a simple process for the fabrication of polymeric nanoparticles. It involves the precipitation of a dissolved material into nanoparticles after exposure to a polymer non-solvent (polymer precipitant) that is miscible with the solvent [20]. The rapid diffusion of the solvent into the non-solvent phase results in the decrease of interfacial tension between the two phases, which increases the surface area and leads to the formation of small droplets of organic solvent [19,21,22]. There are three stages of the formulation of nanoparticles: nucleation, growth by condensation, and growth by coagulation, which leads to the formation of polymer nanoparticles or aggregates [8,23]. The rate of each step determines the particle size, and the ratio of polymer concentration over the solubility of the polymer in the solvent and non-solvent mixture is the driving force of these phenomena. The critical factor for uniform particle formation is separating the nucleation and the growth stages [24,25]. Preferably, operating conditions should allow a high nucleation rate strongly dependent on supersaturation and a low growth rate [22]. Nanoprecipitation has been a widely recognized and established approach in the realm of nanoparticle formulation. Recently, it has been incorporated into the emerging concept of nanoarchitectonics [26–28]. However, even with the development of cutting-edge methods, in some cases, the procedure is still carried out in the same manner as it was two decades ago. Despite all the challenges, nanoprecipitation is a simple, fast, and reproducible method still widely used to prepare NPs [29–34].

Nevertheless, the successful adaptation of nanoparticle formulations still confronts numerous challenges, such as low production efficacy, high batch-to-batch variations, shorter residence time, and the substandard scale-up feasibility of the manufacturing process. Therefore, an approach that can formulate nanoparticles with desired physicochemical characteristics in a high-throughput and reproducible manner is strongly desirable [35].

The conventional methods typically used for polymeric NP preparation (i.e., dropwise method) usually exhibit a broad size distribution due to the lack of precise control over the mixing process. Alternatively, the microfluidics technique can be used for NP fabrication since it allows for rapid mixing and the precise control of different streams to achieve control over NP size and distribution [16,36–39].

Microfluidic technology has been used to formulate polymeric NPs with a high degree of control over particle size, shape, and composition. The main advantage of this technology is the conduction of physical or chemical processes in a small volume with mostly diffusive control of mass transport phenomena, which leads to a repeatable and controlled process as compared to large reactors and mixers. This approach offers a scalable manufacturing process for these materials, with potential applications in drug delivery, diagnostics, and sensing [25,39–41]. Microfluidic devices are used to provide accurate control over the size distribution, agitation, and shear forces [30]. Furthermore, microfluidic devices offer the advantage of reusability. Microfluidic devices typically consist of various components designed to control the fluid flow and enable efficient nanoparticle synthesis. The key components include fluidic inlets and reservoirs, microchannels, junctions, and mixers. A microfluidic device consists of channels typically ranging from 1 to 100 μm in hydraulic diameters [29]. These channels have walls with different geometries that create capillary action. This creates a narrow channel with a small cross-section that is well suited for the suspension of small particles. Narrow channels are more efficient at maintaining shear forces and particle positions during flow, which facilitates the formation of highly efficient dispersions. Additionally, channel dimensions less than 100 μm in diameter allow for high channel pressure without losing flow altogether. Polymeric nanoparticles can be formulated with microfluidic assistance to improve their properties [41].

Microfluidic techniques are widely used for the preparation of colloidal suspensions; however, they can also be used for other applications such as homogenization, heterogenization, and stirring homogeneous dispersions in inorganic materials. Since microfluidic chips can be designed to be very efficient mixers, the mixing rate between a solvent and non-solvent can shift the precipitation rate toward the high nucleation stage rather than the growth stage [22]. Moreover, using precise external equipment (e.g., syringe pumps or gear pumps), it is possible to reproduce synthesis protocol without variety in nanoparticle characteristics. Particle sizes can be precisely controlled using microfluidic assistance, which makes these formulas ideal for formulation purposes such as cosmetics or pharmaceutical formulation development [40,42].

This work compares a novel microfluidic strategy for fabricating polymeric NPs with the traditional nanoprecipitation method in a vessel. The microfluidic chip brings two co-flowing streams into contact in a specially designed flow-focusing device to enhance mixing. The main objectives of this study were to investigate the effect of operating parameters, system geometry, and polymer concentration on the final particle size distribution. Moreover, the cytotoxicity and dye-releasing behavior of synthesized polymeric NPs against L929 and MG63 cells were thoroughly examined to uncover their potential in cancer therapy.

2. Materials and Methods

2.1. Materials

ε-Polycaprolactone (PCL) with a weight average molar mass of 14,000 g/mol, Pluronic®F-127, PBS (phosphate-buffered saline) tablets, fluorescein, and 3-(4,5-dimethylthiazol-2-yl)-2,5-diphenyl tetrazolium bromide (MTT) were purchased from Sigma-Aldrich/Merck (Poznań, Poland), and dimethyl sulfoxide (DMSO) and tetrahydrofuran (THF) (HPLC grade, purity 99.9%) were obtained from Chempur (Piekary Śląskie, Poland). All the other reagents were of pharmaceutical grade and were used without further purification. The antisolvent phase was ultrapure water produced by reverse osmosis (Milli-Q®, Millipore, Burlington, MA, USA) with the addition of surfactant (Pluronic®F-127).

2.2. Microfluidic Device

The fluidic device was designed in Blender 3.0 software (Figure 1A). The inner channel geometry was saved as a .stl file and 3D-printed using a ZMorph VX printer (ZMorph, Wrocław, Poland). The printing material used was acetonitrile butadiene styrene (ABS) 1.75 mm filament (ZMorph, Wrocław, Poland). The printed model was then placed in a rectangular form, followed by polydimethylsiloxane resin—Sylgard 184 Silicone Elastomer (Dow Chemical, Midland, MI, USA). The mixed resin was degassed under the vacuum before application. After 15 min of curing at 90 °C, the 3Dprinted model was removed, and the hollow space left by the model was covered by the second flat piece of partially cured PDMS and a 1 kg weight to press the pieces together. The sandwiched device was further cured at 90 °C overnight. Silicone tubing with 3 mm diameter was installed in the PDMS chip inlets and outlet and sealed with a small amount of silicone glue.

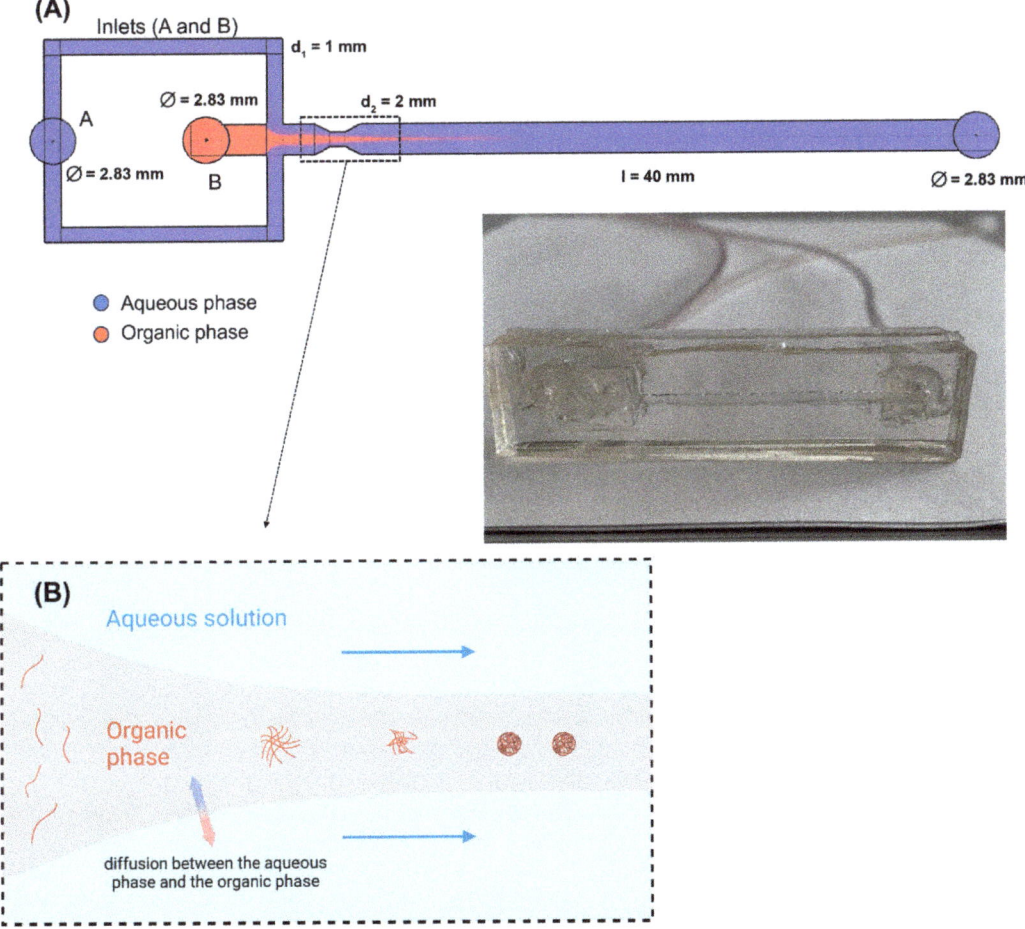

Figure 1. Flow-focusing microfluidic device for NP formulation. (**A**) Schematic of the microfluidic device. (**B**) Proposed mechanism of NP formulation. Created with BioRender.com, accessed on 17 October 2023.

Figure 1 shows the design of our flow-focusing microfluidic device. It has two inlets and one outlet. The organic dispersed phase was introduced from the central channel (inlet B), and the continuous phase (aqueous solution) was introduced through inlet A. The organic and aqueous phases were pumped using KD Scientific and Ascor AP-14 syringe pumps, respectively. Syringes (Beckton Dickinson, Warsaw, Poland) were used for water and polymer/THF solutions. PTFE and silicon tubing (Cole Parmer, Vernon Hills, IL, USA) were used to connect the syringe and the microfluidic device.

2.3. Methods

2.3.1. Preparation of Polymeric Nanoparticles

Preparation of Solutions

To prepare the organic phase for blank emulsions, different amounts of PCL were dissolved in 10 mL of THF to form an organic phase with various polymer concentrations (0.1, 0.5, 1.0, 2.0, 5.0% w/v). For dye-loaded polymeric NPs, the organic phase was prepared by dissolving different amounts of PCL in 10 mL of THF until complete dissolution. Then, fluorescein (1/100 of used PCL amount w/w) was added with continuous stirring until complete dissolution.

The antisolvent phase was an aqueous Pluronics®F-127 surfactant solution (0.6% w/v). The surfactant's role was to stabilize the NPs of the dispersed phase after mixing both phases and prevent agglomeration, coalescence, and imperfect surface formation, as well as facilitate NP size focusing.

Formulation of Nanoparticles

For the classic nanoprecipitation method in a beaker, the organic phase was added dropwise via the syringe pump (LEGATO 210; KD Scientific Inc., Holliston, MA, USA) at a constant rate (0.15 mL/min) to the aqueous phase (ultrapure water MilliQ, Millipore, Burlington, MA, USA) containing surfactant under magnetic stirring (1000 rpm) at room temperature. NPs were formed and the appearance of a milky colloidal suspension was observed. The obtained suspension was stirred magnetically for 10 min. Solvent evaporation was subsequently carried out under magnetic stirring for 72 h at room temperature. The obtained suspension was subjected to sonication and filtration on the syringe filter (0.45 μm).

For the NP formulation using the microfluidic device, the syringes containing organic and aqueous phases were placed in the syringe pumps and connected to the module. A concept of flow rate ratio (R) parameter was introduced, based on hydrodynamic-focusing research in the synthesis of polymeric nanoparticles [43].

$$R = \frac{V_{aq}}{V_{org}} [-]$$

Here:

V_{aq}—flow rate of aqueous phase [mL/h]
V_{org}—flow rate of organic phase [mL/h]

The precipitation process described above was carried out for six R values: 10, 20, 50, 100, 150, and 200 at room temperature. After examination of preliminary results, the flow rate ratio was set at 200. After completion of the process, solvent evaporation was carried out under magnetic stirring for 72 h at room temperature. Then, the obtained suspension was subjected to filtration on the syringe filter (0.45 μm).

2.3.2. Particle Size and Zeta Potential Analysis

The average particle sizes and polydispersity index (PDI) were measured using dynamic light scattering (DLS) using Malvern Zetasizer (Nano ZS, Malvern Instruments, Malvern, UK), equipped with a detector to measure the intensity of the scattered light at 173° to the incident beam. The zeta potential (Z-potential) of the aqueous dispersions was also determined using Zetasizer Nano ZS at 25 °C. All measurements were replicated at least three times and presented as mean values with standard deviations.

2.3.3. Scanning Electron Microscopy (SEM)

The NP morphologies and surface characteristics were investigated based on the images from a scanning electron microscope (SEM, SU8230, Hitachi, Chiyoda City, Tokyo, Japan). First, the NP suspensions were prepared. Following particle size and zeta potential analysis in Section 2.3.2, the samples were diluted 1000 times with ultrapure water and prepared for imaging. The NP diluted suspensions (10 µL) were placed on the surface of the silicon wafer, which was first glued with carbon tape to the aluminum stub. The suspensions on the surfaces of silicon wafers were left for evaporation overnight at ambient temperature (about 22 °C). Such preparation allowed the separate NPs to be imaged as single particles on the surface of the silicon wafer. Images were collected with a 10 kV accelerating voltage, at about 10 mm working distance, using an upper detector of scatted electrons (SE(U)).

2.3.4. FTIR Analysis

Analysis of the chemical interactions of the freeze-dried blank NPs and the loaded NPs was performed using Fourier Transform Infrared Spectroscopy (FTIR) using Nicolet 6700 FTIR (ThermoFisher Scientific®, Waltham, MA, USA). The samples were prepared by mixing the sample fine powder obtained after the lyophilization with IR-grade KBr and subsequent pressing. The scanning range was 4000–500 cm^{-1}.

2.3.5. In Vitro Release Studies

In vitro release of fluorescein from nanoparticles was investigated for three selected formulations from each method (dropwise and microfluidic methods). NP fractions of 1% PCL were chosen as they were characterized by the smallest mean diameter and PDI value. The release of fluorescein from PCL NPs was investigated using a dialysis membrane tubing—regenerated cellulose with a molecular cut-off of 12,000–14,000 Dalton (Spectra/Por Membranes, Spectrum Laboratories, Inc., Rancho Dominguez, CA, USA). Membrane tubes were filled with 3 mL of chosen formulations, sealed with dialysis clips, and then placed in a glass beaker containing 200 mL of PBS and dimethyl sulfoxide (DMSO) solution (4:1 v/v). Experiments were carried out at room temperature for 24 h, and sink conditions were maintained during the analysis. At 0 h, 1 h, 2 h, 3 h, 4 h, and 24 h, 1 mL of the receptor medium was withdrawn, replaced with the same volume of a fresh solution of PBS and DMSO, and then immediately analyzed using a UV-Vis spectrophotometer at 490 nm (BMG, Labtechnologies, Offenberg, Germany). Each measurement was performed in triplicates. Fluorescein solutions ranging from 0.001 to 0.1 mg/mL were prepared to construct a calibration curve that was used to quantify the payload released from the NPs, according to the following equation:

$$y = 102.17x + 0.1401$$

where x represents the fluorescein concentration (mg/mL) and y is the UV/Vis absorbance (nm). The R^2 value was 0.9912. No interference was observed at fluorescein λ_{max} of 490 nm from other components of the formulation. The dye encapsulation efficiency ($EE_\%$) was calculated as the ratio between the unbound dye and the total dye concentration in nanosuspension, according to the following equation:

$$EE_\% = \frac{m_{dye\ total} - m_{dye\ unbound}}{m_{dye\ total}} * 100$$

The drug loading ($DL_\%$) was calculated as the mass fraction of dye in the NPs, according to the following equation:

$$DL_\% = \frac{m_{dye\ total} - m_{dye\ unbound}}{m_{NPs}} * 100$$

2.3.6. Cellular Uptake and Cytotoxicity Assay

To determine the controlled release of the payload from fluorescent NPs, the NPs were incubated with mammalian cells. The fluorescein-loaded NPs were added into the culture of the L929 mouse fibroblast cell line and MG63 human osteosarcoma cells with fibroblast morphology. Fibroblasts are known to be pivotal in contributing to the progression of several malignancies, including endometrial cancer, and therefore represent a possible target for nanoparticle-based therapeutic approaches.

Cytotoxicity Assay

MG63 and L929 cell viability was determined using the standard 3-(4,5-dimethylthiazol-2-yl)-2,5-diphenyl tetrazolium bromide (MTT) assay 24 h after exposure to NPs. Cells suspended in culture medium at the density of 1×10^5/mL (100 µL per well) were seeded onto 96-well plates and cultured for 24 h at 37 °C and 5% CO_2 to adhere. Next, NPs at a series of dilutions—0.01, 0.1, 1.0, and 10 mg/mL—were added (12 wells per variant). Untreated cells served as a positive control of viability. Cells were incubated under standard conditions for 24 h. Then, the NP-containing medium was removed, cells were rinsed three times with PBS, and MTT solution in medium (final MTT concentration 1 mg/mL) was added and incubated (3 h, 37 °C). The MTT solution was removed without disturbing cells, 100 µL/well DMSO was added, the plates were shaken gently (5 min) to dissolve formazan crystals, and the absorbance was read on a microplate reader at 570 nm and 650 nm.

Confocal Microscopy

MG63 and L929 cell lines suspended in culture medium at the density 1×10^5 cell/mL were seeded onto a 24-well plate with a glass plate in each well and cultured for 24 h at 37 °C and 5% CO_2 to adhere. Next, cells were treated with NPs at the concentrations of 1 mg/mL and 0.1 mg/mL (1 mL per well) for 1 h, 2 h, and 24 h under standard conditions. After stimulation, cells were fixed with 4% paraformaldehyde for 15 min, and after washing with PBS, cells were exposed to DAPI (4′,6-diamidino-2-phenylindole) to stain cell DNA. The cells were visualized with a confocal microscope (Zeiss) at the appropriate wavelengths for fluorescein (excitation 470 nm, emission 519 nm) and for DAPI (358 nm excitation, 461 nm emission) at magnification of 20×. Four independent repetitions for each experimental variant were performed.

3. Results and Discussion

The particle sizes of NPs formulated dropwise (DNPs) were found to be from 106 to 185 nm, with a polydispersity index (PDI) of 0.154–0.653 (Figure 2A) for dye-loaded DNPs and 0.090–0.252 for blank DNPs (see Supporting Information). The values of PDI were higher for DNPs compared to the NPs formulated with the use of the microfluidic device (Figure 2B). The PDI values of blank microfluidic NPs (MNPs) were between 0.060 and 0.150 (Table S1. Supporting Information), while the PDI of dye-loaded MNPs was between 0.146 and 0.214, with the NPs' mean diameter from 127 to 193 nm, which was higher than that of the DNPs. The minimum dye-loaded particle size of 106 ± 2 was achieved for DNPs while the smallest polydispersity index (0.146 ± 0.013) was achieved for MNPs. The particles formulated with the microfluidic device were more uniform in size even when higher polymer concentrations were used. Although the size distribution of MNPs was slightly higher than that of DNPs for the analyzed polymer concentration range, the sizes of MNPs were smaller for PCL concentrations of 0.5%, 1%, and 2%. This suggests that the microfluidics method was able to produce NPs with the desired size and PDI over a wider range of polymer concentrations than the dropwise method.

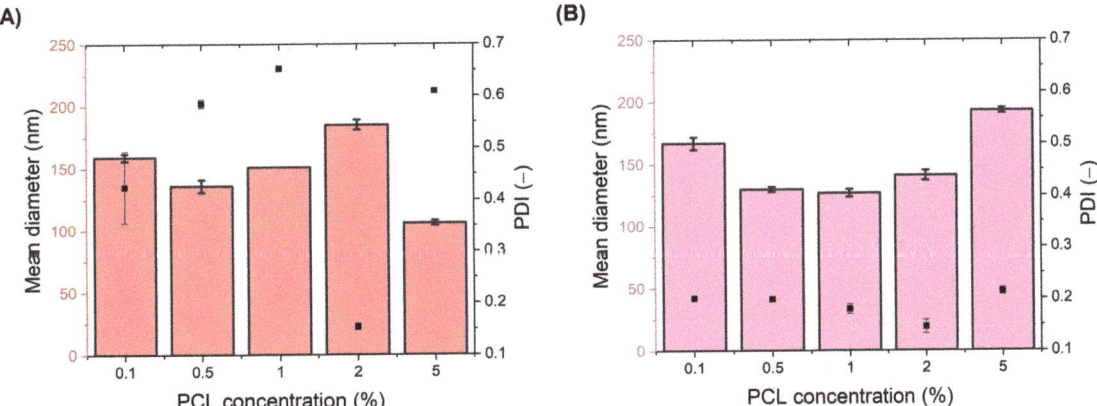

Figure 2. Comparison of the microfluidic device with the dropwise nanoprecipitation method—mean diameter of dye-loaded NPs and PDI (black dot). Dropwise DNPs (**A**), microfluidic device MNPs (**B**).

When the polymer content in the organic phase is increased, the viscosity rises due to higher mass transfer resistance, resulting in larger NP formation. This observation is consistent for 0.5, 1, and 2 polymer mass percentages in the dropwise addition method. The only discrepancy is present in the case of the 0.1 and the 5 mass percentages of the polymer (Figure 2A), where contradictory results were observed. This may be justified by the fact that at the highest polymer concentration, more nuclei were obtained to formulate NPs with small sizes. The PDI value (>0.2) supports uneven nanoparticle growth phenomena, which may be associated, as well, with aggregation tendency.

Formulated NPs have a negative surface charge, which can be a result of the carbonyl group of the PCL polymer present at the surface of the nanoparticle structure [44]. The zeta potential values ranged from −14.6 to −21.0 mV across all formulations, and the method of formulation and variation in polymer concentration did not impact these values. As a result, it can be inferred that the NPs will remain physically stable [45].

Since strict control over a nanoparticle's physical properties is pivotal for nanomedicine usage, the NPs with the smallest mean diameter and PDI were chosen for further research (1% PCL).

3.1. SEM Analysis

Fluorescein-loaded NPs' morphology was visualized using SEM and prepared NPs were assessed in terms of size, shape, and smoothness. Representative images of 1% PCL DNPs and MNPs are presented in Figure 3. The prepared NPs have smooth surfaces. Notably, the MNPs (Figure 3B) have a smaller mean diameter and a narrower size distribution than the DNPs (Figure 3A), which is coherent with values obtained using the DLS technique. DNPs have elongated shapes while MNPs are almost spherical as a result of a controlled formation process. The microfluidic device has a short mixing time and residence time, which prevents the excessive growth of particles. Additionally, the soluble molecules move uniformly in all directions towards the nucleus surface, resulting in spherical particles that minimize surface energy and maintain stability [46]. On the other hand, in the dropwise method, the nucleation and growth mechanisms of NPs cannot be separated due to the lack control of precise control over mixing conditions, leading to excessive particle growth that affects particle size and shape [29,47,48]. The morphology of the NPs appears to be slightly distorted, which could be attributed to the facile dissolution of surfactant on the

NPs' outer surface. This phenomenon leads to the exposure of PCL and may also result in the gradual dissolution of the exposed PCL over time [49].

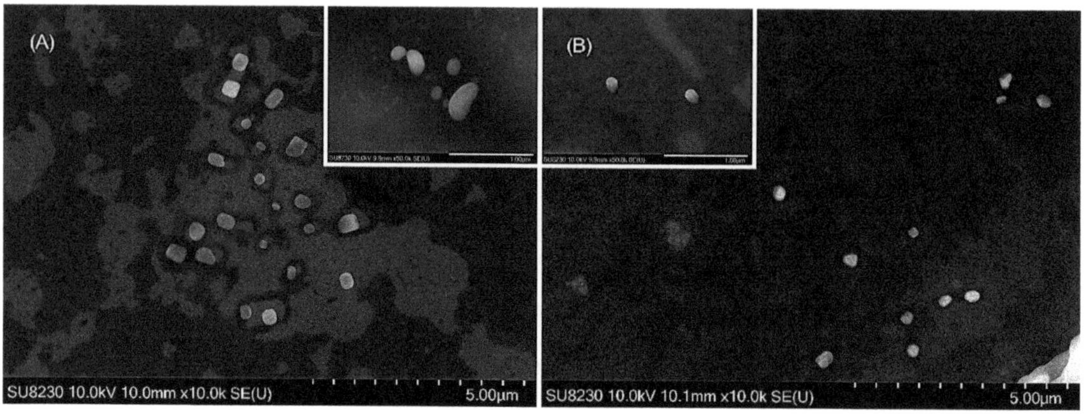

Figure 3. Scanning electron microscopy (SEM) images of (**A**) DNPs and (**B**) MNPs. Inserts show the smallest particles within the investigated samples. Insert scale bars represent 1 μm.

3.2. FTIR Analysis

To confirm nanoparticle chemical composition, FTIR analysis was carried out on pure fluorescein, blank MNPs, and MNPs loaded with fluorescein (Figure 4). All compounds presented their characteristic bands. The MNPs showed a band with asymmetric and symmetric stretching of C-H$_2$ at 2940 cm^{-1} and 2869 cm^{-1}, respectively. Moreover, a peak corresponding to carbonyl stretching at 1722 cm^{-1} was observed as well. The band at 2867 cm^{-1} was related to the stretching vibration of the C–H bond from Pluronic®F-127; at 1187 cm^{-1}, a band of the C–O bond stretching appeared [50]. For fluorescein, the C=C stretching was observed within the range of 1643–1465 cm^{-1} [51]. Increasing PCL concentration in NP formation process does not change the shape of the spectrum; it only decreases the peak intensity (Figure S1, Supporting Information). This report supported the successful encapsulation of fluorescein into the formulated nanoparticles.

3.3. In Vitro Release Studies

The fluorescent labeling of the NPs was performed to study their biological localization. We prepared different formulations of NPs with various amounts of polymer used and proportional amounts of dye used. Based on the data presented in Figure 5, it can be observed that the fluorescence intensity of fluorescein is noticeably augmented as the concentration of fluorescein-loaded DNPs and MNPs increases. Notably, the fluorescence intensity values for DNPs are higher compared to those for MNPs.

The in vitro dye release from the fluorescein-loaded DNPs and MNPs was conducted in PBS (pH 7.4) containing DMSO (4:1 v/v). DMSO was used to enhance the solubilization of the hydrophobic dye. Because of the very low aqueous solubility of fluorescein, the addition of this solubility-enhancing component was necessitated to ensure the sink conditions and to achieve detectable UV/VIS concentrations during the release studies [52]. The dye release mechanisms are important in these formulations because of the proposed applications in sustained drug delivery.

Figure 6 illustrates that the dye-loaded PCL NPs manifested a standard biphasic dye release pattern from the nanoparticle matrix. The dye release profile displayed an initial burst release of 44% for DNPs and 49% for MNPs within the first 4 h, followed by a sustained dye release from the polymer matrix for 24 h. The initial phase of dye release is mainly attributed to the desorption or diffusion of the dye located on the large surface area of the nanoparticles or loosely bound to the polymer matrix. The remaining unreleased dye is thought to be tightly associated with PCL molecules and/or well trapped in the nanoparticle matrix and originates primarily from the diffusion or erosion of the matrix under sink conditions. If the diffusion of the dye is faster than matrix erosion, the release mechanism is largely controlled by the diffusion process [29,53–55].

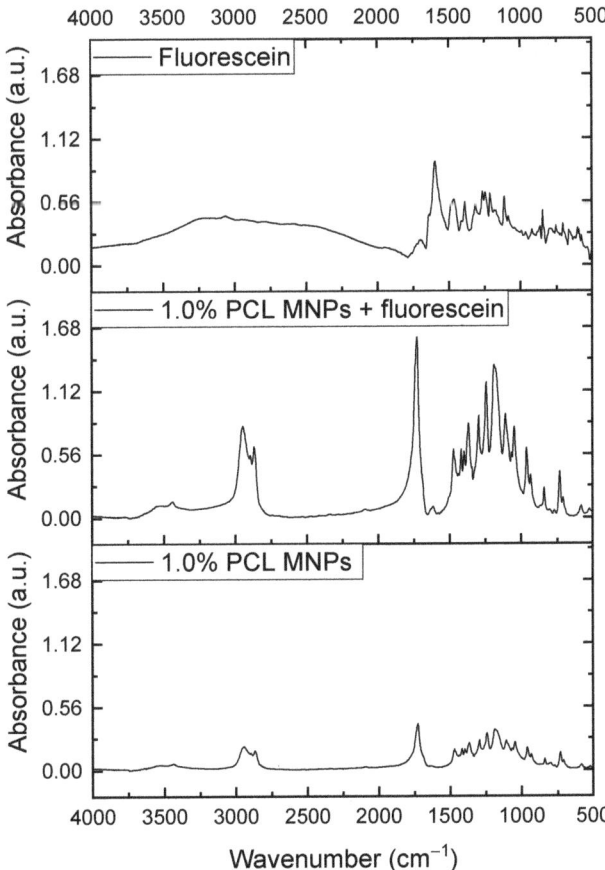

Figure 4. FTIR spectrum of pure fluorescein (**top**) and 1.0% PCL MNPs with (**middle**) and without (**bottom**) fluorescein.

In this study, both groups had relatively small mean diameters—DNPs at 151 nm and MNPs at 127 nm. During the experiment, it was noticed that DNPs exhibited a lower cumulative dye release of 48% compared to MNPs, which showed a relatively higher cumulative release of 55%. This difference could be explained by the fact that smaller nanoparticles possess a larger surface area, which results in a higher concentration of dye molecules at the surface of the NPs, ultimately leading to a faster dye release [53,54,56]. This observation has significant implications in the field of drug delivery as it provides a deeper understanding of the impact of particle size on drug release kinetics.

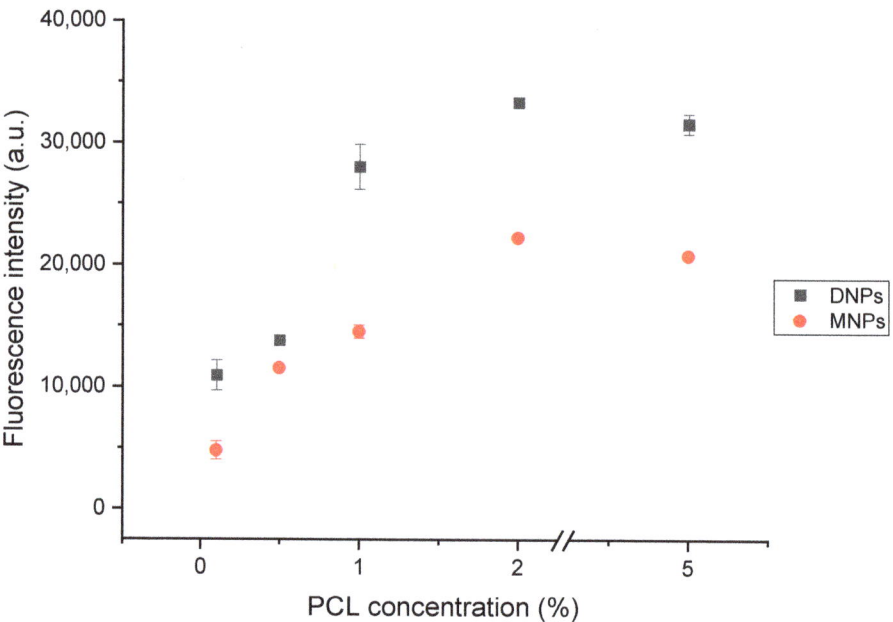

Figure 5. The relation between fluorescence intensity and increasing concentration of fluorescein-loaded DNPs and MNPs (1% PCL = 10 mg/mL).

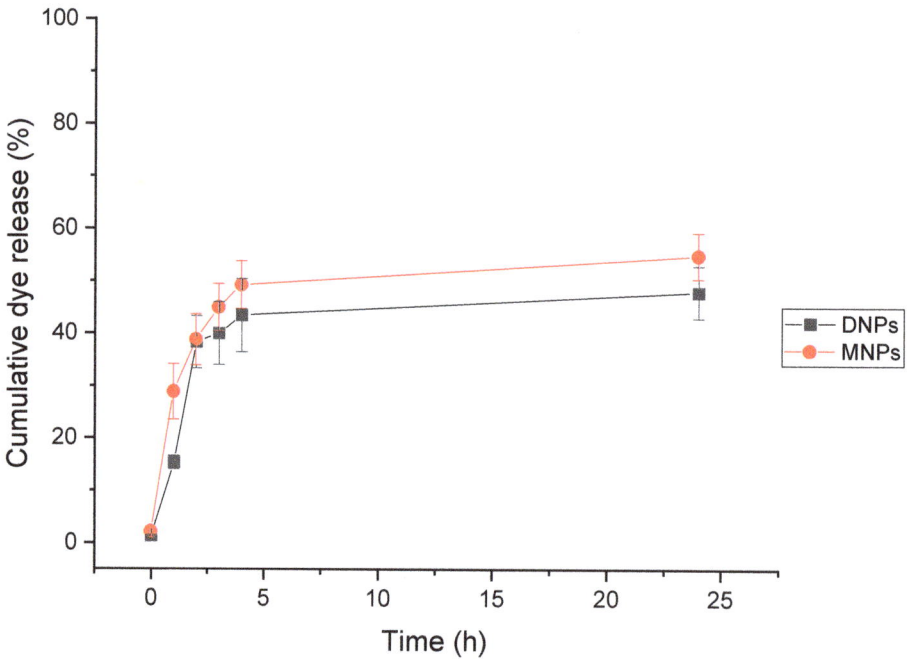

Figure 6. The cumulative release profiles of fluorescein from 1% PCL DNPs and MNPs of three independent measurements. The samples were measured at 0 h, 1 h, 2 h, 3 h, 4 h, and 24 h. Data are presented as means ± SDs.

The $EE_\%$ and $DL_\%$ of the fluorescein DNPs and MNPs are presented in Figure 7. The percentages of fluorescein loaded in the DNPs and MNPs were 4.53% and 6.48%, respectively. The encapsulation of the DNPs was 95.17%, and that of the MNPs was 97.22% (n = 3). It could be seen that both parameters reached higher values for MNPs.

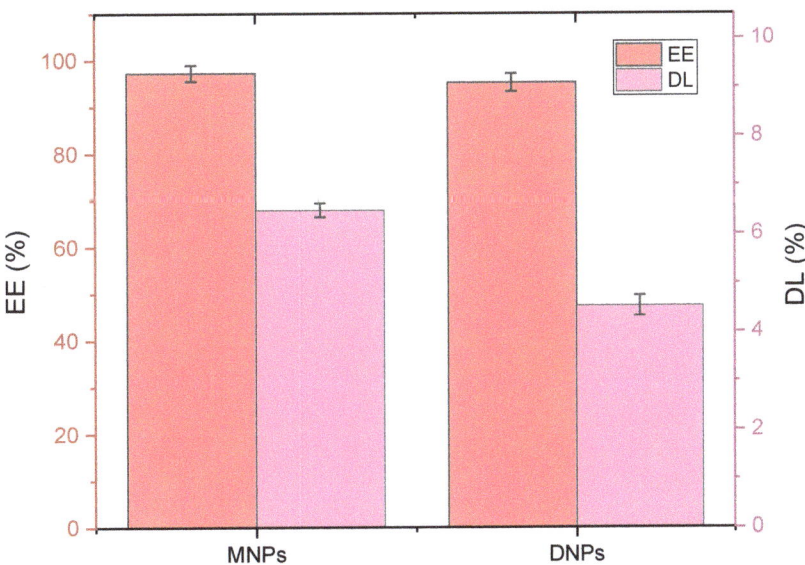

Figure 7. Drug loading and encapsulation efficiency [%] of fluorescein-loaded 1% PCL MNPs and DNPs.

3.4. In Vitro Cytotoxicity and Cellular Uptake Studies

MNPs are the preferred choice when conducting cell experiments due to their beneficial characteristics. Compared to DNPs, they have a smaller mean diameter and PDI value, along with higher $EE_\%$ and $DL_\%$ values. Additionally, MNPs demonstrate a controlled cumulative release profile, making them an ideal choice for such experiments. The formulated nanoparticles were examined for their potential cytotoxicity. Different concentrations of PCL NPs were added into the cell cultures (MG63 cell line and L929 cell line) and tested for their toxicity under 24 h of incubation. The results are presented in Figure 8. For MG63 cells, dosages of 0–10 mg/mL nanoparticles were examined for cytotoxicity, and in these examinations, PCL NPs showed no noticeable toxicity. This non-toxic trend can be observed until 10 mg/mL (cell viability then dropped to 86 ± 14%). In comparison, L929 cells showed elevated sensitivity as the dosage of NPs over 0.1 mg/mL started to lower the cell viability to 81 ± 11%, and to 79 ± 5% for the dosage of 10 mg/mL of NPs. Therefore, there are slightly different toxicity profiles for each cell line from the same nanoparticles.

Finally, we performed confocal microscopy to study the effect of dye-loaded NPs composed of PCL in L929 and MG63 cells. Importantly, our results (Figure 9) supported data obtained via MTT assay, as shown in Figure 8. The dye-loaded NPs seemed to be effective in the delivery of payload in only an hour to both normal and cancerous cells. Figure 9 presents images of cells treated with NPs at the concentration of 1 mg/mL. For 0.1 mg/mL, the results were similar.

To evaluate the penetration of the NPs into the cells and the targeting effects of the loaded NPs, cellular uptakes of the MNPs were performed using L929 and MG63 cell lines. The lipophilic dye, fluorescein, was chosen in this study because the water solubility of fluorescein is poor and it could be easily encapsulated into the hydrophobic cores of the NPs [57]. The internalization of fluorescein-loaded NPs incubated for 1, 2, and 24 h was visualized using confocal laser scanning microscopy (CLSM). To observe the cellular distributions of NPs, we visualized green fluorescence from fluorescein and blue fluorescence from DAPI nuclei labeling. The fact that nanoparticle uptake by L929 and MG63 cells was significantly higher at 24 h compared to 1 or 2 h of incubation highlights the time-dependent accumulation of these nanoparticles within the cells. It was demonstrated that particle size determines both the mechanism and rate of intracellular uptake as well as the ability of a particle to permeate through tissue [58,59]. Indeed, the size of a particle can affect the efficiency and pathway of its cellular uptake by influencing its adhesion and interaction with cells [60]. A qualitative approach to cellular fluorescence revealed a gradual increase over 24 h, with evidence of a significant increase in payload release over time (Figure 9). Moreover, after 24 h, we could observe that the dye was cumulating in regions around the nucleus (Figure 9E,F). These results indicate that NPs loaded with a high fluorescein concentration release their cargo intracellularly. We suggest that lysosomal acid hydrolases may facilitate release in our experiments after the initial distribution of NPs in early endosomes and lysosomes due to the presence of PCL, a polyester highly sensitive to hydrolases [61]. The hydrolysis of the polyester leads to the destabilization of the vesicle structure, possibly leading to the accelerated release of fluorescein at physiological temperature (~37 °C) compared to the storage temperature (room temperature) of the NP formulation. Further high-resolution imaging studies will allow the better description of this process.

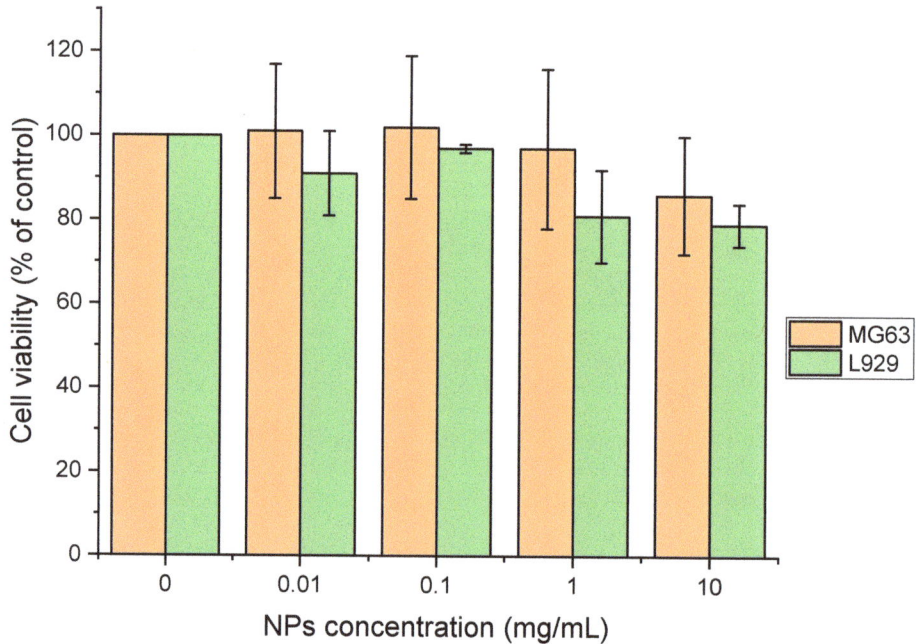

Figure 8. Cytotoxicity assessment of 1.0% PCL MNPs in two different cell lines, osteosarcoma cells MG63, and mouse fibroblast cell line L929. Cell viability (%) was calculated as (B)/A * 100, where A and B are the absorbances of control and treated cells, respectively. Values represent means ± SDs (n = 6).

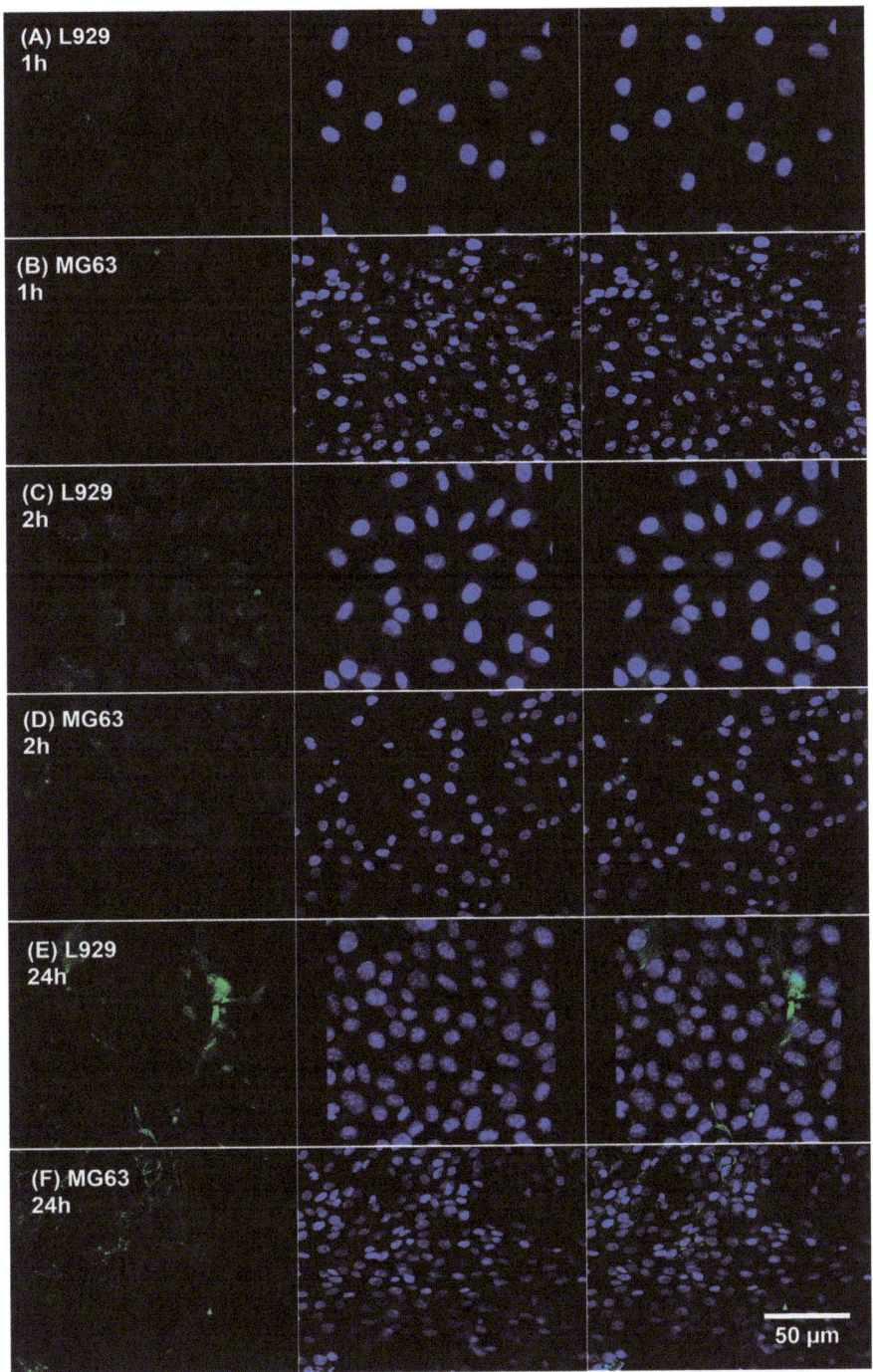

Figure 9. Confocal microscopy images of L929 and MG63 cells after 1, 2, and 24 h incubation at 37 °C with PCL MNPs with 120 nm in diameter (NPs concentration 1 mg/mL). Fluorescein (**left**), DAPI (**middle**), combined (**right**). Magnification: 20×; 1.0% PCL.

Controlling ordered polymeric nanoparticles' pore size, morphology, and particle size is significant for biomedical applications. Nano-sized particles (<200 nm) are better carriers than bulky polymeric materials with sizes greater than 1 μm as they exhibit rapid mass transfer, efficient adhesion to substrates, and good suspension in solution. Nanoparticles bigger than human albumin and smaller than approximately 200 nm have a better chance of staying in the circulatory system for a longer period [62]. Therefore, in this paper, polymeric nanoparticles with diameters below 200 nm are suggested to be well sized for cellular uptake. Size effects on cellular uptake are expected to result in size-dependent biochemical responses [58,63]. However, the particular responses of downstream cells require further investigation.

4. Discussion and Conclusions

Advances in nanomedicine require developing more robust and controllable procedures for manufacturing nanoparticles. Conventional methods rely heavily on bulk mixtures with poor batch-to-batch reproducibility, making it difficult to rapidly screen and optimize the properties of nanomaterials. There are several limitations associated with conventional methods of producing NPs. One such limitation is the need for additional chemical and physical processes such as freeze–thaw, high-pressure homogenization, and extrusion [64]. These processes not only require extra resources but also affect the quality and efficacy of the final product. Another drawback is insufficient macro-mixing, which can result in the uneven distribution of active ingredients. Additionally, there is a risk of potential contamination during the production process. These limitations pose a significant challenge in translating NPs from the laboratory to clinical use. Consequently, exploring and developing methodologies that can effectively tackle the obstacles and produce nanomaterials with utmost accuracy and precision is imperative. This critical aspect demands considerable attention and innovation from researchers and industry practitioners alike. The utilization of microfluidic devices presents a viable solution to the limitations of bulk mixing and top–down approaches given these devices' microscale dimensions and ability to precisely control flow parameters, particle size tunability, and reproducibility. In bottom-up approaches, bulk mixing is influenced by various factors, such as temperature, precursor concentration, time, and pH. However, microfluidic NPs are further impacted by total flow rate, flow rate ratio, and residence time due to the continuous flow operation of microfluidics. These additional factors impact the physicochemical properties of NPs and the aforementioned parameters [65]. The miniature size of microfluidic tools also offers low power consumption, precise laminar flow due to small Reynolds numbers where viscous forces dominate, and improved mass and heat transfer due to large surface area. Aside from advantages, the main disadvantage is that these tools rely on novelty and not thoroughly investigated ongoing processes. Hence, microscale reactions become more complex and dependent on the process parameters than macroscale reactions. Microfluidic emulsion droplets are produced at tens to hundreds of microliters per minute, so emulsification is a lengthy process with low throughput per hour. Some groups have tried to overcome this situation by parallelizing the droplet generators [66], which can significantly increase a plant's cost.

In this study, size-tunable PCL NPs were prepared using the conventional nanoprecipitation method and a specially designed microfluidic device. For the microfluidic method, we tested six different R values, and after a thorough data analysis, the flow rate ratio was set at 200. The increased amount of water reduces the tendency of the particles to agglomerate because the particles collide and stick together less frequently. The particle buildup is most likely to occur near the orifice, with high localized particle concentrations. In addition, higher flow rate ratios ensure faster mixing in microfluidic systems [67]. With a faster mixing process, the critical supersaturation required for nucleation is reached more quickly, resulting in the formation of more nuclei whose growth is limited by the amount of polymer available in the liquid phase. Therefore, the higher the number of nuclei is, the smaller the nanoparticle size is. Laoini et al. also observed smaller particle sizes at higher *R* ratios in the production of liposomes and polymeric micelles in membrane contractors [68,69]. Jahn et al. observed these in the formation of liposomes in a planar flow-focusing microfluidic mixer [70], and Othman et al. did the same in the formulation of polymeric NPs using glass capillary micromixers [16]. A high flow rate allowed small-sized NPs to be produced with a low polydispersity index. In addition, a microfluidic device can prepare PCL-based dye-loaded NPs with a wide size range (127–193 nm) and lower PDI (0.146–0.214) than those prepared using the conventional bulk method. The Z-potential is a valuable parameter for predicting colloidal dispersions' storage stability, as reported by Thode et al. [71]. When the zeta potential exhibits high negative values, it indicates the presence of electrostatic repulsion between particles, which prevents aggregation and stabilizes the nanoparticulate dispersion [18,45,72]. Moreover, negatively charged NPs have low toxicity to cells [73]. From the SEM images, it is clear that the prepared NPs had smooth surfaces. The sizes of blank and dye-loaded NPs formulated with both methods were within a range of 106–193 nm. When administered intravenously, NPs should be small enough (100–300 nm) to passively cross the tumor endothelial barrier and remain in the tumor bed for a prolonged period. This is due to reduced lymphatic drainage, which is also known as the enhanced permeability and retention (EPR) effect [16,74,75].

The release of fluorescein from both DNPs and MNPs followed a two-step pattern, with an initial rapid burst release observed during the first 4 h followed by a slower, sustained release over 24 h. Implementing a controlled and sustained release system is highly desirable for effectively maintaining optimal therapeutic drug levels over an extended period [76,77]. MNPs, which are smaller in size and have higher $EE_\%$ and $DL_\%$ than DNPs, may be better suited for applications requiring faster drug release rates. Further research in this area has the potential to lead to the development of more effective drug delivery systems with greater therapeutic efficacy.

The cytotoxicity assay revealed that although the cell viability gradually declined as the concentration of NPs increased from 0.01 to 10 mg/mL, both tested cell lines tolerated NPs well. We found that 120 nm-sized fluorescein-loaded NPs successfully entered the L929 and MG63 cells only after 1 h and accumulated in the cytosol, and it is plausible that some of the dye released may have diffused into the cell nuclei. In time, as in the fluorescent dye release assay, the dye was released and gathered in cells for 24 h. The next step should include drug encapsulation into PCL NPs and determining encapsulation efficiency. The further investigation of this method will allow the formulation of loaded NPs that could accumulate at the tumor sites via the EPR effect [74,75]. We believe that polymer-based microfluidic NP production will provide substantial opportunities for future clinical applications of size-controlled nanomedicines. A proposed extension of this work is to introduce other functional additives to tailor the properties of the nanoparticles for specific applications and to evaluate the effect of these substances on the NP precipitation process. The geometry of the mixing device can be easily modified to achieve better mixing and reduce the sizes and polydispersity of the nanoparticles obtained. The construction of the mixer device is simple, and it can be manufactured using a consumer-grade 3D printer and open-source software. The aspect ratio of the channels, defined as the width-to-height ratio, can influence the flow dynamics and NP formulation. A higher aspect ratio can promote better mixing while a lower aspect ratio can favor laminar flow conditions and enable easier control of NP size. The design of microfluidic channels can also incorporate various features such as bends, converging or diverging sections, and junctions to enhance mixing and achieve desired nanoparticle characteristics. It is worth noting that these dimensions are not fixed and can be tailored based on the specific requirements of the polymeric NP formulation process and the target application. Factors such as the viscosity of the polymer solution, the desired nanoparticle size range, and the intended drug loading or release characteristics can influence the selection of the microfluidic channel dimensions. The current study is the basis for further experiments addressing more specific applications of nanoparticles and improving mixing devices.

Supplementary Materials: The following supporting information can be downloaded at: https://www.mdpi.com/article/10.3390/polym15224375/s1, Table S1. Mean diameter and PDI of blank NPs. Size and polydispersity index (PDI) of PCL NPs measured by dynamic light scattering (DLS). Table S2. Comparison of dye-loaded NP characteristics formulated with analyzed methods. Size and polydispersity index (PDI) of PCL NPs with dye measured by dynamic light scattering (DLS). Figure S1. FTIR spectrum of dye-loaded NPs formulated with different amounts of PCL.

Author Contributions: Conceptualization, E.R., J.T. and T.C.; methodology, E.R. and P.K.; validation, J.T. and T.C., data curation, E.R.; funding acquisition, S.C.-Ś. and T.C.; investigation, E.R. and M.W.; supervision, J.T., S.C.-Ś. and T.C.; writing—original draft preparation, E.R. and P.K.; writing—review and editing, J.T., T.C. and M.W.; design and 3D printing of microfluidic device, P.K. All authors have read and agreed to the published version of the manuscript.

Funding: The research was supported by the National Centre for Research and Development (NCBR) within the EuroNanoMed III project (acronym: TARBRAINFECT, Project Contract No. 16/EuroNanoMed/2019).

Institutional Review Board Statement: Not applicable.

Data Availability Statement: The data presented in this study are available on request from the corresponding author.

Acknowledgments: The authors express their sincere thanks to Rafał Podgórski and Michał Wojasiński for invaluable consultations with cell sample preparation and the confocal laser scanning microscopic imaging of cells.

Conflicts of Interest: The authors declare that they have no known competing financial interest or personal relationship that could have appeared to influence the work reported in this paper.

References

1. Wang, Y.; Li, P.; Tran, T.T.-D.; Zhang, J.; Kong, L. Manufacturing Techniques and Surface Engineering of Polymer Based Nanoparticles for Targeted Drug Delivery to Cancer. *Nanomaterials* **2016**, *6*, 26. [CrossRef] [PubMed]
2. Gonçalves, C.; Pereira, P.; Gama, M. Self-assembled hydrogel nanoparticles for drug delivery applications. *Materials* **2010**, *3*, 1420–1460. [CrossRef]
3. Gui, R.; Wang, Y.; Sun, J. Encapsulating magnetic and fluorescent mesoporous silica into thermosensitive chitosan microspheres for cell imaging and controlled drug release in vitro. *Colloids Surf. B Biointerfaces* **2014**, *113*, 1–9. [CrossRef] [PubMed]
4. Wilhelm, S.; Tavares, A.J.; Dai, Q.; Ohta, S.; Audet, J.; Dvorak, H.F.; Chan, W.C.W. Analysis of nanoparticle delivery to tumours. *Nat. Rev. Mater.* **2016**, *1*, 16014. [CrossRef]
5. Yang, Y.; Wang, Y.; Powell, R.; Chan, P. Frontiers in Research Review: Cutting-Edge Molecular Approaches to Therapeutics Polymeric Core-Shell Nanoparticles For Therapeutics. *Clin. Exp. Pharmacol. Physiol.* **2006**, *33*, 557–562. [CrossRef]
6. Pathak, Y.; Thassu, D. *Drug Delivery Nanoparticles Formulation and Characterization*; CRC Press: Boca Raton, FL, USA, 2016; Volume 191.
7. Nehilla, B.J.; Bergkvist, M.; Popat, K.C.; Desai, T.A. Purified and surfactant-free coenzyme Q10-loaded biodegradable nanoparticles. *Int. J. Pharm.* **2008**, *348*, 107–114. [CrossRef]
8. Rao, J.P.; Geckeler, K.E. Polymer nanoparticles: Preparation techniques and size-control parameters. *Prog. Polym. Sci.* **2011**, *36*, 887–913. [CrossRef]
9. Seyler, I.; Appel, M.; Devissaguet, J.-P.; Legrand, P.; Barratt, G. Macrophage activation by a lipophilic derivative of muramyldipeptide within nanocapsules: Investigation of the mechanism of drug delivery. *J. Nanoparticle Res.* **1999**, *1*, 91–97. [CrossRef]
10. Legrand, P.; Lesieur, S.; Bochot, A.; Gref, R.; Raatjes, W.; Barratt, G.; Vauthier, C. Influence of polymer behaviour in organic solution on the production of polylactide nanoparticles by nanoprecipitation. *Int. J. Pharm.* **2007**, *344*, 33–43. [CrossRef]
11. Tavares, M.R.; de Menezes, L.R.; Filho, J.C.D.; Cabral, L.M.; Tavares, M.I.B. Surface-coated polycaprolactone nanoparticles with pharmaceutical application: Structural and molecular mobility evaluation by TD-NMR. *Polym. Test.* **2017**, *60*, 39–48. [CrossRef]
12. Dash, T.K.; Konkimalla, V.B. Poly-ε-caprolactone based formulations for drug delivery and tissue engineering: A review. *J. Control. Release* **2012**, *158*, 15–33. [CrossRef] [PubMed]
13. Karuppuswamy, P.; Reddy, J. Polycaprolactone nano fibers for the controlled release of tetracycline hydrochloride. *Mater. Lett.* **2015**, *141*, 180–186. [CrossRef]
14. Bilensoy, E.; Sarisozen, C.; Esendağlı, G.; Doğan, A.L.; Aktaş, Y.; Şen, M.; Mungan, N.A. Intravesical cationic nanoparticles of chitosan and polycaprolactone for the delivery of Mitomycin C to bladder tumors. *Int. J. Pharm.* **2009**, *371*, 170–176. [CrossRef] [PubMed]

15. Payyappilly, S.S.; Panja, S.; Mandal, P.; Dhara, S.; Chattopadhyay, S. Organic Solvent-Free Low Temperature Method of Preparation for Self Assembled Amphiphilic Poly(ε-Caprolactone)–Poly(Ethylene Glycol) Block Copolymer Based Nanocarriers for Protein Delivery. *Colloids Surf. B Biointerfaces* **2015**, *135*, 510–517. [CrossRef]
16. Othman, R.; Vladisavljević, G.T.; Nagy, Z.K. Preparation of biodegradable polymeric nanoparticles for pharmaceutical applications using glass capillary microfluidics. *Chem. Eng. Sci.* **2015**, *137*, 119–130. [CrossRef]
17. Badilescu, S.; Packirisamy, M. Microfluidics-nano-integration for synthesis and sensing. *Polymers* **2012**, *4*, 1278–1310. [CrossRef]
18. Kumar, A.; Sawant, K. Encapsulation of exemestane in polycaprolactone nanoparticles: Optimization, characterization, and release kinetics. *Cancer Nanotechnol.* **2013**, *4*, 57–71. [CrossRef]
19. Fessi, H.; Puisieux, F.; Devissaguet, J.P.; Ammoury, N.; Benita, S. Nanocapsule formation by interfacial polymer deposition following solvent displacement. *Int. J. Pharm.* **1989**, *55*, R1–R4. [CrossRef]
20. Karnik, R.; Gu, F.; Basto, P.; Cannizzaro, C.; Dean, L.; Kyei-Manu, W.; Langer, R.; Farokhzad, O. Microfluidic platform for controlled synthesis of polymeric nanoparticles. *Nano Lett.* **2008**, *8*, 2906–2912. [CrossRef]
21. Mishra, B.; Patel, B.B.; Tiwari, S. Colloidal nanocarriers: A review on formulation technology, types and applications toward targeted drug delivery. *Nanomed. Nanotechnol. Biol. Med.* **2010**, *6*, 9–24. [CrossRef]
22. Tao, J.; Chow, S.F.; Zheng, Y. Application of flash nanoprecipitation to fabricate poorly water-soluble drug nanoparticles. *Acta Pharm. Sin. B* **2018**, *9*, 4–18. [CrossRef] [PubMed]
23. Marchisio, D.L.; Barresi, A.A.; Garbero, M. Nucleation, growth, and agglomeration in barium sulfate turbulent precipitation. *AIChE J.* **2002**, *48*, 2039–2050. [CrossRef]
24. Johnson, B.K.; Prud'Homme, R.K. Flash NanoPrecipitation of Organic Actives and Block Copolymers using a Confined Impinging Jets Mixer. *Aust. J. Chem.* **2003**, *56*, 1021–1024. [CrossRef]
25. Hamdallah, S.I.; Zoqlam, R.; Erfle, P.; Blyth, M.; Alkilany, A.M.; Dietzel, A.; Qi, S. Microfluidics for pharmaceutical nanoparticle fabrication: The truth and the myth. *Int. J. Pharm.* **2020**, *584*, 119408. [CrossRef] [PubMed]
26. Ferhan, A.R.; Park, S.; Park, H.; Tae, H.; Jackman, J.A.; Cho, N. Lipid Nanoparticle Technologies for Nucleic Acid Delivery: A Nanoarchitectonics Perspective. *Adv. Funct. Mater.* **2022**, *32*, 2203669. [CrossRef]
27. Ariga, K. Materials Nanoarchitectonics for Advanced Physics Research. *Adv. Phys. Res.* **2023**, *2*, 2200113. [CrossRef]
28. Ariga, K. Materials nanoarchitectonics in a two-dimensional world within a nanoscale distance from the liquid phase. *Nanoscale* **2022**, *14*, 10610–10629. [CrossRef]
29. Heshmatnezhad, F.; Nazar, A.R.S.; Aghaei, H.; Varshosaz, J. Production of doxorubicin-loaded PCL nanoparticles through a flow-focusing microfluidic device: Encapsulation efficacy and drug release. *Soft Matter* **2021**, *17*, 10675–10682. [CrossRef]
30. Bramosanti, M.; Chronopoulou, L.; Grillo, F.; Valletta, A.; Palocci, C. Microfluidic-assisted nanoprecipitation of antiviral-loaded polymeric nanoparticles. *Colloids Surf. A Physicochem. Eng. Asp.* **2017**, *532*, 369–376. [CrossRef]
31. Jamwal, S.; Ram, B.; Ranote, S.; Dharela, R.; Chauhan, G.S. New glucose oxidase-immobilized stimuli-responsive dextran nanoparticles for insulin delivery. *Int. J. Biol. Macromol.* **2018**, *123*, 968–978. [CrossRef]
32. Yan, X.; Bernard, J.; Ganachaud, F. Nanoprecipitation as a simple and straightforward process to create complex polymeric colloidal morphologies. *Adv. Colloid Interface Sci.* **2021**, *294*, 102474. [CrossRef] [PubMed]
33. Javaid, S.; Ahmad, N.M.; Mahmood, A.; Nasir, H.; Iqbal, M.; Ahmad, N.; Irshad, S. Cefotaxime loaded polycaprolactone based polymeric nanoparticles with antifouling properties for in-vitro drug release applications. *Polymers* **2021**, *13*, 2180. [CrossRef] [PubMed]
34. Saqib, M.; Ali Bhatti, A.S.; Ahmad, N.M.; Ahmed, N.; Shahnaz, G.; Lebaz, N.; Elaissari, A. Amphotericin B loaded polymeric nanoparticles for treatment of *Leishmania* infections. *Nanomaterials* **2020**, *10*, 1152. [CrossRef] [PubMed]
35. Martins, J.P.; Torrieri, G.; Santos, H.A. The importance of microfluidics for the preparation of nanoparticles as advanced drug delivery systems. *Expert Opin. Drug Deliv.* **2018**, *15*, 469–479. [CrossRef] [PubMed]
36. Soleimani, S.; Hasani-Sadrabadi, M.M.; Majedi, F.S.; Dashtimoghadam, E.; Tondar, M.; Jacob, K.I. Understanding biophysical behaviours of microfluidic-synthesized nanoparticles at nano-biointerface. *Colloids Surf. B Biointerfaces* **2016**, *145*, 802–811. [CrossRef]
37. Michelon, M.; Oliveira, D.R.B.; de Figueiredo Furtado, G.; De La Torre, L.G.; Cunha, R.L. High-throughput continuous production of liposomes using hydrodynamic flow-focusing microfluidic devices. *Colloids Surf. B Biointerfaces* **2017**, *156*, 349–357. [CrossRef]
38. Lari, A.S.; Khatibi, A.; Zahedi, P.; Ghourchian, H. Microfluidic-assisted production of poly(ε-caprolactone) and cellulose acetate nanoparticles: Effects of polymers, surfactants, and flow rate ratios. *Polym. Bull.* **2021**, *78*, 5449–5466. [CrossRef]
39. Liu, Y.; Yang, G.; Hui, Y.; Ranaweera, S.; Zhao, C. Microfluidic Nanoparticles for Drug Delivery. *Small* **2022**, *18*, e2106580. [CrossRef]
40. Niculescu, A.-G.; Chircov, C.; Bîrcă, A.C.; Grumezescu, A.M. Nanomaterials synthesis through microfluidic methods: An updated overview. *Nanomaterials* **2021**, *11*, 864. [CrossRef]
41. Li, W.; Chen, Q.; Baby, T.; Jin, S.; Liu, Y.; Yang, G.; Zhao, C.-X. Insight into drug encapsulation in polymeric nanoparticles using microfluidic nanoprecipitation. *Chem. Eng. Sci.* **2021**, *235*, 116468. [CrossRef]
42. Bendre, A.; Bhat, M.P.; Lee, K.-H.; Altalhi, T.; Alruqi, M.A.; Kurkuri, M. Recent developments in microfluidic technology for synthesis and toxicity-efficiency studies of biomedical nanomaterials. *Mater. Today Adv.* **2022**, *13*, 100205. [CrossRef]
43. Kowalczyk, P.; Wojasiński, M.; Wasiak, I.; Ciach, T. Investigation of controlled solvent exchange precipitation of fluorescent organic nanocrystals. *Colloids Surf. A Physicochem. Eng. Asp.* **2018**, *545*, 86–92. [CrossRef]

44. Mahmoudi, M.; Saeidian, H.; Mirjafary, Z.; Mokhtari, J. Preparation and characterization of memantine loaded polycaprolactone nanocapsules for Alzheimer's disease. *J. Porous Mater.* **2020**, *28*, 205–212. [CrossRef]
45. Badran, M.M.; Alanazi, A.E.; Ibrahim, M.A.; Alshora, D.H.; Taha, E.; HAlomrani, A. Optimization of Bromocriptine-Mesylate-Loaded Polycaprolactone Nanoparticles Coated with Chitosan for Nose-to-Brain Delivery: In Vitro and In Vivo Studies. *Polymers* **2023**, *15*, 3890. [CrossRef]
46. Tapia-Guerrero, Y.S.; Del Prado-Audelo, M.L.; Borbolla-Jiménez, F.V.; Gomez, D.M.G.; García-Aguirre, I.; Colín-Castro, C.A.; Morales-González, J.A.; Leyva-Gómez, G.; Magaña, J.J. Effect of UV and gamma irradiation sterilization processes in the properties of different polymeric nanoparticles for biomedical applications. *Materials* **2020**, *13*, 1090. [CrossRef]
47. Boken, J.; Soni, S.K.; Kumar, D. Microfluidic Synthesis of Nanoparticles and their Biosensing Applications. *Crit. Rev. Anal. Chem.* **2016**, *46*, 538–561. [CrossRef]
48. Heshmatnezhad, F.; Nazar, A.R.S. On-chip controlled synthesis of polycaprolactone nanoparticles using continuous-flow microfluidic devices. *J. Flow Chem.* **2020**, *10*, 533–543. [CrossRef]
49. Nair, B.P.; Vaikkath, D.; Mohan, D.S.; Nair, P.D. Fabrication of a microvesicles-incorporated fibrous membrane for controlled delivery applications in tissue engineering. *Biofabrication* **2014**, *6*, 045008. [CrossRef]
50. Elzein, T.; Nasser-Eddine, M.; Delaite, C.; Bistac, S.; Dumas, P. FTIR study of polycaprolactone chain organization at interfaces. *J. Colloid Interface Sci.* **2004**, *273*, 381–387. [CrossRef]
51. Gupta, P.; Sharma, S.; Godara, S.K.; Kaur, V. Synthesis of fluorescein dye using microwave radiations and its application on textile substrates. *Fibres Text. East. Eur.* **2021**, *29*, 100–105. [CrossRef]
52. Duse, L.; Agel, M.R.; Pinnapireddy, S.R.; Schäfer, J.; Selo, M.A.; Ehrhardt, C.; Bakowsky, U. Photodynamic therapy of ovarian carcinoma cells with curcumin-loaded biodegradable polymeric nanoparticles. *Pharmaceutics* **2019**, *11*, 282. [CrossRef]
53. Ali, S.; Amin, M.U.; Tariq, I.; Sohail, M.F.; Ali, M.Y.; Preis, E.; Ambreen, G.; Pinnapireddy, S.R.; Jedelská, J.; Schäfer, J.; et al. Lipoparticles for synergistic chemo-photodynamic therapy to ovarian carcinoma cells: In vitro and in vivo assessments. *Int. J. Nanomed.* **2021**, *16*, 951–976. [CrossRef]
54. Dinarvand, R.; Moghadam, S.H.; Mohammadyari-Fard, L.; Atyabi, F. Preparation of biodegradable microspheres and matrix devices containing naltrexone. *AAPS PharmSciTech* **2003**, *4*, 45–54. [CrossRef] [PubMed]
55. Jiang, Y.; Zhou, Y.; Zhang, C.Y.; Fang, T. Co-delivery of paclitaxel and doxorubicin by ph-responsive prodrug micelles for cancer therapy. *Int. J. Nanomed.* **2020**, *15*, 3319–3331. [CrossRef] [PubMed]
56. Dayanandan, A.P.; Cho, W.J.; Kang, H.; Bello, A.B.; Kim, B.J.; Arai, Y.; Lee, S.-H. Emerging nano-scale delivery systems for the treatment of osteoporosis. *Biomater. Res.* **2023**, *27*, 68. [CrossRef] [PubMed]
57. Scarpa, E.; Bailey, J.L.; Janeczek, A.A.; Stumpf, P.S.; Johnston, A.H.; Oreffo, R.O.C.; Woo, Y.L.; Cheong, Y.C.; Evans, N.D.; Newman, T.A. Quantification of intracellular payload release from polymersome nanoparticles. *Sci. Rep.* **2016**, *6*, 29460. [CrossRef] [PubMed]
58. Guo, H.; Qian, H.; Sun, S.; Sun, D.; Yin, H.; Cai, X.; Liu, Z.; Wu, J.; Jiang, T.; Liu, X. Hollow mesoporous silica nanoparticles for intracellular delivery of fluorescent dye. *Chem. Central J.* **2011**, *5*, 1. [CrossRef]
59. Smith, A.M.; Duan, H.; Mohs, A.M.; Nie, S. Bioconjugated quantum dots for in vivo molecular and cellular imaging. *Adv. Drug Deliv. Rev.* **2008**, *60*, 1226–1240. [CrossRef]
60. Lee, K.D.; Nir, S.; Papahadjopoulos, D. Quantitative Analysis of Liposome-Cell Interactions In Vitro: Rate Constants of Binding and Endocytosis with Suspension and Adherent J774 Cells and Human Monocytes. *Biochemistry* **1993**, *32*, 889–899. [CrossRef]
61. Li, S.; Molina, I.; Martinez, M.B.; Vert, M. Hydrolytic and enzymatic degradations of physically crosslinked hydrogels prepared from PLA/PEO/PLA triblock copolymers. *J. Mater. Sci. Maed. Med.* **2002**, *13*, 81–86. [CrossRef]
62. Hoshyar, N.; Gray, S.; Han, H.; Bao, G. The effect of nanoparticle size on in vivo pharmacokinetics and cellular interaction. *Nanomedicine* **2016**, *11*, 673–692. [CrossRef] [PubMed]
63. Zhang, S.; Li, J.; Lykotrafitis, G.; Bao, G.; Suresh, S. Size-dependent endocytosis of nanoparticles. *Adv. Mater.* **2009**, *21*, 419–424. [CrossRef] [PubMed]
64. Liu, Y.; Jiang, X. Why microfluidics? Merits and trends in chemical synthesis. *Lab Chip* **2017**, *17*, 3960–3978. [CrossRef] [PubMed]
65. Agha, A.; Waheed, W.; Stiharu, I.; Nerguizian, V.; Destgeer, G.; Abu-Nada, E.; Alazzam, A. A review on microfluidic-assisted nanoparticle synthesis, and their applications using multiscale simulation methods. *Discover Nano* **2023**, *18*, 18. [CrossRef]
66. Khan, I.U.; Serra, C.A.; Anton, N.; Vandamme, T.F. Production of nanoparticle drug delivery systems with microfluidics tools. *Expert. Opin. Drug Deliv.* **2015**, *12*, 547–562. [CrossRef]
67. Génot, V.; Desportes, S.; Croushore, C.; Lefèvre, J.-P.; Pansu, R.B.; Delaire, J.A.; von Rohr, P.R. Synthesis of organic nanoparticles in a 3D flow focusing microreactor. *Chem. Eng. J.* **2010**, *161*, 234–239. [CrossRef]
68. Laouini, A.; Charcosset, C.; Fessi, H.; Holdich, R.; Vladisavljević, G. Preparation of liposomes: A novel application of microengineered membranes—Investigation of the process parameters and application to the encapsulation of vitamin E. *RSC Adv.* **2013**, *3*, 4985–4994. [CrossRef]
69. Laouini, A.; Charcosset, C.; Fessi, H.; Holdich, R.; Vladisavljević, G. Preparation of liposomes: A novel application of microengineered membranes—From laboratory scale to large scale. *Colloids Surf. B Biointerfaces* **2013**, *112*, 272–278. [CrossRef]
70. Jahn, A.; Stavis, S.M.; Hong, J.S.; Vreeland, W.N.; DeVoe, D.L.; Gaitan, M. Microfluidic Mixing and the Formation of Nanoscale Lipid Vesicles. *ACS Nano* **2010**, *4*, 2077–2087. [CrossRef]

71. Thode, K.; Müller, R.H.; Kresse, M. Two-time window and multiangle photon correlation spectroscopy size and zeta potential analysis—Highly sensitive rapid assay for dispersion stability. *J. Pharm. Sci.* **2000**, *89*, 1317–1324. [CrossRef]
72. Feng, S.-S.; Huang, G. Effects of emulsifiers on the controlled release of paclitaxel (Taxol®) from nanospheres of biodegradable polymers. *J. Control Release* **2001**, *71*, 53–69. [CrossRef] [PubMed]
73. Shao, X.R.; Wei, X.Q.; Song, X.; Hao, L.Y.; Cai, X.X.; Zhang, Z.R.; Peng, Q.; Lin, Y.F. Independent effect of polymeric nanoparticle zeta potential/surface charge, on their cytotoxicity and affinity to cells. *Cell Prolif.* **2015**, *48*, 465–474. [CrossRef] [PubMed]
74. Brzeziński, M.; Kost, B.; Gonciarz, W.; Krupa, A.; Socka, M.; Rogala, M. Nanocarriers based on block copolymers of L-proline and lactide: The effect of core crosslinking versus its pH-sensitivity on their cellular uptake. *Eur. Polym. J.* **2021**, *156*, 110572. [CrossRef]
75. Wu, J. The enhanced permeability and retention (EPR) effect: The significance of the concept and methods to enhance its application. *J. Pers. Med.* **2021**, *11*, 771. [CrossRef]
76. Kusumasari, F.C.; Samada, L.H.; Budianto, E. Preparation, Characterization and In Vitro Release Study of Microcapsule Simvastatin Using Biodegradable Polymeric Blend of Poly(L-Lactic Acid) and Poly(ε-Caprolactone) with Double Emulsifier. *Mater. Sci. Forum* **2020**, *977*, 178–183. [CrossRef]
77. Nozal, V.; Rojas-Prats, E.; Maestro, I.; Gil, C.; Perez, D.I.; Martinez, A. Improved controlled release and brain penetration of the small molecule S14 using PLGA nanoparticles. *Int. J. Mol. Sci.* **2021**, *22*, 3206. [CrossRef]

Disclaimer/Publisher's Note: The statements, opinions and data contained in all publications are solely those of the individual author(s) and contributor(s) and not of MDPI and/or the editor(s). MDPI and/or the editor(s) disclaim responsibility for any injury to people or property resulting from any ideas, methods, instructions or products referred to in the content.

Review

Self-Assembled Block Copolymers as a Facile Pathway to Create Functional Nanobiosensor and Nanobiomaterial Surfaces

Marion Ryan C. Sytu [1], David H. Cho [2] and Jong-in Hahm [1,*]

1 Department of Chemistry, Georgetown University, 37th & O Sts. NW., Washington, DC 20057, USA
2 National Institute of Biomedical Imaging and Bioengineering, National Institutes of Health, 9000 Rockville Pike, Bethesda, MD 20892, USA; dc1210@georgetown.edu
* Correspondence: jh583@georgetown.edu

Citation: Sytu, M.R.C.; Cho, D.H.; Hahm, J.-i. Self-Assembled Block Copolymers as a Facile Pathway to Create Functional Nanobiosensor and Nanobiomaterial Surfaces. *Polymers* **2024**, *16*, 1267. https://doi.org/10.3390/polym16091267

Academic Editors: Arash Moeini, Pierfrancesco Cerruti and Gabriella Santagata

Received: 27 March 2024
Revised: 24 April 2024
Accepted: 26 April 2024
Published: 1 May 2024

Copyright: © 2024 by the authors. Licensee MDPI, Basel, Switzerland. This article is an open access article distributed under the terms and conditions of the Creative Commons Attribution (CC BY) license (https://creativecommons.org/licenses/by/4.0/).

Abstract: Block copolymer (BCP) surfaces permit an exquisite level of nanoscale control in biomolecular assemblies solely based on self-assembly. Owing to this, BCP-based biomolecular assembly represents a much-needed, new paradigm for creating nanobiosensors and nanobiomaterials without the need for costly and time-consuming fabrication steps. Research endeavors in the BCP nanobiotechnology field have led to stimulating results that can promote our current understanding of biomolecular interactions at a solid interface to the never-explored size regimes comparable to individual biomolecules. Encouraging research outcomes have also been reported for the stability and activity of biomolecules bound on BCP thin film surfaces. A wide range of single and multicomponent biomolecules and BCP systems has been assessed to substantiate the potential utility in practical applications as next-generation nanobiosensors, nanobiodevices, and biomaterials. To this end, this Review highlights pioneering research efforts made in the BCP nanobiotechnology area. The discussions will be focused on those works particularly pertaining to nanoscale surface assembly of functional biomolecules, biomolecular interaction properties unique to nanoscale polymer interfaces, functionality of nanoscale surface-bound biomolecules, and specific examples in biosensing. Systems involving the incorporation of biomolecules as one of the blocks in BCPs, i.e., DNA–BCP hybrids, protein–BCP conjugates, and isolated BCP micelles of bioligand carriers used in drug delivery, are outside of the scope of this Review. Looking ahead, there awaits plenty of exciting research opportunities to advance the research field of BCP nanobiotechnology by capitalizing on the fundamental groundwork laid so far for the biomolecular interactions on BCP surfaces. In order to better guide the path forward, key fundamental questions yet to be addressed by the field are identified. In addition, future research directions of BCP nanobiotechnology are contemplated in the concluding section of this Review.

Keywords: BCP nanobiotechnology; BCP self-assembly; BCP thin films; self-assembled BCP nanopatterns; protein arrays; nanobiosensors; biomaterials; proteins; cells

1. Introduction

The development of functional biosensors has long drawn considerable research interests across many different disciplines in fundamental science, biotechnology, and medicine [1–11]. One of the notable trends in recent efforts for biosensor development involves high miniaturization [6,8,12–14] and flexibility/wearability [3,5,7,9,15,16]. Advances in nanoscience continuously propel such a drive to create flexible and miniaturized biosensors, permitting high-throughput detection of bioanalytes that are held on an array of nanometer-sized sensor surfaces in a flexible setting. The majority of conventional biosensors are fabricated by top-down approaches such as photolithography, soft (microcontact printing) lithography, and inkjet printing, which can be costly and time-consuming [2,13,17–25]. Fabrication techniques relying on conventional lithographic procedures also present limitations in the size of the smallest possible sensor unit that can be individually addressed. This is due to the optical diffraction limit of light-based lithographic tools commonly used in the fabrication process. Although there exist lithographic tools of higher spatial resolution such as electron-beam lithography [26–28], scanning probe-based lithography [29,30] and nanoimprint lithography [31,32], the involvement of these procedures can lead to an even slower fabrication process and a higher production cost. Hence, alternative approaches based on self-assembly have emerged to create nanoscopic patterns of individually addressable biosensor surfaces and nanoscale bioreactors in simple steps [17,33–39].

The remarkable self-assembly behaviors of block copolymers (BCPs) have been well-recognized as one of the most versatile and convenient mechanisms to exploit a bottom-up assembly approach in organizing nanoscale features [40–48]. BCPs can be synthesized from a rich selection of monomers, whose chemical compositions can be tuned to match the desired functionalities for their applications [43,49–54]. It is also straightforward to create BCPs into thin structures that can be flexible and wearable. Furthermore, there exists a wealth of theoretical, computational, and experimental works performed to understand the phase separation behaviors of BCPs [40,45–47,49,55–65]. Owing to these efforts, nanoscale features resulting from BCPs' phase separation processes and their two-/three-dimensional (2D/3D) periodicities have been well-characterized. The size and shape of these nanopatterns that can be controlled thermodynamically and kinetically have also been mapped out for many BCPs. As such, nanoscale BCP surface patterns have extensively been utilized as templates to organize inorganic nanomaterials in BCP lithography [33,34,42,49,66–74]. The first attempt to use BCP nanopatterns for assembling biomolecules such as proteins was undertaken in mid 2000s [75]. Many ensuing endeavors have since been made in the field of BCP nanobiotechnology as represented in Table 1. The various BCP–biomolecule systems in Table 1 summarize the stimulating research endeavors and findings that will be discussed in this Review. All these efforts have successfully demonstrated the application of underlying BCP nanopatterns in controlling the spatial density, large area assembly, adsorption/desorption dynamics, biofunctionality, and other important interfacial characteristics of biologically relevant molecules such as proteins, peptides, biomineral nanocrystals, cell adhesive molecules, and cells. The focus of this Review is to provide a comprehensive and detailed overview of those research efforts pertaining to nanoscale surface assembly of functional biomolecules, biomolecular interaction properties at nanoscale polymer interfaces, functionality of nanoscale surface-bound biomolecules, as well as specific examples in biosensing.

Table 1. Various BCPs and biosystems demonstrated for controlling key characteristics of biomolecules via BCP nanopatterns.

Biomolecule Name (Abbreviation)	BCP Nanotemplate Used with Biomolecules	Section Covered	Ref.
Proteins and Peptides			
Immunoglobulin G (IgG)	Polystyrene-block-polymethylmethacrylate (PS-b-PMMA)	3.1.1. 3.1.2. 3.1.3. 3.1.4.	[75] [76,77] [78] [79]
	Poly(styrene-co-4-bromostyrene)-block-polyethylene oxide (P(S-co-BrS)-b-PEO)	3.1.3.	[80]
	Poly(2-methacryloyloxyethyl phosphorylcholine)-block-poly(dimethylsiloxane) (PMPC-b-PDMS)	3.1.1.	[81]
S-layer protein (SbpA)	Polystyrene-block-polyethylene oxide (PS-b-PEO)	3.1.1.&3.1.2.	[82]
	Polystyrene-block-poly(2-vinylpyridine) (PS-b-P2VP)	3.1.2.	[82]

Table 1. Cont.

Biomolecule Name (Abbreviation)	BCP Nanotemplate Used with Biomolecules	Section Covered	Ref.
Amelogenin (Amel)	PS-b-PMMA	3.1.1. 3.2.	[83]
Fibrinogen (Fg)	PS-b-PMMA	3.1.1. 3.1.2. 3.1.3. 3.1.4. 3.2.	[84,85] [86] [78,87] [85] [85]
	Polystyrene-block-poly(2-hydroxyethyl methacrylate) (PS-b-PHEMA)	3.1.2. 3.1.4.	[88,89] [89]
γ-globulin	PS-b-PMMA	3.1.2.	[86]
Fibronectin (FN)	PS-b-PMMA	3.1.2.	[86]
	PMPC-b-PDMS	3.1.2.	[81]
	Polymethylmethacrylate-block-polyacrylic acid (PMMA-b-PAA) Polymethylmethacrylate-block-poly(2-hydroxyethyl methacrylate) (PMMA-b-PHEMA) Polyacrylic acid-block-polymethylmethacrylate-block-polyacrylic acid (PAA-b-PMMA-b-PAA) Polymethylmethacrylate-block-poly(2-hydroxyethyl methacrylate)-block-polymethylmethacrylate (PMMA-b-PHEMA-b-PMMA)	3.1.2.	[90]
	PS-b-PEO	3.1.2.	[91]
	Polystyrene-block-polyisoprene (PS-b-PI)	3.3.	[92]
	PMPC-block-poly(3-methacryloyloxy propyltris(trimethylsilyloxy) silane) (PMPTSSi)	3.3.	[93]
Thrombomodulin (TM)	PS-b-PMMA	3.1.2.	[86]
Type I collagen (Col I)	PS-b-PMMA	3.1.2.	[86]
Collagen fibrils	PS-b-PEO	3.4.	[94]
Human/bovine serum albumin (HSA/BSA)	PS-b-PMMA	3.1.3.	[78,87]
	PS-b-PI	3.1.2.	[95]
Ovalbumin (OVA)	Poly(acrylic acid)-block-poly(N-isopropyl acrylamide) (PAA-b-PNIPAM)	3.1.2.	[96]
Streptavidin (SAv)	Polyethylene glycol-block-polystyrene (PEG-b-PS)	3.1.2.	[97]
Myoglobin (Mb)	Polystyrene-block-poly(2-hydroxyethyl methacrylate) (PS-b-PHEMA)	3.1.2.&3.1.4.	[89]
	PS-b-PEO	3.1.2.	[98]
Lysozyme (LZM)	PS-b-PHEMA	3.1.2.&3.1.4.	[89]
	PS-b-PEO	3.1.2.	[99]
Green fluorescent protein (GFP)	PS-b-PEO	3.1.2.	[91]
Arginine-Glycine-Aspartate (RGD) peptide motifs	PS-b-PEO	3.1.2. 3.3.	[91] [91,100]
	Polyacrylamide/bis-acrylamide-block-poly(acrylic acid) (PAAm/bisAAm-b-PAA)	3.3. 3.4.	[101]
TAT peptide	PS-b-PEO	3.1.2.	[99]
Coiled-coil α-helix bundle (heme-binding motif)	PS-b-PEO	3.1.2.	[98]
Lsmα protein	PS-b-PEO	3.1.2.	[102]

Table 1. Cont.

Biomolecule Name (Abbreviation)	BCP Nanotemplate Used with Biomolecules	Section Covered	Ref.
	PS-b-PMMA	3.1.4.	[79,103]
Horseradish peroxidase (HRP)	Polystyrene-block-polyethylene oxide/polystyrene-block-poly(l-lactide) (PS-b-PEO/ PS-b-PLLA)	3.1.2. 3.1.4.	[104]
avß3 integrin receptor of c(-RGDfK-)	Polystyrene-block-poly(2-vinylpyridine) (PS-b-P2VP)	3.1.2. 3.3.	[105,106]
Tyrosinase	PS-b-PMMA	3.1.4.	[79]
Nucleic Acids			
DNA origami	PS-b-PMMA	3.1.2.	[68,107]
	PS-b-P2VP	3.1.2.	[108]
Cells			
Chinese Hamster ovary cells (CHO)	PS-b-PI	3.3.	[92]
MC3T3-osteoblasts	PS-b-P2VP	3.3.	[105,106]
B16-melanocytes	PS-b-P2VP	3.3.	[106]
REF52-fibroblasts	PS-b-P2VP	3.3.	[106]
	PS-b-P2VP	3.3.	[106]
3T3 and NIH-3T3 fibroblasts	PS-b-PEO	3.3. 3.4.	[91,100] [94]
	Polyacrylamide/bis-acrylamide-block-poly(acrylic acid) (PAAm/bisAAm-b-PAA)	3.4.	[101]
L929 fibroblasts	PMPC-block-poly(3-methacryloyloxy propyltris(trimethylsilyloxy) silane) (PMPC-b-PMPTSSi)	3.3.	[93]
	PMPC-b-PDMS-PMPC	3.4.	[81]
Escherichia coli (*E.coli*)	PS-b-P2VP	3.4.	[109]
Staphylococcus aureus (*S.aureus*)	PS-b-P2VP	3.4.	[109]
Bone marrow mesenchymal stem cells (BMMSC), Mesenchymal precursor cells	PS-b-P2VP	3.4.	[110]
	PS-b-P2VP Polystyrene-block-poly(4-vinylpyridine) (PS-b-P4VP)	3.4.	[111]
Osteosarcoma cells (SaOS-2)	PS-b-P2VP	3.4.	[110]
Dermal fibroblasts	PS-b-P2VP PS-b-P4VP	3.4.	[111]
Mouse preosteoblasts (MC3T3-E1)	Polystyrene-block-poly(ethylene oxide)/dodecylbenzenesulfonic acid (PS-b-PEO/DBSA)	3.4.	[112]
Pancreatic tumor cells, PaTu 8988t	PAAm/bisAAm-b-PAA	3.4.	[101]
Endothelial cells (ECs)	Polystyrene-block-poly(ethylene-co-butylene)-block-polystyrene (SEBS)	3.4.	[113]
Biomineral Nanocrystals			
Calcium phosphate (CaP), Hydroxy-apatite (HAP), Triple CaP (TCP)	PS-b-PMMA	3.2.	[83,85]
Biosensors			
rop B gene	PS-b-P4VP	4.4.&4.5.	[114]
Glucose oxidase (GOx)/Glucose	PS-b-P4VP	4.5.	[115]
Choline oxidase (ChO)/Choline	Poly(n-butylmethacrylate)-block-poly(N,N-dimethylaminoethyl methacrylate) (PnBMA-b-PDMAEMA)	4.5.	[116]
Dopamine (DA)	PS-b-P4VP	4.5.	[117]

109

2. Block Copolymers as Nanoscale Templates

2.1. Block Copolymer Nanostructures in Bulk

BCPs are synthesized by covalently linking two or more, chemically distinct, polymer blocks via methods such as atom transfer free radical polymerization (ATRP) and reversible addition fragmentation chain transfer (RAFT) [43,49–54]. In bulk, BCPs self-assemble into various nanostructures with a tunable periodicity typically in the range of 5–100 nm through a process known as microphase separation. The phase separation processes of BCPs occur as a direct consequence of self-assembly driven by chemically incompatible polymer segments in a given BCP to maximize (minimize) the spatial contact between similar (dissimilar) blocks. However, these forces driving phase separation are countered by the entropic forces of polymer chain mixing since the different blocks of the BCP are covalently bonded together. The BCP microphase separation is ultimately achieved by a balance between forces associated with separating and mixing and hence, the process is thermodynamically driven by enthalpic and entropic parameters. The enthalpic term of the process is defined by the Flory–Huggins interaction parameter (χ) which is related to the free energy cost between the different polymer blocks. The interaction parameter of χ is inversely proportional to temperature (T). The entropic term of mixing is affected by the degree of polymerization (N) and the relative composition fraction of polymer blocks in terms of volume fraction (f).

The phase separation behaviors of different BCP systems have been extensively studied both theoretically and experimentally [40,45–47,49,55–65]. BCP phase diagrams obtained by a self-consistent mean-field and other related theories provide the exact relationship between χN and f which, in turn, dictates the spatial configuration and packing nature of the polymer nanostructures for a given BCP system. For a simple linear A-B diblock whose χN value is greater than ~10.5, ordered nanostructures that range from spheres (body-centered cubic A spheres in a B matrix), to cylinders (hexagonally packed A cylinders in a B matrix), to bicontinuous gyroids (two interpenetrating networks of A and B), and to lamellae (alternating planes of A and B) can be formed depending on the volume fraction and the immiscibility of the polymer blocks. Even a larger collection of nanostructures is available in BCPs composed of triblocks or polymer blocks with higher architectural complexities [56,60,118–123]. Figure 1A displays representative phase diagrams obtained by a self-consistent field theory (SCFT) for (a) AB-type diblock and (b) symmetric ABA triblock copolymer systems. For these two-component systems, the BCP morphologies predicted for the ordered state include body-centered cubic spheres (S), hexagonally close-packed spheres (S_{cp}), cylinders (C), gyroids (G), lamellae (L), and F_{ddd} (O^{70}) [124]. Figure 1B further shows exemplar nanostructures associated with various phases of AB diblock copolymers that were identified by a theory and/or experiment in the literature [125].

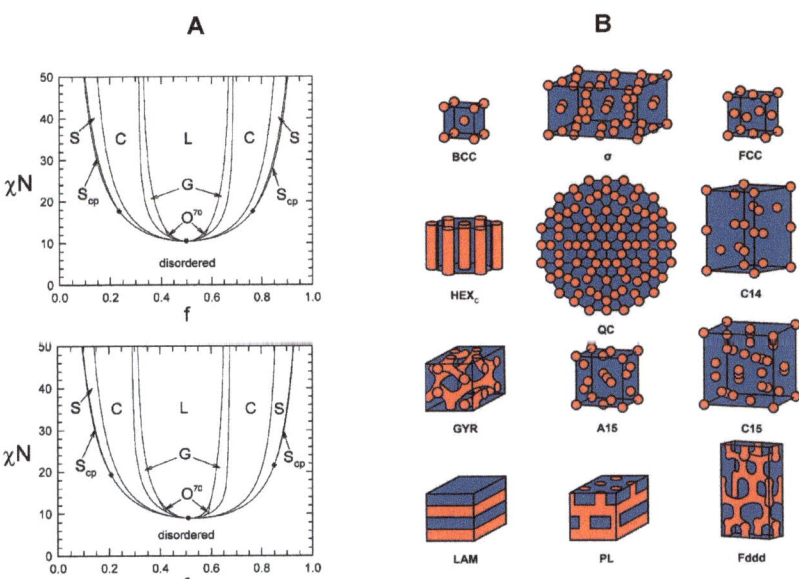

Figure 1. (**A**) The plots display representative phase diagrams of BCP systems determined by the SCFT. The top and bottom plots belong to the phase diagrams of AB diblock and ABA triblock copolymers, respectively. χ is the Flory–Huggins interaction parameter, N is the degree of polymerization, and f is the volume fraction. The symbols of S, S_{cp}, C, G, L, and O^{70} denote for body-centered cubic spheres, hexagonally close-packed spheres, cylinders, gyroids, lamellae, and F_{ddd}, respectively. Reproduced with permission from Ref. [124] Copyright (2012) American Chemical Society. (**B**) Nanostructures that were determined for linear AB diblock copolymers are schematically depicted. These structures include BCC (body-centered cubic), σ (Frank–Kasper sigma phase), FCC (face-centered cubic), HEX_c (hexagonally packed cylinders), QC (dodecagonal quasi-crystal), C_{14} (Frank–Kasper AB_2 Laves phase), GYR (double gyroid), A_{15} (Frank–Kasper AB_3 phase), C_{15} (Frank–Kasper AB_2 Laves phase), LAM (lamellae), PL (perforated lamellae), and F_{ddd} (O^{70} network). Reproduced with permission from Ref. [125] Copyright (2020) American Chemical Society.

One of the crucial aspects of phase diagrams is the characteristic phase separation behaviors of BCPs and therefore, the resulting BCP nanostructures are highly predictable and tunable. The size and shape of BCP domains along with the periodicity between the polymer domains can be readily controlled at the nanoscale level by simply changing experimental variables such as the composition, molecular weight, and volume fraction of the BCP blocks. Moreover, nanostructures that are not thermodynamically stable can be experimentally achieved in certain cases [40,126]. These morphologies arise from additional experimental constraints of kinetic or chemical factors that are applied during the synthesis and fabrication of BCPs. Examples of these factors include heterogeneities in the molecular weight and structure as well as interactions of solvent vapors to select polymer blocks. More detailed discussions will follow in the next section.

2.2. Block Copolymer Nanostructures in Thin Films

Surface energetics and confinement effects become extremely important for predicting the phase separation behaviors of BCPs in thin films [41,44,47,62,127–130]. Therefore, when BCPs are prepared on a solid support, wetting energies associated with polymer–air and polymer–solid interactions are often considered in addition to the thermodynamic parameters previously discussed for bulk phase diagrams. The enhanced role of surface and interfacial energetics as well as the interplay between the BCP film thickness and the equilibrium period of microphase separation can drive a richer array of nanomorphologies than what can be obtained from their bulk counterparts. For example, the formation of BCP nanodomains in thin films can take place in different orientations with respect to the substrate surface. Balancing the energetics between the polymer blocks and the material interfaces above and below the BCP is used to create nanopatterns that are not expected in bulk, specifically in the direction perpendicular to the underlying substrate or in combinations of perpendicular and parallel orientations along the thickness axis. The enthalpic contributions from selective interactions at the top and bottom interfaces are minimized under which condition BCP nanodomains are aligned perpendicular to the substrate due to the entropic contributions of better chain stretching in this direction. Similarly, variations in the polymer–air interfacial energies for the different blocks of a BCP can be used to create chemically alternating nanopatterns at the polymer–air boundary. Nanostructures normal to the underlying substrate can exist partially into the depth of a BCP film. For example, an ultrathin diblock film of polystyrene-block-polymethylmethacrylate (PS-b-PMMA) can be prepared to produce periodic nanopatterns on the film surface where the small difference in interaction energy between PS–air and PMMA–air causes both the PS and PMMA blocks to be exposed at the polymer–air boundary. BCP nanopatterns organized this way can offer distinct chemical properties whose length scale varies at nanoscopic dimensions.

More examples of nanomorphologies can be found in amphiphilic BCP systems, whose structures are first self-assembled in solution and subsequently transferred to a solid substrate [67,126,131–140]. In these systems, the polymer nanoarchitectures are controlled by the interactions not only between the different polymer segments of a given BCP but also between each polymer segment and the solvent. Amphiphilic BCPs form micellar assemblies in a solution above a critical polymer concentration. For a typical BCP–solvent system, the volume fraction of the polymer segments plays the largest role in determining the morphology of the nanoassemblies. However, their exact sizes and structures can be additionally adjusted by changing the solution properties and environmental parameters such as the pH, temperature, and ionic strength of the solvent as well as the chemical composition, length, and relative solubility of the polymer segments. The most common nanomorphologies found from diblock BCPs are micelles of a spherical and cylindrical (worm-like) shape, although vesicle-shaped micelles (polymersomes) analogous to naturally occurring liposomes are also formed. In triblock BCP as well as nonlinear BCP systems involving hyperbranched polymers, much more complex morphologies such as toroids, helices, and multicompartment micelles have been obtained [131,132,137].

The formation of micelles in BCP systems of polystyrene-block-polyacrylic acid (PS-b-PAA), polystyrene-block-polyethylene oxide (PS-b-PEO), polystyrene-block-poly(2-vinylpyridine) (PS-b-P2VP), and polystyrene-block-poly(4-vinylpyridine) (PS-b-P4VP) has been extensively examined [131,137,141–145]. In these systems, it is well-known that additional BCP nanostructures beyond those predicted by thermodynamic considerations can be kinetically isolated via solvent vapor annealing (SVA). This method employs a solvent vapor selective to a particular polymer block which then induces preferential interactions of the selective block to the solvent vapor. A swollen and mobile BCP thin film, upon exposure to the solvent vapor, can result in well-ordered nanostructures on the BCP surface even at a temperature that is well below the glass transition temperature of the polymer blocks. Hence, this method has been widely used to kinetically trap various non-equilibrium, but air-stable micellar nanomorphologies as an avenue to producing periodic nanopatterns on a solid surface. For example, hexagonal micelles of PS-b-P4VP were formed in toluene and prepared into a thin film on a Si support. The PS-b-P4VP thin film was subsequently annealed under the vapor of chloroform. In addition to the original micellar spheres, additional nanostructures that include holes, reformed spheres, embedded spheres, enlarged spheres, cylinder precursors, and cylinders were produced during the solvent annealing process [126,146]. BCP nanostructures that can be generated by controlling solvent vapor annealing have also been identified by computer simulations [147]. For producing different BCP nanostructures, solvent vapor annealing provides additional degrees of freedom such as the fraction of different solvents incorporated, their selectivity with respect to each block, and solvent effects on the surface and interface energies. The effects of these factors on the final nanodomain morphologies of BCPs were examined in 2D simulation studies, as depicted in Figure 2. The simulation results were then compared with the experimental outcomes of polystyrene-block-polydimethylsiloxane (PS-b-PDMS) annealed under block-selective solvents of toluene and heptane [147].

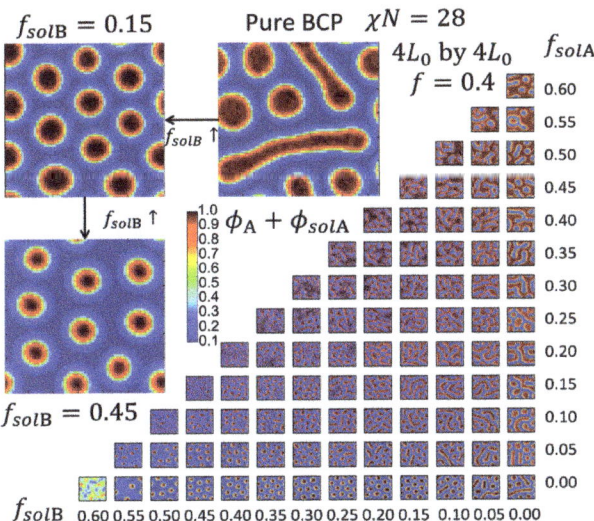

Figure 2. Explicit simulations for solvent vapor annealing were performed in 2D with an AB-type of BCP with $f_A = 0.4$ and a fixed $\chi N = 28$ that were exposed to different amounts of block A- and block B-selective solvents with the respective volume fraction of f_{solA} and f_{solB}. Various BCP phase regions observed in simulations are then illustrated for various values of f_{solA} and f_{solB}. ϕ denotes the local density. L_0 is the spacing of a set of metastable hexagonally packed cylinders. The 2D bulk morphology with no solvent included nanostructures of circles corresponding to through-plane cylinders and lines corresponding to lamellae of block A. As more f_{solA} was added to the system, the line structures became more dominant and eventually transitioned to a perforated A network surrounding B. As f_{solB} increased, the morphology changed to hexagonally close-packed circles that eventually solvated A-rich micelles. Ordered structures were lost with the increase in both f_{solA} and f_{solB}. Reproduced with permission from Ref. [147] Copyright (2015) Royal Society of Chemistry.

3. Block Copolymer Surfaces Interfacing Biomolecules

As discussed so far, BCP self-assembly enables a straightforward and convenient means to obtain patterned nanostructures with high precision and controllability without the need for sophisticated nanofabrication techniques such as extreme UV and electron-beam lithography. Further, self-assembled BCP nanostructures can serve as a powerful and viable platform to position different nanomaterials of interest onto a solid surface with excellent scalability and exquisite nanoscale spatial precision. As such, BCP nanodomains have been well-recognized and utilized as nanotemplates for seeding inorganic nanoparticles (NPs) and as lithographic masks for large-area chemical patterning [42,69,70,144,145,148–150].

Another area in which self-assembly plays a crucial role is found in biological systems. Many biological processes and functions rely on the precise positioning and assembly of biomolecules. The spatial positioning and high-level organization of proteins and nucleic acids, for example, inside virus capsids, collagen matrices, and cell membranes occur with nanoscale precision via self-assembly-driven processes. The exact nanoscale arrangements of intricate molecular structures are vital for their proper functions in those cases. It is also imperative to control the assembly of biomolecules in biomaterials for their use in engineered bioplatforms such as medical implant devices, artificial tissue scaffolds, and antibacterial coatings. Likewise, the self-assembly dynamics pertinent to nanoscale organizations of biomolecules onto various surfaces, including those on heterogeneous templates, are critical to controlling protein crystallization, protein printing onto a surface, and timed protein release from a surface. These aspects have direct and important consequences for biosensing and biocharacterization applications.

In comparison to BCP applications in inorganic NP assembly and nanolithography masks, their use in nanoscale bioassembly has not been realized until later [35–37,72,75,151–155]. Comparatively speaking, limited work has been undertaken to exploit BCP self-assembly in nanoscale spatial partitioning of biomolecules and their extended assembly on a solid surface. Yet, many intriguing and encouraging discoveries have been put forward so far in this field. The first endeavor in this regard was made by Kumar et al. [75], whose work demonstrated the possibility of creating well-organized protein nanoarrays. Owing to this and ensuing research efforts, it is now well-understood that the spatial assembly of proteins can be faithfully guided not only by the size and periodicity of the nanostructures formed on an underlying BCP surface upon microphase separation, but also by preferential interactions between the different BCP nanodomains and a given protein. This section highlights those research endeavors that successfully utilized self-assembled nanostructures of BCPs as surface guides to derive simple and hierarchical ordering of biomolecules during which processes biomolecules themselves were also organized on the BCP nanostructures via self-assembly.

3.1. Proteins

3.1.1. BCP Nanodomains for Proteins: Single-Component Systems

Protein Interactions on BCP Thin Films. The spearheading study of protein nanoarrays guided by an underlying BCP surface of PS-b-PMMA demonstrated that individual protein molecules self-assemble on BCP nanodomains via preferential protein–PS interactions [75]. A model protein of immunoglobulin G (IgG) was successfully ordered on the PS nanodomain areas of PS-b-PMMA. Figure 3A displays such exclusive interaction behavior of IgG with the PS nanodomains that was unambiguously resolved at the individual protein level on the BCP nanodomain surface. It is clear from the atomic force microscopy (AFM) data that the surface partitioning of IgG molecules was entirely exclusive to the PS nanodomains, leaving the PMMA nanodomains completely free of IgG. This was due to the preferential interaction of IgG with the more hydrophobic block of PS relative to PMMA. The different degree of IgG loading on the BCP surface in Figure 3A was controlled by adjusting the bulk solution concentration of IgG and incubation time on the surface. When the loading condition was tuned to a monolayer-forming coverage, all available PS sites were packed densely with a single layer of adsorbed IgG. The packed protein layer on the PS nanodomains contained two IgG molecules along the short axis of the nanodomain direction, as shown in the rightmost panels of Figure 3A. This was because the width of the underlying PS domains used for the study was commensurate with approximately two IgG molecules assembled side by side along the short nanodomain axis. The study was the first demonstration of a BCP thin film-based approach for achieving nanopatterned proteins on a solid surface, while solely relying on the self-assembly processes of the BCP as well as the biomolecules.

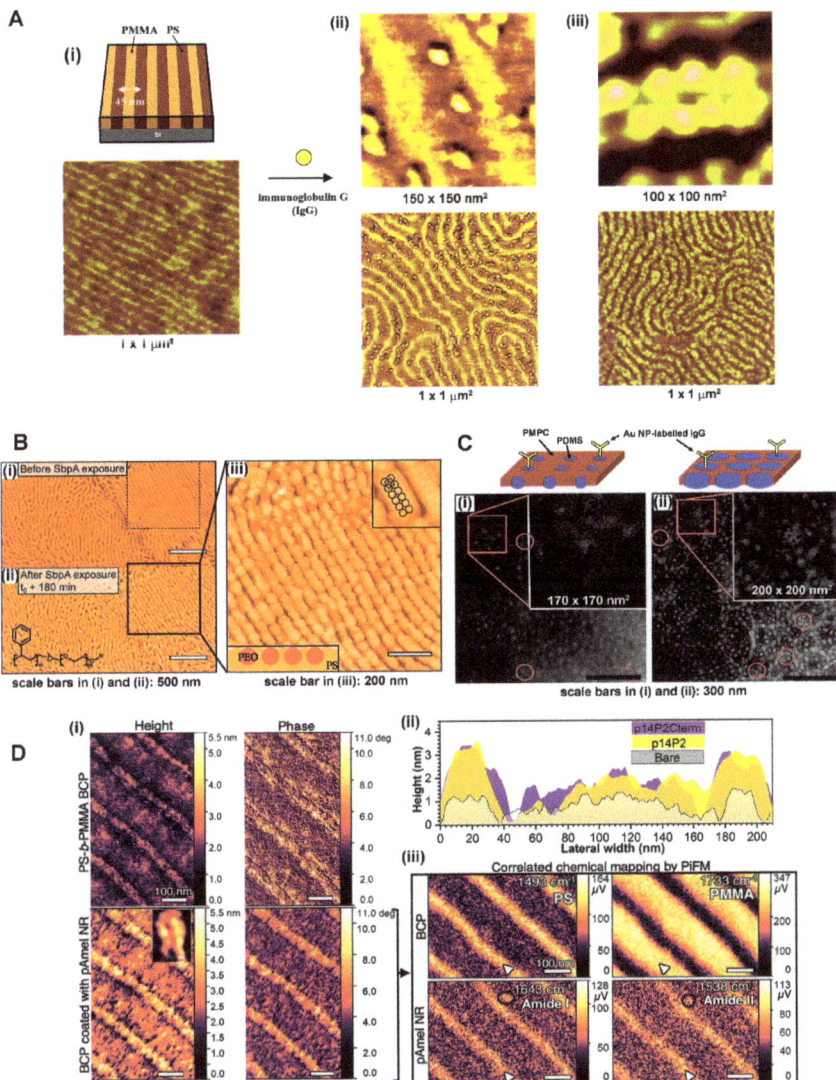

Figure 3. (**A**) The schematic diagram and the AFM panel in (**i**) correspond to the alternating PS (orange) and PMMA (yellow) nanodomains formed on a PS-b-PMMA thin film surface. The repeat spacing of the nanostripes (PS to PS nanodomains) is 45 nm. The AFM images in (**ii**,**iii**) display the exclusive interaction of IgG molecules (appearing as spheres) with the more hydrophobic PS block of PS-b-PMMA under different protein loading conditions. In all cases, the distribution of IgG molecules was consistently observed to be segregated only on the PS nanodomain areas of the BCP surface. The AFM data in (**iii**) belong to an IgG monolayer-forming condition under which all available PS nanodomains were fully occupied by densely packed IgG molecules. Two IgG molecules assembled along the short axis of the PS nanodomains at maximum due to the protein size with respect to the width of the underlying PS nanodomain. The assembly of IgG molecules on the BCP surface resembled the packing nature found in a 2D protein crystal. Adapted with permission from Ref. [75] Copyright (2005) American Chemical Society. (**B**) A blank template of nanostriped PS-b-PEO thin film is displayed in (**i**) onto which SbpA was incubated. The treatment resulted in the formation of S-layer crystals confined to the PS regions of the PS-b-PEO surface, as shown in the AFM panel of

(ii). The internal crystal structure of the S-layer is shown in the magnified image of (iii). Adapted with permission from Ref. [82] Copyright (2019) American Chemical Society. (**C**) The TEM data display the surface of PMPC-b-PDMS after treating the surface with a solution of AuNP-labelled IgG. The PMPC-b-PDMS surfaces used in (**i,ii**) contained a PDMS monomer unit composition of 40.7% and 55.4%, respectively. Cylindrical PDMS domains were produced in a PMPC matrix with different domain sizes as schematically shown in (**i,ii**). Small dark dots inside the red circles in the TEM panels correspond to the AuNP-labelled IgG molecules segregated on the more hydrophobic PDMS regions of the BCP surface. Adapted with permission from Ref. [81] Copyright (2009) Elsevier. (**D**) The AFM results in (**i,ii**) and PiFM data in (**iii**) show the peptide analogues of pAmel NRs assembled on 50 nm PS stripes of PS-b-PMMA. The top and bottom panels in (**i**) correspond to the blank template of PS-b-PMMA and after incubation with p14P2, where the inset in the bottom left panel (scale bar of 10 nm) displays the morphology of two pAmel NRs on PS. Height profiles measured perpendicular to the stripes are compared among the cases of the bare BCP, p14P2-coated BCP, and p14P2Cterm-coated BCP in (**ii**). The PiFM surface maps of p14P2-coated BCP in (**iii**) were obtained at the excitation wavelength specific to the BCP as well as to the β-sheet pAmel NRs. The specific wavelength used is marked in each map and the arrows point to the region of excitation. The strong signals appearing as bright stripes in the PiFM data are from pAmel NRs with a β-sheet conformation which were assembled on the PS regions. The relatively low signal of the PMMA areas in the Amide I and II maps indicated the lack of β-sheet NRs. Adapted with permission from Ref. [83] Copyright (2023) American Chemical Society.

In many other stimulating studies following this work, other proteins, largely globular in shape, were able to be similarly assembled into nanopatterns [35,37,79,81,82,86,95,151,156,157]. Proteins and peptides such as human and bovine serum albumins (HSA and BSA), horseradish peroxidase (HRP), mushroom tyrosinase (MT), green fluorescent protein (GFP), protein G (PG), and amelogenin (Amel) behaved similarly as IgG on a PS-b-PMMA thin film. Much like the data presented in Figure 3A, the assembled protein patterns under a monolayer forming condition faithfully followed the size and shape of the more hydrophobic BCP nanodomains [35,37,79,81–83,86,95,151,156,157]. The AFM, transmission electron microscopy (TEM), and photo-induced force microscopy (PiFM) data in Figure 3B through 3D display some of these examples. As shown in Figure 3B, S-layer protein (SbpA) treated on a nanostriped BCP thin film of polystyrene-block-polyethylene oxide (PS-b-PEO) yielded the formation of S-layer crystals confined to the more hydrophobic PS block of the BCP surface. The TEM data in Figure 3C correspond to gold nanoparticle (AuNP)-labelled IgG that segregated into the more hydrophobic PDMS block on the thin film surface of poly(2-methacryloyloxyethyl phosphorylcholine)-block-poly(dimethylsiloxane) (PMPC-b-PDMS). The AFM and PiFM results in Figure 3D present the assembly of the peptide analogues of phosphorylated amelogenin (pAmel) nanorods (NRs) on the PS stripes of PS-b-PMMA when the BCP surface was incubated with p14P2. The pAmel NRs on the PS-b-PMMA thin film were formed via the self-assembly of p14P2 (GHPGYINF p(S) YEVLT) and p14P2Cterm (GHPGYINF p(S) YEVT DKTKREEVD) on the more hydrophobic PS block.

Elongated Protein Interactions on BCP Thin Films. It was also demonstrated that the size and periodicity of the BCP nanodomains could be further utilized to modulate the surface partitioning of elongated proteins into nanoscopic patterns upon their self-assembly on the BCP surface [84]. It was revealed that, for a system of an elongated protein of fibrinogen (Fg) on PS-b-PMMA, the interaction differences between the two polymer blocks as well as those between the D, E, and αC subunits within a Fg molecule can lead to protein concentration-dependent and protein subunit-specific behaviors of Fg partitioning on the BCP surface. Individual Fg molecules show high aspect ratios of ~10 (length to width) and ~25 (length to height). Unlike the globular protein case discussed above, the interaction of Fg to the PS nanodomains was less exclusive where, depending on the protein concentration, Fg showed a more neutral tendency for shared interactions with both blocks of PS and PMMA. The interaction forces governing Fg were found to arise from not only hydrophobic but also electrostatic in nature. This is different from the globular protein interactions discussed earlier which were dominated by the hydrophobic interactions.

Figure 4A presents such complex interaction behaviors observed from the elongated protein of Fg on a PS-b-PMMA surface. Mixed populations of Fg molecules with TP and SP configurations were observed on the BCP surface that contained unaligned nanodomains with a repeat spacing of 25 nm. TP and SP stand for the configuration of Fg on the BCP surface, where the entire length of a Fg molecule (~48 nm in length) lies across the PS and PMMA nanodomain areas (two phases, TP) versus only within the PS nanodomain areas (single phase, SP). The study also revealed surface-specific Fg conformations on the BCP as well as on the homopolymer surface consisting of PS or PMMA. In addition, BCP surface-driven topological changes of single proteins were experimentally resolved for the first time at the sub-biomolecule level. Figure 4B presents such BCP surface-driven effect on the assembly of Fg molecules. Compared to the data on an unaligned nanodomain template in Figure 4A, Figure 4B displays Fg molecules on fully aligned nanodomains of a PS-b-PMMA substrate. The aligned BCP template used for Figure 4B had a repeat spacing comparable to the unaligned sample of Figure 4A. All populations of Fg molecules on the aligned BCP exhibited the SP configuration.

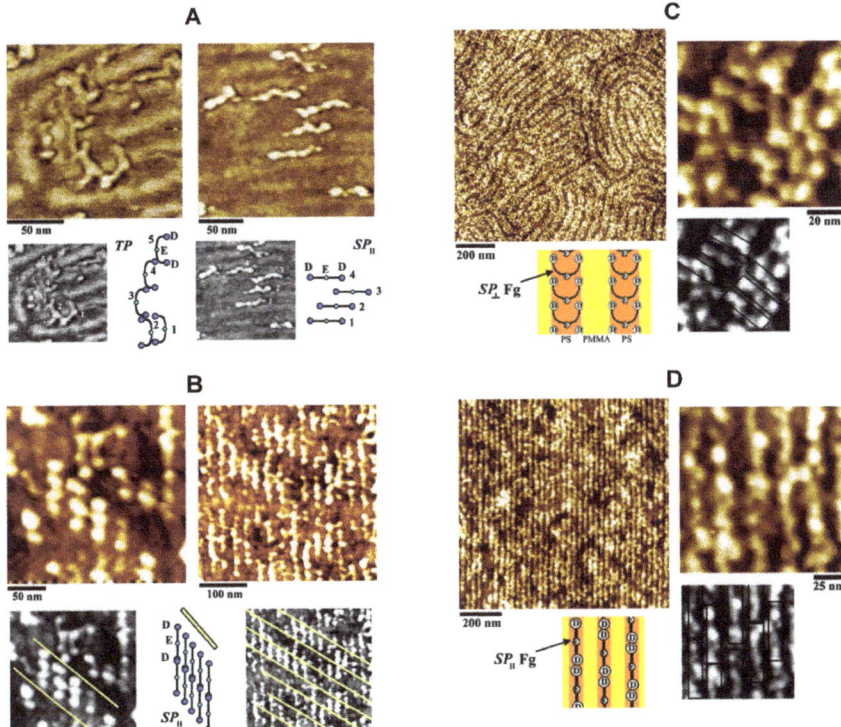

Figure 4. The AFM panels display different Fg configurations on various PS-b-PMMA surfaces. TP and SP refer to the Fg configuration for which the backbone of a Fg molecule lies both on the PS and PMMA nanodomain areas (TP) and only on the PS regions (SP). $SP_\parallel$ and $SP_\perp$ denote the backbone of a Fg molecule lying along the long ($SP_\parallel$) and short ($SP_\perp$) axis of the PS nanodomains. The exact Fg configuration and large-area assembly on PS-b-PMMA were dependent not only on the protein concentration but also on the periodicity and alignment degree of the underlying BCP nanotemplate. (**A**) Mixed populations of Fg molecules with TP and $SP_\parallel$ were found on unaligned PS-b-PMMA nanodomains of 25 nm in repeat spacing. The cartoons inserted next to each AFM panel depict the inter- and intra-molecule arrangements of the different Fg subunits of D and E. (**B**) On a PS-b-PMMA surface with fully aligned nanodomain of 28 nm in periodicity, all Fg molecules assembled on the PS

nanodomain areas in the direction parallel to the long axis of the PS nanodomains ($SP_\parallel$). The yellow lines show the characteristic slope formed by the neighboring Fg molecules. (**C**) Fg molecules assembled on the PS areas in the orientation parallel to the short axis of the PS nanodomains ($SP_\perp$ configuration, side-on packing) when a PS-b-PMMA surface with 45 nm in periodicity was used to form a monolayer of Fg molecules. Black boxes mark individual Fg molecules. (**D**) On a PS-b-PMMA surface with fully aligned nanodomains of 28 nm in periodicity, Fg molecules under a monolayer forming condition occupied the PS areas in the orientation parallel to the long axis of the PS nanodomains ($SP_\parallel$ configuration, end-on packing). Reproduced with permission from Ref. [85] Copyright (2016) American Chemical Society.

In a later study, the effects of BCP periodicities and alignments on Fg interactions were scrutinized [85]. Different PS-b-PMMA substrates that contained fully aligned or randomly oriented nanodomain of varying sizes were employed. The length scale of the nanodomains between the samples was varied to exhibit a dimension that was much larger, comparable to, and much smaller than the length of Fg. The adsorption behaviors of several Fg molecules in isolation as well as the assembly of many Fg molecules in large-area surface packing were further investigated on the different BCP substrates. The study reported that the periodicity and orientation of the chemically alternating BCP nanodomains can be exploited to manipulate the packing configuration of Fg molecules on the BCP surface. For example, an end-on (side-on) packing geometry, where the backbone of Fg is parallel (and perpendicular) to the long axis of the PS nanodomain, can be achieved by providing a BCP template with a nanodomain periodicity much smaller than (compatible to) the length of the protein. Figure 4C,D summarize these results. The application of a PS-b-PMMA template with a repeat spacing of 45 nm (comparable to the length of a Fg molecule) versus 28 nm (much smaller than the Fg length) for the protein assembly was able to induce side-on (Figure 4C) versus end-on (Figure 4D) packing of Fg molecules on the PS nanodomain areas. Highly oriented nanostructures formed on melt-drawn, ultrahigh molecular weight polyethylene (UHMWPE) surfaces were also shown to induce Fg assembly [158]. Similar to the nanostructures on the BCP surfaces, the nanocrystalline lamellae on the UHMWPE surface were able to control the conformation and aggregation of human plasma Fg. The lateral orientational order of proteins on the polymer surface was dependent on multiple parameters such as nanoscale topography, chemistry, crystallinity, and molecular chain anisotropy of the UHMWPE surfaces.

These works showed that the structural and chemical features of BCP and related polymer surfaces could be effectively used to control not only the spot size and periodicity of the assembled proteins at the nanometer range, but also the orientation and packing geometry in the large-area organization of elongated proteins. Controlling the spatial arrangement of Fg molecules on solid surfaces has important biomedical relevance to blood clotting and wound healing since a specific arrangement of Fg molecules is required in these processes [159]. As evidenced in Figure 4B,D, the BCP-generated Fg nanoassemblies produced Fg molecules in a half-staggered manner, yielding protofibrils of Fg molecules arranged in different PS nanodomains. The half-staggered packing of Fg molecules enabled contact points for the D-E subunits between neighboring Fg molecules on adjacent PS nanodomains. This intermolecular assembly pattern of Fg molecules is similar to the natural process of fibrin assembly in blood clotting. Such an aspect will be important for future applications of BCP-based protein nanoassemblies in developing biomaterials.

3.1.2. BCP-Guided Protein Assembly on Extended Systems Involving Various BCP Thin Films and Proteins

Other studies have since demonstrated that a BCP-based method can be effectively used to attain a large-scale surface organization of biomolecular nanopatterns in a controllable and predictable manner [78,81,85–88,90,91,95–97,103,126,140]. Diblock and triblock copolymer systems used for protein assembly have been extended to include a range of different BCP blocks such as polystyrene-block-polyisoprene (PS-b-PI), polyethylene glycol-block-polystyrene (PEG-b-PS), poly(2-methacryloyloxyethyl phosphorylcholine-block-poly(dimethylsiloxane) (PMPC-b-PDMS), polystyrene-block-poly(2-hydroxyethyl methacrylate) (PS-b-PHEMA), poly(acrylic acid)-block-poly(N-isopropyl acrylamide) (PAA-b-PNIPAM), PS-b-PEO, PMMA-b-PHEMA-b-PMMA, and PAA-b-PMMA-b-PAA. In addition, a diverse system of whole proteins, protein fragments, protein coats, peptides, and extracellular matrix (ECM) fragments has been employed as a model biomolecule for BCP-based self-assembly. Regardless of the BCP and protein model systems used, it was possible to effectively modulate selective adsorption, morphology, orientation, and alignment of proteins on the BCPs. This was achieved by tuning the underlying BCP nanostructures to favorably recognize the different physicochemical properties of the proteins.

Preferential Protein Interaction with the Hydrophobic BCP Domains. Many studies reported that proteins and protein coats preferentially interact with the more hydrophobic segments of BCPs [81,82,86,95]. An example of this can be found in a study that employed highly oriented lamellar nanopatterns of PS-b-PMMA as a platform to assemble nanopatterns of various serum, antithrombogenic, as well as cell adhesive proteins such as γ-globulin, Fg, fibronectin (FN), thrombomodulin (TM), and type I collagen (Col I) [86]. The preparation of the BCP thin film was formulated to have a perpendicularly oriented, lamellar morphology of alternating PS and PMMA regions on the surface. The lamellar structures were then aligned along the thickness gradient for producing unidirectional protein nanopatterns of γ-globulin molecules, FN, and TM on the hydrophobic PS areas. Unlike these proteins, Col I molecules did not show any particular orientation on the BCP template, but almost all of the adsorbed parts of Col I were reported to interact with the PS domains. In a different study involving a polymer blend surface, the important roles that the size and surface coverage of polymer heterogeneities play in modulating the diameter and length of Col I assemblies have been identified [160]. When a blend thin film consisting of PS and PMMA was employed, the organization of Col I was reported to be affected by the size of the PS areas in the blend film. The study also confirmed that the amount of Col I adsorbed on the surface was linearly correlated with the PS surface fraction of the blend film.

The general tendency of proteins favoring the more hydrophobic domain of BCPs was further confirmed by examining the adsorption behaviors of BSA, FN, and crystalline surface layers (S-layer crystals) on the BCP surfaces of PS-b-PI, PMPC-b-PDMS, and PS-b-PEO [81,82,92,95]. Various nanopatterns and surface chemistry of solvent-annealed PS-b-PEO were found to efficiently steer the formation of crystalline S-layers from monomeric SbpA confined to the PS nanodomains of the BCP surface [82]. A study using well-ordered, nonequilibrium nanostructures of PS-b-PI also reported that BSA as well as FN tended to bind selectively on the PS domains of PS-b-PI [92,95]. The overall patterns produced by the BSA or FN molecules were found to closely resemble the nanopattern shape of the underlying PS-b-PI nanostructures. Phase-separated BCP surfaces composed of PMPC-b-PDMS were shown to exhibit selective binding of FN molecules to the hydrophobic PDMS domains as well [81].

Topographical versus Chemical Contrast on a BCP Surface for Protein Assembly. PS-b-P2VP and PS-b-PEO were employed as model BCP systems in a research effort to discern the BCP effects of a structural (i.e., topographic) versus chemical origin on protein assembly [82]. Figure 5A displays the preparation processes of these BCP thin films for the assessment of structural versus chemical effects. The two BCP surfaces were prepared to have a topographic contrast similar to each other, while presenting different chemical contrasts in terms of alternating hydrophobicity and hydrophilicity. The PS-b-P2VP surfaces, uniformly hydrophilic relative to PS-b-PEO in terms of their chemical contrast, resulted in no confinement of the S-layer crystals to any specific nanodomains. Figure 5B schematically illustrates the different formation processes of the S-layer due to the structural and chemical variations associated with the underlying polymer substrates. It was concluded that the presence of a chemical contrast on the PS-b-PEO template played a critical role in the spatially confined assembly of the crystalline S-layers [82].

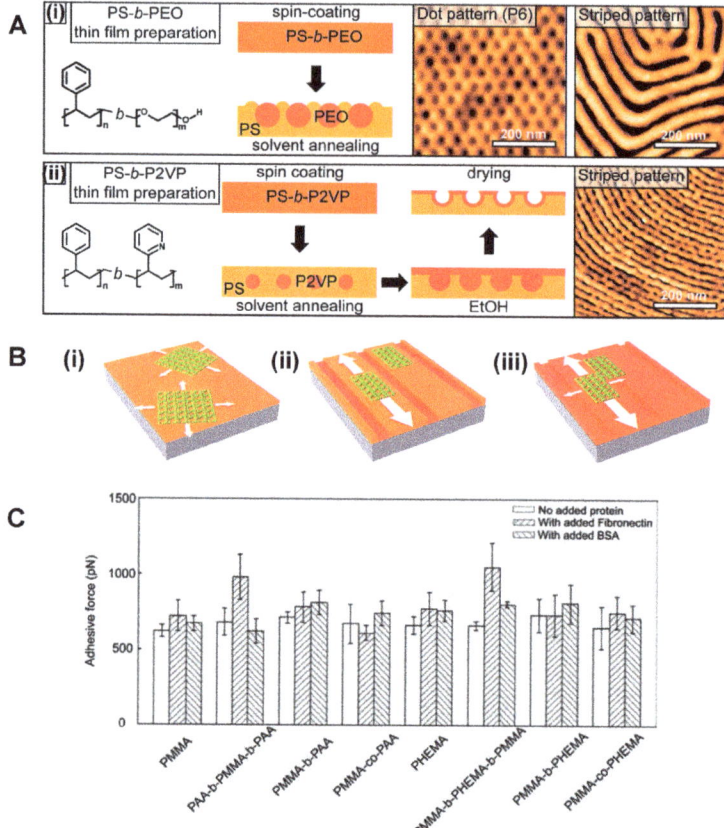

Figure 5. (**A**) The BCP surfaces of (**i**) PS-b-PEO and (**ii**) PS-b-P2VP prepared to examine the effects of structural and chemical contrasts on protein organization are displayed. Relative to the PS-b-PEO surface, the patterned PS-b-P2VP presented a comparable physical contrast but lacked a chemical contrast in terms of alternating hydrophobicity and hydrophilicity. (**B**) The illustrations depict different S-layer formation processes due to the surface effects of structural and chemical contrasts. Isotropic nucleation and growth are expected for S-layers on a uniform surface in (**i**). On a nanopatterned surface with alternating hydrophobic and hydrophilic domains in (**ii**), S-layers nucleate and preferentially grow on the hydrophobic regions only. Lastly, on a nanopatterned surface with no chemical contrast in (**iii**), S-layers nucleate equally on both nanodomains although its growth rate is faster along the long axis of the nanodomain. (**A**,**B**) Adapted with permission from Ref. [82] Copyright (2019) American Chemical Society. (**C**) The bar diagram summarizes measured adhesive forces between antibody-functionalized tips and various polymer surfaces with and without added protein. In both series of BCP templates containing the more hydrophilic (PMMA and PAA) and the less hydrophilic (PMMA and PHEMA) blocks, it was the triblock copolymers that exhibited the highest adhesive force. Adapted with permission from Ref. [90] Copyright (2012) Wiley Periodicals, Inc.

The effects of BCPs' structural and chemical contrasts on protein assembly have been further examined not only on diblock but also on triblock and other related polymer systems. The different roles of the BCP's structural and chemical effects have been studied by measuring the adhesion forces of proteins on surfaces. Diblock and triblock as well as random copolymers consisting of PMMA, PAA, and PHEMA were evaluated for their differences in FN interaction [90]. While keeping PMMA as one of the blocks, the other two polymers were used as varying segments. Different diblock and triblock surfaces such as PMMA-b-PAA, PMMA-b-PHEMA, PMMA-b-PHEMA-b-PMMA, and PAA-b-PMMA-b-PAA were prepared this way. It was found that the surface distribution of FN molecules was dictated by both the chemical effect stemming from the interactions between FN and the polymer chain of PMMA, PAA, or PHEMA, and the topographic effect due to the nanoscale dimension and spacing of the polymer domains. The conformation and orientation of FN were determined by the surface chemistry as well as the nanomorphology of the BCP templates. However, the study pointed out that the adhesion forces between FN on the BCP surfaces and FN antibody hanging from a probe tip did not depend either on the chemistry, charges, or wettability of the BCP surfaces. Rather, the adhesion forces between FN-FN antibodies were governed by the BCP nanomorphology. In general, higher adhesion was monitored for the triblock surfaces that presented a larger domain size relative to the diblock samples. This tendency was consistently observed whether the sample surfaces contained the more hydrophilic (PMMA and PAA blocks) or the less hydrophilic (PMMA and PHEMA blocks) polymer segments. Figure 5C summarizes the measured adhesion forces between the different polymer surfaces and antibody-functionalized tips.

Stimuli-Responsive BCP Segments for Protein Interactions. Protein behaviors at surfaces have been successfully tuned with the aid of stimuli-responsive segments in BCPs. A BCP thin film of PAA-b-PNIPAM, assembled in a layer-by-layer manner, was employed to study the adsorption behaviors of ovalbumin (OVA) while varying the temperature and the pH of the protein solution [96]. The pH-responsiveness of the PAA block and the thermo-responsiveness of the PNIPAM block provided a dual sensitivity to modulate protein interactions at the PAA-b-PNIPAM surface. It was found that OVA adsorption to the BCP surface was dependent on the temperature. The BCP film exhibited high OVA adsorption at 50 °C whereas the same BCP surface strongly repelled the protein at 20 °C.

Micellar BCP Inversion in Protein Assembly. Tuning protein behaviors at the surface has been attempted by flipping the spatial arrangement of the polymer segments belonging to the core and matrix (corona) portions of BCP nanostructures as well. Heterogeneous nanopatterns assembled from an amphiphilic BCP of PS-b-PHEMA were used for Fg adsorption [88]. The study found that the protein-adhesive/-resistant property of the underlying surface can be tuned by switching out the core and matrix polymer components of the heterogeneous nanopatterns. When the PS-b-PHEMA surface was processed to yield PHEMA (PS) domains to occupy the majority (minority) of the surface, the film became strongly protein-repulsive. In contrast, the opposite distribution of majority PS and minority PHEMA domains on the film surface led to protein adsorption.

Chemical Modifications of BCPs and Proteins for Specific Interactions. Strategies to chemically modify the nanodomains of a specific BCP block as well as proteins of interest have been used to induce exclusive polymer block–protein interactions on a BCP surface. In one study, a biotinylated BCP surface of PEG-b-PS was prepared into cylindrical nanostructures by mixing a small amount (4 mol%) of biotin-functionalized BCP into non-functionalized BCP [97]. Upon subsequent incubation with streptavidin (SAv) on the biotinylated BCP surface, the strong interaction between biotin and SAv led to the immobilization of SAv to the PEG-b-PS thin film. The protein immobilization was controlled by varying the amount of biotinylated PEG-b-PS used for mixing. In another study, alkyne-functionalized BCP nanopatterns of PS-b-PHEMA were demonstrated for linking azide-tagged protein molecules of Fg, myoglobin (Mb), and lysozyme (LZM) [89]. The azide-tagged protein molecules bound to the alkyne-functionalized PS nanodomains of PS-b-PHEMA. The approach was able to conveniently produce nanoarrays containing individual protein molecules per spot via specific protein binding to the PS nanodomains and eliminated any issues in protein quantification that might arise from nonspecific protein adsorption. In addition, a chemically modified BCP of PS-b-PEO was self-assembled to produce the functional group of maleimides on the PEO nanodomains while controlling the size, number density, and lateral spacing of the nanodomains [91]. The maleimide group was then employed to bind proteins and extracellular matrix (ECM) fragments such as

GFP, FN fragments, and arginine-glycine-aspartate (RGD)-containing peptides on the PEO domains. The study also showed that the same maleimide-functionalized BCP templates were applicable for linking other biomolecules such as poly-histidine tagged proteins and Zn-chelating peptide sequences.

Protein Embedded in BCP Thin Films. Attempts to create hierarchically structured, functional biomaterials have been made by directing co-assembly of BCP thin films and biomolecules of proteins or peptides. These platforms were also used to carry out a quantitative examination of the release kinetics of biomolecular cargos within BCP thin films. PS-b-PEO thin films prepared into different thicknesses were co-assembled with cargo proteins and peptides such as LZM and a peptide of TAT [99]. This process led to the distribution of protein or peptide cargos within the hexagonally packed PEO nanodomains of PS-b-PEO. In a different study, the co-assembly method was extended to build assembled structures of a greater hierarchy [98]. Structures consisting of PS-b-PEO and a bio-motif were designed for a simultaneous co-assembly scheme. Bio-motifs such as horse-heart Mb and a heme-binding protein were used in the co-assembly to produce protein/cofactor complexes as well as catalytically active enzymes within the BCP thin film. In another endeavor, a solvent-induced film of hexagonally packed PS-b-PEO nanostructures was processed into vertically arranged, cylindrical nanoscaffolds for the assembly of Lsmα [102]. Lsmα is a protein that self-organizes into stackable, doughnut-shaped, heptameric structures whose pore size can be tuned for encapsulation molecules of interest. Upon co-assembly of PEGylated Lsmα (LsmαPEG) with PS-b-PEO in a solvent mixture composed of water, methanol, and benzene, the protein molecules were able to form into a regular array. The assembled array structure contained doughnut-shaped tunnels of Lsmα. The work showed that BCP templates can effectively guide even a coordinated assembly of hierarchical protein nanostructures into BCP nanodomains, beyond what has been demonstrated for the assembly of simple proteins on BCP nanodomains. All these efforts will be crucial for the future applications of biocargo-loaded BCP films in cell culture and mechanotransduction studies as well as in biocatalytic reactions and biosensing.

Nanoporous BCP Thin Films for Protein Assembly. Nanoscale protein interactions with BCPs have been extended to those with nanoporous thin film structures [104,161–163]. Nanopores in self-assembled BCP thin films are typically produced by selectively removing a polymer block from a phase-separated BCP film [163]. Methods used for the selective segment removal include ultraviolet (UV) degradation, reactive ion etching (RIE), ozonolysis, and chemical etching [164–167]. The resulting size and shape of the nanopores in the BCP templates are governed by the original nanostructures formed during the BCP's phase separation process and, thereby, the nanoporous structures can be controlled by the same experimental parameters that are used to modulate the BCP nanodomains according to their phase diagrams. In a study using the nanopore approach, a thin film of nanometric channels was fabricated from a BCP mixture of polystyrene-block-poly(l-lactide) (PS-b-PLLA) and PS-b-PEO [104]. Nanoporous structures with elongated nanopores of ~20 nm in width were generated after the selective removal of PLLA from the phase-separated BCP mixture. The resulting nanochannels contained PEO chains pending from PS walls. The BCP nanopore thin film was then successfully utilized for the immobilization of HRP molecules.

Indirect BCP–Protein Interactions via Inorganic Nanoparticles. Research efforts have been made to assemble biomolecules at BCP surfaces through a mediating layer of inorganic NPs using a process known as BCP micelle nanolithography, instead of having BCP nanopatterns directly interface with proteins on the polymer surfaces [68,105–108]. In these works, inorganic NPs such as gold NPs (AuNPs) of 1–15 nm in size were preassembled on BCP nanoguides with a tunable lateral spacing of 15–250 nm through a preferential metal–polymer segment interaction. Well-defined patterns of AuNPs were subsequently produced after subjecting BCP thin films to a plasma process to remove the BCP from the substrate, leaving only the AuNPs. As the lateral spacing in the BCP template can be controlled by the BCP molecular weight, the periodic spacing between AuNP dots in the array can be adjusted accordingly. Fabrication processes similar to those depicted in the schematics of Figure 6A are typically used to create inorganic NP-linked templates via the BCP micelle nanolithography. The NP-containing BCP surfaces can then be used for assembling DNA, peptides, or proteins.

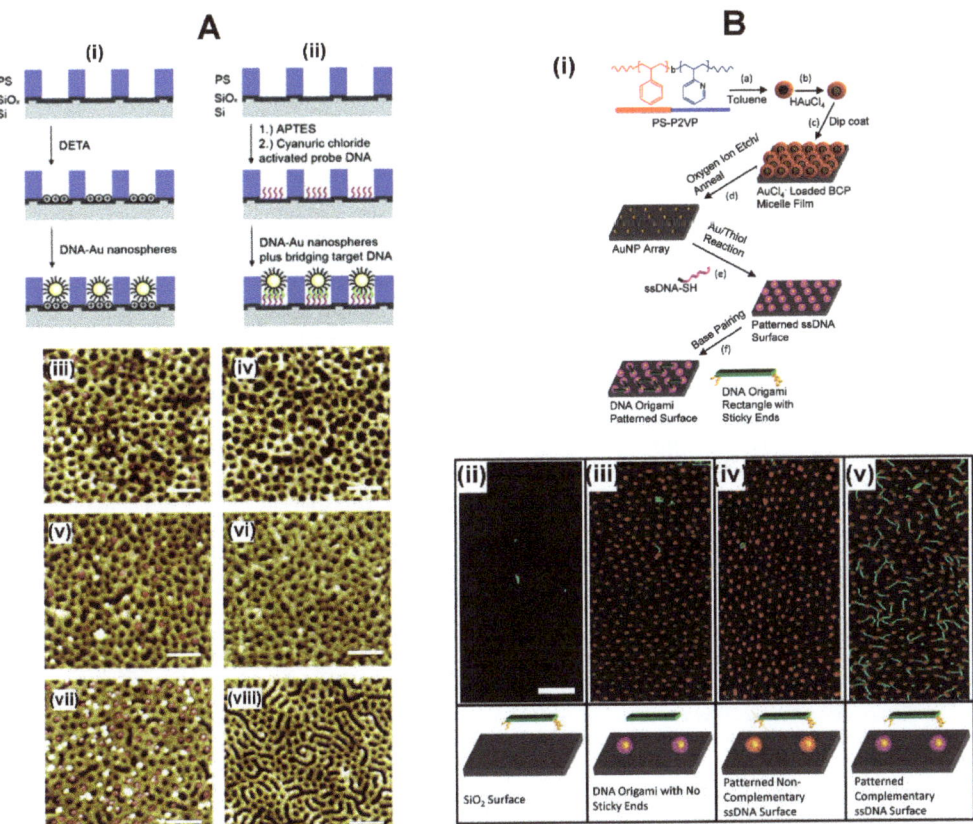

Figure 6. (**A**) The schematic representations show the processes for Au-DNA functionalization via (**i**) DETA and (**ii**) APTES on nanoporous PS-b-PMMA thin films. DETA and APTES denote (3-trimethoxysilylpropyl)-diethylenetriamine and (3-aminopropyl)-trimethoxysilane, respectively. The AFM images correspond to the DETA-functionalized (**iii,v,vii**) and unfunctionalized (**iv,vi,viii**) nanoporous BCP templates exposed to DNA-AuNPs. Red circles inserted in the images are the DNA-AuNPs that were deposited into the DETA-functionalized nanopores. All scale bars are 200 nm in size. Adapted with permission from Ref. [107] Copyright (2006) American Chemical Society. (**B**) The schematic in (**i**) displays the patterning process of AuNPs on PS-b-P2VP micelles for the adsorption of DNA origami on the BCP surface. The AFM panels show (**ii**) sticky end-modified, DNA origami placed on a clean SiO$_2$ surface, (**iii**) non-modified DNA origami on a patterned, single-strand DNA (ssDNA) surface, (**iv**) sticky end-modified DNA origami on a patterned, noncomplementary ssDNA surface, and (**v**) modified DNA origami on a patterned, complementary ssDNA surface. All scale bars are 200 nm in size. Adapted with permission from Ref. [108] Copyright (2011) American Chemical Society.

In studies using the micelle nanolithography method, AuNPs with tunable sizes were generated into a quasi-hexagonal pattern by employing a sacrificial PS-b-P2VP template [105,106]. The AuNP array was further used to assemble thiolated αvβ3 integrin receptor of c(-RGDfK-) functionalized through thiol–Au interactions. Nanoporous structures fabricated from PS-b-PMMA were also utilized for the deposition of DNA-conjugated AuNPs [107]. As illustrated in Figure 6A, nanopatterns on the PS-b-PMMA thin film were first exposed to UV radiation to cross-link the PS chains while degrading PMMA. Nanopores were then created on the PS-b-PMMA surface by rinsing away the degraded PMMA with a solvent. The resulting nanopores on the BCP surface were able to serve as nanocontainers for the AuNPs whose NP surfaces were pre-conjugated with oligonucleotides. Similarly, PS-b-PMMA and PS-b-P2VP surfaces were fabricated to produce AuNP and Au nanorod (AuNR) arrays of various diameters and center-to-center distances [68,108]. The Au-modified BCP templates were used afterwards to selectively place DNA origami at directed surface locations. An example of such efforts involves AuNPs and AuNRs formed on the hexagonal array of PS-b-P2VP micelles [108]. The Au-containing BCP surface was functionalized with thiol-modified, single-stranded DNA (ssDNA-SH). DNA origami created with sticky ends was then attached to the surface by extending appropriate staple strands on each end. The modified staple strands connected to DNA origami subsequently pair up with the ssDNA-SH. The AFM results from the AuNP-modified PS-b-P2VP micelles compared to those of control surfaces are shown in Figure 6B.

Protein Adsorption and Release Kinetics on BCP Thin Films. It has been revealed that the time-dependent adsorption behaviors of proteins differ on nanoscale BCP surfaces when compared with those on the surfaces of homopolymer counterparts. So far, investigations of proteins on nanoscale polymer surfaces have been largely centered on static instead of time-dependent behaviors. This is mainly due to the experimental challenges associated with directly attaining single biomolecule imaging and kinetic data. Being able to experimentally identify key kinetic segments that can be substantiated by corresponding topological data will be critical. However, the measurement process becomes especially difficult for the very early stage of protein adsorption, where ensemble-averaged measurement techniques may not be adequate for correctly rendering the kinetics associated with single biomolecule behaviors at nanoscale surfaces.

Despite these hurdles, the exact adsorption pathways and kinetics of IgG were determined successfully by tracking individual IgG molecules on the striped nanodomains of PS-b-PMMA [76]. Owing to the direct measurements of the same IgG molecules over time, it was possible to establish meaningful correlations between various topological states of the IgG assembly and specific adsorption kinetic regimes on the BCP surface. These characteristics were then compared to those data similarly acquired on a PS homopolymer surface. A distinct adsorption pathway of a single to double-file IgG assembly was revealed on the BCP surface. Additionally, unique adsorption characteristics such as the presence of two Langmuir-like segments and an undulating nonmonotonic regime were identified on the nanoscale BCP surface [76]. On the control surfaces of PS and PMMA homopolymers, the kinetic profile of IgG adsorption exhibited a single Langmuir-like segment with no undulating regime [76,77]. The IgG adsorption kinetics on the BCP surface of PS-b-PMMA versus on the homopolymer surfaces of PS and PMMA are presented in Figure 7A,B. Data in Figure 7A were collected from single protein tracking by AFM in a time-lapse manner, and those in Figure 7B were obtained by surface plasmon resonance (SPR) spectroscopy.

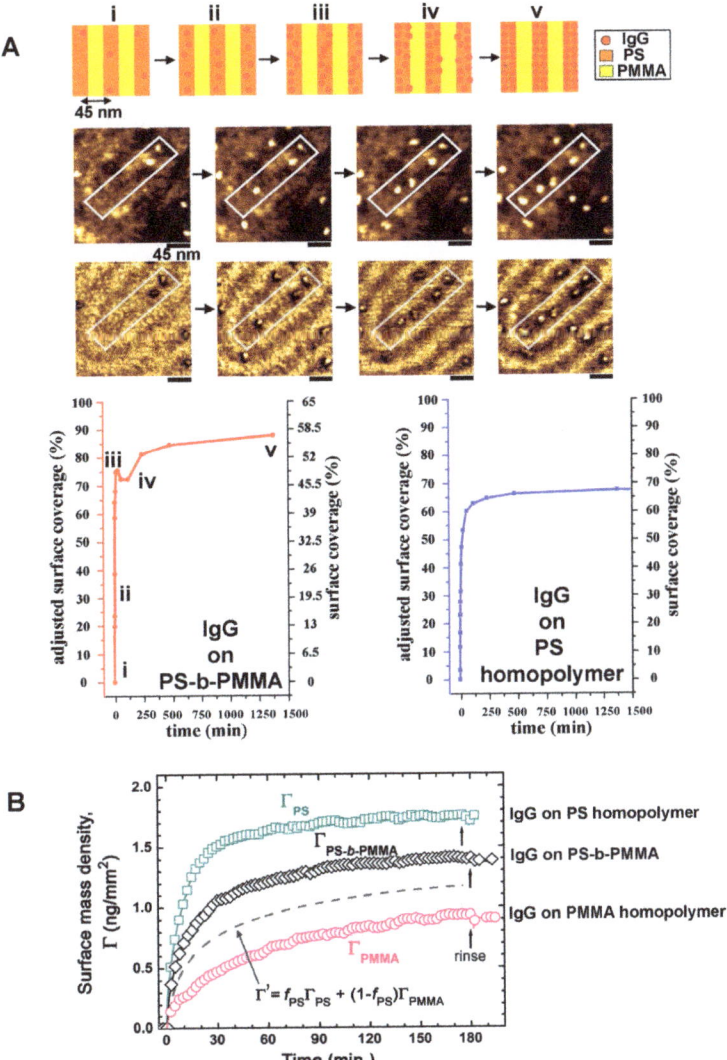

Figure 7. (**A**) The AFM data and IgG adsorption isotherms were obtained by time-lapse imaging of the same PS-b-PMMA surface areas while tracking individual IgG molecules. As a guide, white boxes are inserted in the time-lapse AFM data to mark the same BCP area. Key kinetic segments identified from the IgG assembly on the BCP surface are linked to the topographic data corresponding to each segment specified as (**i–v**). Unlike the IgG behavior on the control surface of PS homopolymer, IgG adsorption isotherms presented two unique features at the nanoscale BCP surface, i.e., the presence of two Langmuir-like segments and the existence of an undulating, nonmonotonic adsorption regime. Adapted with permission from Ref. [76] Copyright (2022) American Chemical Society. (**B**) The plot of surface mass density versus time corresponds to IgG adsorption on the BCP surface of PS-b-PMMA as well as on the homopolymer control surfaces of PS and PMMA. The data were obtained by SPR spectroscopy. The dashed line shows the hypothetical amount of adsorbed IgG, which was calculated from the weighted average of PS and PMMA homopolymer data while considering the volume fraction of the two polymer blocks in the BCP. Adapted with permission from Ref. [77] Copyright (2009) American Chemical Society.

In addition to adsorption kinetics, the release kinetics of single-component proteins from BCP surfaces were also examined [99]. The release kinetics of fluorescein isothiocyanate isomer (FITC)-coupled protein of LZM as well as FITC-coupled peptide of TAT from PS-b-PEO thin films were measured by spectrofluorometry [99]. By taking advantage of the fact that BCP film thickness can be easily modulated during the spin coating process of the film preparation, PS-b-PEO samples of 45–60 nm in thickness were prepared. The thinnest (thickest) BCP film yielded the least (greatest) amount of released protein cargo. Quantitatively, 20–80 ng cm^{-2} of cargo was reported to be released from PS-b-PEO films, where the larger (smaller) molecule of LZM (TAT peptide) was released over a longer (shorter) period. As for the release kinetics of the biomolecules, the study confirmed that an initial burst release of the protein or the peptide was followed by either a gradual or a steady-state release depending on the cargo. The study was able to demonstrate that the released quantity of the biomolecular cargo can be effectively controlled simply by altering the thickness of the BCP thin film.

3.1.3. BCP Nanodomains for Proteins: Multicomponent Systems

Biomedical applications in many practical settings are expected to involve multiple protein components, rather than single protein species. However, the interaction dynamics and kinetics of multicomponent proteins on solid surfaces are understood much less than single protein component systems in general, let alone for those polymer surfaces of nanoscale topology and chemical variability. Insights from single-component protein studies may not be applicable to adequately explaining more complex, multicomponent protein behaviors on nanoscopic material surfaces. On macroscopic solid surfaces, a protein exchange process known as the Vroman effect has been commonly observed from the competitive interactions of multicomponent proteins. The effect has been extensively documented in the areas of hemostasis, thrombosis, and biomaterials [168–174]. The Vroman process describes a phenomenon in which proteins, preferentially bound on a solid surface at early times, are displaced by other proteins in the bulk solution over time. Fast-diffusing protein species of lower molecular weights with lower surface affinity tend to arrive at the solid surface at earlier times. These species are replaced later in time by other slow-diffusing protein species of higher molecular weights and higher surface affinity.

Unlike the cases for macroscopic surfaces, not much insight into protein behaviors on nanoscale surfaces can be currently drawn from the literature. There is currently a lack of definitive experimental data at the single biomolecule level for unambiguously revealing multiprotein protein interaction processes on nanoscale surfaces. Despite this, it is crucial to determine the exact molecular mechanism underlying competitive protein–surface interactions on nanoscale surfaces and to reveal the precise compositions of adsorbed proteins at a given time. Furthermore, it is imperative to acquire such experimental evidence and move beyond the present stage of the field where existing postulations deduced from ensemble-averaged measurements are used to speculate on possible kinetics and mechanisms. Considering all these situations, there still is plenty of room to explore competitive protein interactions on nanoscale polymer templates, particularly those attributes examined at the individual protein level. Research efforts have begun to be put forward for the multicomponent protein systems on nanoscale BCP surfaces, leading to important discoveries on protein behaviors that are exclusive to those interaction mechanisms and dynamics at the nanoscopic interfaces.

Multicomponent Protein Assembly on BCPs. Nanostructures formed from a PS-b-PEO derivative, P(S-co-BrS)-b-PEO, have been used for the fabrication of multicomponent biomolecular arrays by combining nonspecific and site-specific interactions between proteins and the BCP nanodomains [80]. The BCP of P(S-co-BrS)-b-PEO contained 5 wt% of 4-bromostyrene (BrS) copolymerized within the PS block for crosslinking with 254 nm light. The PEO segment of the BCP was biotinylated. Various nanopatterns of lines and dots with parallel or perpendicular PEO cylinders with respect to the substrate were generated by using combinations of preparation protocols such as solvent annealing and shadow-mask irradiation. The PS nanodomain areas of the BCP surface were passivated by BSA in order to prevent nonspecific protein adsorption. The biotinylated PEO areas then served as a modular template to pattern neutravidin and biotinylated IgG. The process relied on the specific interaction between biotin and neutravidin to form a complex of biotinylated IgG-neutravidin-biotinylated PEO. A general approach that can be similarly used to pattern multicomponent proteins to the different nanodomain areas of a BCP surface is schematically depicted in Figure 8A.

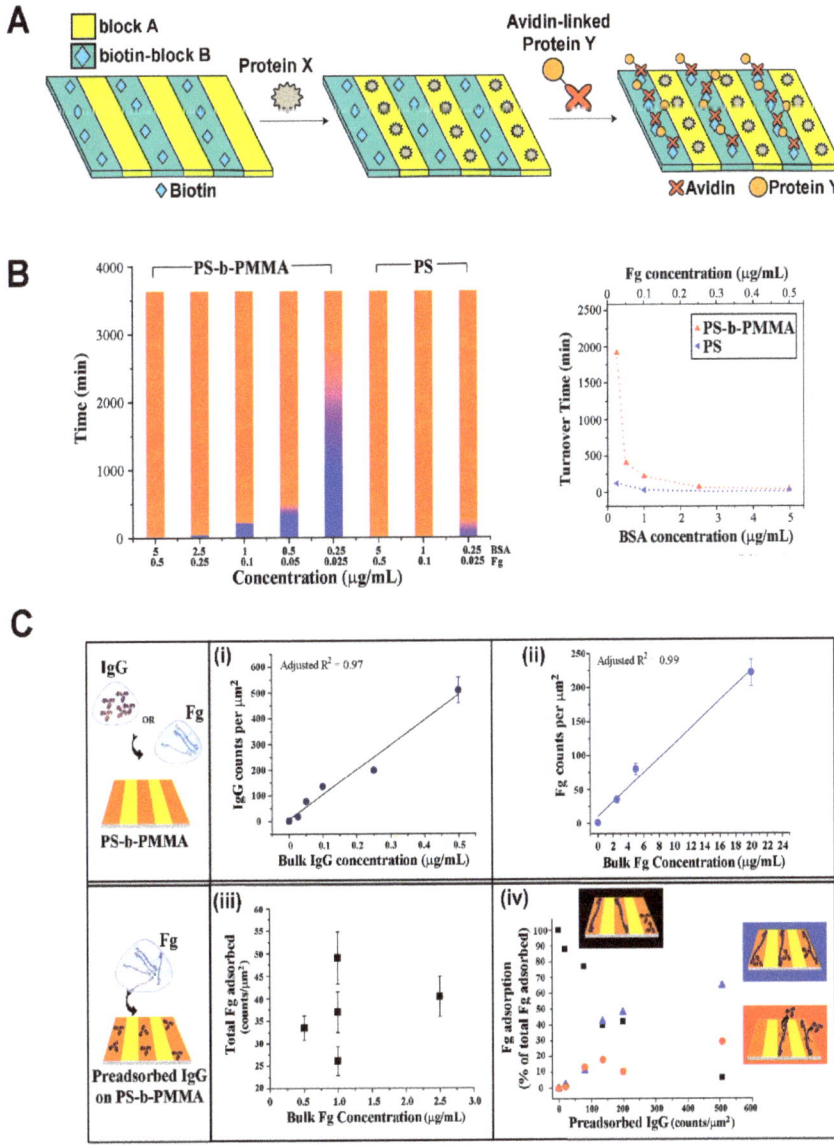

Figure 8. (**A**) The schematic illustration displays an experimental approach that can be used for the spatial control of multicomponent proteins onto self-assembled nanopatterns of an AB-type BCP. The

BCP surface contains nanoscale patches of alternating blocks of A and B, where A is more hydrophobic than B and B is pre-functionalized with biotin. Protein X, exhibiting a preferential interaction with the more hydrophobic block A, is first deposited to coat the block A nanodomains. Subsequently, protein Y conjugated with avidin is deposited into the nanodomain areas of block B by way of biotin–avidin interactions. (**B**) Time-dependent behaviors of multicomponent proteins were examined for simultaneous competitive adsorption. The model system involved BSA and Fg simultaneously exposed to the surface of PS-b-PMMA as well as to that of PS homopolymer. At earlier times, BSA was the dominant protein species assembled on the BCP surface. Over time, Fg replaced the BSA molecules on the surface and became dominant. The colored bar graphs display the time-dependent transition between BSA (blue shaded portion) and Fg (orange shaded portion) on the PS-b-PMMA surface as well as on the PS homopolymer template. The different transition stages of BSA-dominant phase, the Fg onset/turnover phase, and the Fg-dominant phase are identified in blue, gradient purple, and orange, respectively. The plot shown in the right panel displays the times corresponding to the turnover point from BSA to Fg for different protein concentrations. Adapted with permission from Ref. [87] Copyright (2016) Royal Society of Chemistry. (**C**) Sequentially occurring, competitive protein behaviors were examined by using the model protein system of IgG and Fg on the surface of PS-b-PMMA. The control data in (**i,ii**) were obtained by examining the case of (**i**) IgG and (**ii**) Fg adsorption onto a clean BCP substrate with no preadsorbed proteins. In both cases, the plots show the adsorbed protein amount is linearly dependent on the bulk protein concentration. The data in (**iii,iv**) correspond to the BCP surface containing preadsorbed IgG proteins from a prior incubation step. Fg molecules were introduced as a subsequent-stage adsorber. The plot of adsorbed Fg versus Fg bulk concentration in (**iii**) shows that the adsorbed Fg amount was no longer dependent on the bulk Fg concentration. Fg adsorption in this case was dependent on preadsorbed IgG amounts on the surface. The plot in (**iv**) shows the occurrence frequencies of Fg on the BCP surface for the case of distal Fg adsorption (black), proximal Fg adsorption (blue), and Fg replacing IgG (red). Adapted with permission from Ref. [78] Copyright (2018) Royal Society of Chemistry.

In a different study, multicomponent protein interactions on PS-b-PMMA were examined for the situation of a simultaneous, rather than sequential, exposure to BSA and Fg [87]. When the protein mixture was applied to the PS-b-PMMA surface, the protein components found on the PS nanodomains of the BCP surface were revealed to be time-dependent. At earlier times, BSA constituted the dominant protein species assembled on the BCP, whereas Fg molecules became the major protein kind that occupied the BCP surface at later times. The data shown in Figure 8B present the change in the dominant protein species on the PS-b-PMMA thin film over time.

Multicomponent Protein Dynamics on BCPs. Experimental and simulation research endeavors have been undertaken jointly to reveal competitive protein adsorption behaviors on nanoscale BCP surfaces. When the BCP platform of PS-b-PMMA was exposed to proteins of different kinds such as BSA, Fg, and IgG, it was confirmed that a protein exchange process similar to those on macroscopic polymer surfaces indeed occurred on the nanoscale BCP surface as well [78,87]. However, protein adsorption occurred exclusively on the PS nanodomains regardless of deposition time. Furthermore, the extent to which the initially bound BSA resists its displacement by Fg was much greater on the nanoscale, chemically varying BCP surface relative to the macroscopic, chemically homogeneous surface of PS homopolymer. This phenomenon can be clearly seen in the data presented in Figure 8B. The protein exchange of BSA by Fg took place much more slowly on the nanoscale BCP relative to the PS homopolymer surface [87]. The results indicated that nanoscale BCP surfaces present a more energetically favorable environment for surface-bound proteins which, in turn, enables prolonged residence time of the initially bound protein species and significant retardation in the onset of the protein exchange process.

In another study, individual protein tracking was successfully carried out for competitive protein adsorption of IgG and Fg that occurred in a sequential manner on PS-b-PMMA [78]. The study was able to provide valuable experimental evidence for the dominant adsorption pathway, occurrence frequency, and directionality in protein exchange, all resolved at the single biomolecule level. In addition, the adsorption profiles of subsequent-stage proteins were proven to be significantly different between those sample surfaces with and without pre-adsorbed proteins from earlier stages. For single-component protein adsorption to a neat PS-b-PMMA surface, the protein amount adsorbed on the BCP surface increased linearly with the bulk protein concentration. However, for the sequential interaction case involving a subsequent-stage protein of Fg introduced to the BCP surface treated with IgG in an earlier step, such a linear relationship was no longer observed. In this case, the adsorbed amount of Fg, the subsequent-stage protein, had no dependence on the Fg solution concentration. Rather, the adsorbed amount of the subsequent-stage protein showed a strong correlation to the amount of the prior-stage protein on the BCP surface. The data shown in Figure 8C summarize these features that are associated with the BCP surface under different, competitive adsorption stages in a sequential deposition scenario.

3.1.4. Protein Functionality on BCP Thin Films

Proteins immobilized on a solid surface may present biological functionalities different from those in their native states. The presence of an underlying surface may restrict necessary changes in protein conformation and protein chain rearrangement for exposing its binding sites toward a ligand molecule, for instance. In fact, conflicting results are found in the literature in terms of protein functionality upon surface immobilization. Some reported reduced activities due to substrate-induced, steric hindrance of protein binding to ligands [79,103]. On the other hand, some reported increased protein activity on a solid platform [175,176]. The disparity can be largely explained by the fact that the former conclusion was drawn for protein systems that were randomly adsorbed onto a surface, whereas the latter case involved protein molecules specifically oriented in space with respect to the surface. Tethering of proteins to the platform in the latter case was typically attained by chemical or biological moieties. For protein reactions in solution that occur without the involvement of a solid surface, Brownian motion related to the stochastic chances of biomolecular collisions dominates the reaction process. On the contrary, biomolecules strategically oriented on a surface can guide more effective ligand binding along a well-defined molecular coordinate and increase protein activity. Nevertheless, very little is yet known about the activity and stability of proteins upon their binding onto BCP nanotemplates. It is important to determine the influence of the nanoscale BCP surface on the biofunctionality as well as the stability of proteins for a diverse system of BCPs and biomolecular reactions.

Protein Activity and Stability on BCP Thin Films. Research endeavors have begun in this regard using an antigen–antibody system on PS-b-PMMA. Antibody binding activities were examined for IgG molecules bound to a PS-b-PMMA surface using an IgG antibody as well as other control proteins with no specificity to IgG [79]. It turned out that the specificity of IgG molecules in antibody recognition remained on the PS-b-PMMA surface. The IgG molecules on the PS-b-PMMA formed paired complexes only when they were reacted with the IgG antibody, but not in control reactions with nonbinding proteins. In a different study, it was reported that the total enzymatic activity and long-term stability of HRP was greater on a nanoporous BCP thin film when compared to those on the macroscopic surfaces of glass and PS [104]. The nanoporous thin film used in the study was fabricated from a mixture of two BCPs, 90 wt% PS-b-PLLA and 10 wt% PS-b-PEO. The nanoporous BCP platforms provided a greater surface area and easier mass-transfer than the control surfaces. This promoted the enzymatic reactions and increased the catalytic activity of HRP. In another study, biological activities in antibody binding were tested for the proteins of Fg, Mb, and LZM immobilized on a PS-b-PHEMA surface [89]. The immunoreactions were carried out on PS-b-PHEMA by using specific antigen–antibody pairs for each protein, i.e., Fg with anti-Fg, Mb with anti-Mb, and LZM with anti-LZM. It was demonstrated that the amounts of adsorbed antibodies were in qualitative agreement with the number density of the protein molecules that were preassembled on the BCP surface.

In addition to the qualitative assessments, quantitative comparisons of enzyme activities have been made for the case of surface-immobilization versus free solution [79,85,103,140]. Enzymatic activities were quantitatively determined for HRP and tyrosinase molecules that were configured to be BCP surface-bound versus freely moving in a solution. It was revealed that, when compared to the same number of HRP molecules in solution, the enzyme molecules bound to the surfaces of PS-b-PMMA and PS-b-P4VP were able to retain approximately 85% and 78% of the free-state activity, respectively [79,103,140]. The HRP molecules on the BCP templates remained stable and catalytically active even after 100 days, when kept at 4 °C. Other biological activities of BCP surface-bound proteins have also been examined. For example, Fg molecules immobilized on PS-b-PMMA were evaluated for their biofunctionality in the activation of microglial cells [85]. It was shown that the surface-bound Fg retained its cell-activating functionality on the BCP template. The outcomes summarized in this section provide encouraging early data of high protein activity and stability upon immobilization to BCP surfaces. These results suggest that protein nanopatterns assembled with the guidance of BCPs can be exploited to fabricate biofunctional constructs for applications in biosensors and biomaterials.

3.2. Biomineral Nanocrystals

The structural anisotropy in various mineralized tissues such as nacre, bone, and dental enamel plays a vital role in their remarkable functionalities [177–179]. For example, the high mechanical properties and chemical stability of enamel are due to the intricate spatial organization of hydroxyapatite (HAP) nanocrystals that are bundled to form thick prisms and interprismatic regions of different orientations [180,181]. As the supramolecular organization of matrix proteins in mineralizing tissues largely regulates the nucleation and growth of minerals, various strategies have been explored to create mineralizing material platforms, especially those based on organic matrices. Thin films of BCPs can offer excellent chemical contrasts of nanoscopic dimensions that can be easily varied by altering the size and confinement direction of the nanodomains. BCPs can further be formulated to assemble into 3D scaffolds, even making the incorporation of 3D printing possible to fabricate tailored biomaterials [182]. Therefore, the BCPs' capability to produce well-controlled 2D and 3D nanopatterns in a facile and rapid manner can present distinctive advantages to the biomineralization field. For instance, protein nanopatterns and peptide nanoassemblies on BCP surfaces can be used to seed mineral filaments and platelets similar to those processes seen in natural biominerals, and further direct a mineralization process with high fidelity. As discussed, spatial control is one of the most critical factors in mineralization since the specific organization of individual nanocrystals and their larger-scale arrangements determine the resulting material's properties. To this end, BCPs can provide exquisite spatial control at the nanometer scale in guiding mineralization.

Calcium Phosphate Nanoparticle Assembly on BCP Thin Films. PS-b-PMMA nanopatterns predecorated with a protein layer have been successfully employed to seed calcium phosphate (CaP) NPs [83,85]. CaP-based materials such as HAP and triple calcium phosphate (TCP) are biomedically important materials that are often used to coat the surface of implant biomaterials. The incorporation of the CaP-based materials increases the biocompatibility of implant materials and accelerates the man-made material's integration with living tissues [183,184]. Aligned PS-b-PMMA has been employed to guide CaP growth after the BCP surface was first patterned by using Amel-derived peptide NRs associated with tooth enamel formation [83]. In the study, two prototypical Amel peptide sequences of p14P2 and p14P2Cterm were used. The peptide NRs bound to the PS domains were able to retain their β-sheet structure and biological activity on the surface and direct the formation of filamentous and plate-shaped minerals of CaP. CaP crystals were mineralized from both an aqueous solution of precursor ions and a polymer-induced liquid-like precursor (PILP). Each mineral was revealed to be a single crystal whose crystalline planes were similar to those of apatite filaments in enamel. By employing nanopattern dimensions ranging from 50 to 150 nm in PS width, it was determined that the width of the apatite crystals was directly dependent on the width of the PS stripes. These results are displayed in Figure 9A. In a different study, nanostriped domains of PS-b-PMMA were used to produce CaP NPs on Fg-covered PS nanodomains. The nanodomain width and orientation were tuned for the alignment of packed Fg molecules in an end-to-end manner parallel to the stripe direction [85]. The BCP template with a densely packed Fg layer was then exploited for nucleating CaP NPs whose results are summarized in Figure 9B. These research efforts have demonstrated that the BCP-directed approach may serve as a highly generalizable platform for nanopatterning of mineral crystals by being able to effectively control the size, number density, and spatial locations of the mineral particles via the underlying BCP nanostructures.

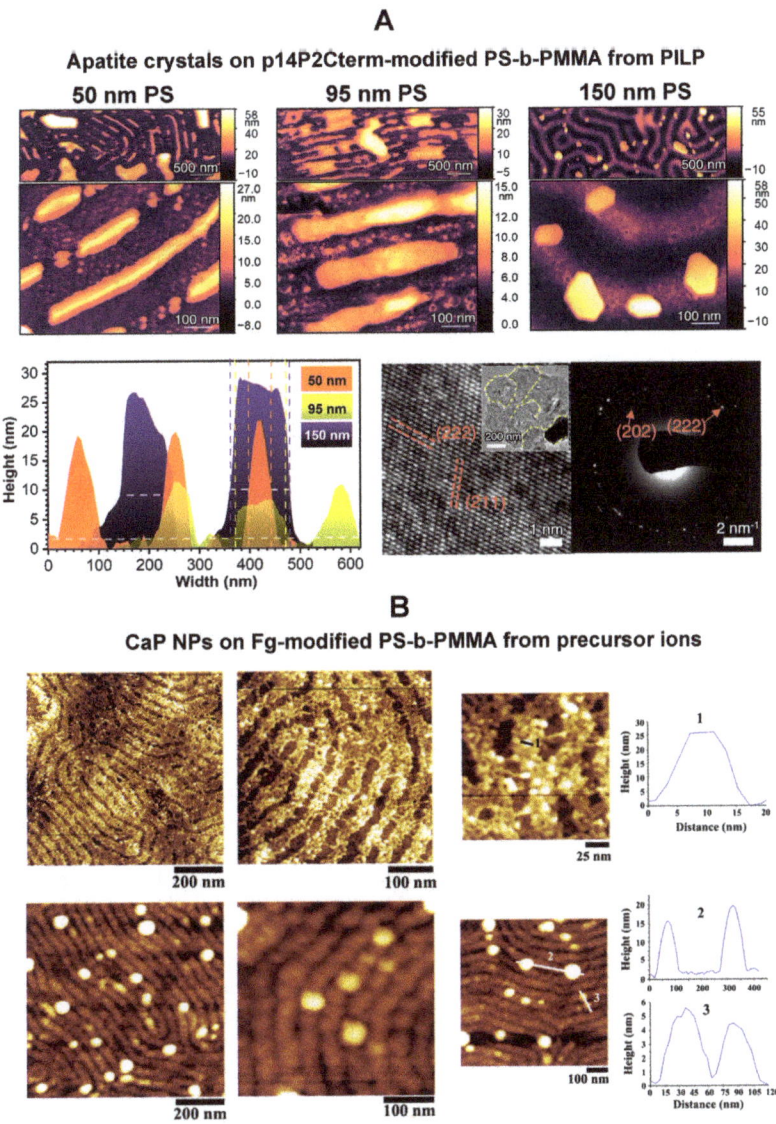

Figure 9. (**A**) The AFM and TEM data were obtained from apatite crystals formed on p14P2Cterm-modified PS stripes of PS-b-PMMA thin films using PILP. The height versus width plot displays

the apatite crystal dimensions formed on the BCP templates with 50 nm, 95 nm, and 150 nm p14P2Cterm-PS stripes. White horizontal dashes indicate the base of the particle (PS stripe) and color-coded vertical dashes indicate the filament width. The high-resolution TEM (HRTEM) and selected area electron diffraction (SAED) data were obtained from the particles extracted from the template of 150 nm PS stripe. The TEM data confirmed that the particles were crystalline with lattice and reflections specific to apatite. The inset displays a low-magnification TEM image and corresponding SAED pattern of aggregated single crystals and their grain boundaries as yellow dashed lines. Adapted with permission from Ref. [83] Copyright (2023) American Chemical Society. (**B**) The AFM panels show CaP NPs preferentially nucleated and grown on the Fg-covered PS nanodomains of PS-b-PMMA after (top row) 5 min and (bottom row) 7 h of incubation time in a precursor solution. The particle sizes measured along the white lines are provided for the two different incubation periods in the line analysis panels. Adapted with permission from Ref. [85] Copyright (2016) American Chemical Society.

3.3. Cell Adhesive Molecules

It is well-understood that proteins organized on macroscale polymer surfaces can function as a cell-mediating layer and ultimately affect cell behaviors of adhesion, proliferation, and differentiation by influencing upstream cytoskeletal dynamics and downstream gene expression. It is also known that the number, morphology, and alignment of cells are affected by the surface density and gradient of cell-corresponding adhesion molecules [185–188]. The initial cell–surface interactions can be regulated by influencing the presence of these cell-corresponding adhesion molecules on a material surface to which cell receptors bind afterwards. The adhesion molecules usually consist of specific peptide sequences or proteins such as RGD-containing epitopes, FN, collagen (Col), and gelatin. The activation of specific transmembrane receptors such as integrins further induces the assembly of adhesion sites known as focal adhesions [189]. Hence, upon initial cell attachment, cell behaviors can be additionally controlled by altering the degree of integrin binding and focal adhesion formation. This is usually achieved by providing different micro- and nano-environments of chemical cues to a material surface in the form of a protein layer. Proteins such as cytokines, growth factors, hormones, and adhesion molecules are used for this purpose [190]. Chemical cues can also take the form of hydrophobic, Coulombic, and van der Waals forces as well as surface energies between the cell membranes and the underlying polymer surfaces [190].

BCP Thin Films for Cell Adhesive Molecules. BCP-based approaches are highly conducive to generating molecular patterns to control the clustering of cell adhesion receptors and structural signaling activities of cell adhesion. This is because protein layers on BCP templates can be exploited as well-defined chemical cues. The spatial organization of protein layers can be achieved spontaneously and instantaneously at nanoscale precision on BCPs for tuning the chemical specificity of adhesive epitopes. In addition, BCP-based methods can be beneficial in regulating the physical features of an epitope-containing platform through adjusting its geometry, rigidity, and spacing. Transplanted cells can recognize and respond to these different nanoscale cues on BCP surfaces in a highly sensitive manner, ultimately affecting the degree of gene expression and tissue formation. Hence, self-assembled BCP nanopatterns show great potential to be utilized for modulating a variety of experimental parameters critical for cell receptor-initiated processes.

BCP surfaces have been engineered to match the bioligand spacing found in cells. BCP templates have also been constructed to provide a nanoscale gradient with varying bioligand spacing in order to monitor changes in cell sensitivity with respect to the spatial distribution of the adhesion ligands. For example, BCP nanotemplates were generated to provide periodic surface sites of ~8 nm in size to match the diameter of integrin in the cell membrane [106]. Nanoscopic topological features were also varied to control the spacing and density of biorecognition molecules for cell receptors which led to substantial changes in cell behaviors [105,106,112]. In one study, a BCP of PS-b-PI with ring-like FN nanopatterns was shown to increase the percent surface coverage and density of Chinese hamster ovary (CHO) cells relative to control substrates [92]. The controls were composed of either a homogeneous FN surface or a PS-b-PI template with striped nanopatterns of FN. It was also found that the PS-b-PI surface with ring-like FN nanopatterns induced more actin fibers, cell spreading, and focal adhesion formation. These results were attributed to a high local FN density on the ring areas, consequently leading to increased integrin clustering and stable focal adhesions.

In another study, an optimal range for the BCP template spacing that is necessary for integrin adhesion and focal adhesion was determined [106]. A BCP surface of PS-b-P2VP was first prepared to produce hexagonally arranged Au dots on each BCP micelle to which c(-RGDfK-) peptides were linked via thiol–Au interactions. When the separation distance between each micelle was greater than 73 nm, cell attachment and spreading as well as the formation of focal adhesions were revealed to be highly restricted. These templates prevented integrin clustering, whose step is important not only for the initial binding of cells to the surface but also for the subsequent cell attachment via stable adhesion sites. The optimal spacing between each micelle was determined as 58–73 nm. This range was considered universal after examining different cells of MC3T3-osteoblasts, B16-melanocytes, REF52-fibroblasts, and 3T3-fibroblasts.

Other BCPs have been employed for nanopatterning RGD peptides as well. For instance, BCP brush samples of polyacrylamide/bis-acrylamide-block-poly(acrylic acid) (PAAm/bisAAm-b-PAA) were prepared with various cross-linking density [101]. The PAA segment of the BCP was then conjugated with an RGD peptide of GRGDS via NHS/EDC chemistry for cell adsorption and spreading. In a different study, the size and spacing of the PEO nanodomains on PS-b-PEO templates were varied as 8–14 nm and 62–44 nm, respectively [91]. These templates were further modified with RGD binding peptides and used for the adhesion of NIH-3T3 fibroblasts. The study identified that the spacing between the PEO nanodomains was crucial for controlling cell spreading on the BCP surfaces. A decrease in the patch spacing for the RGD binding peptides led to an increase in the spreading of NIH-3T3 cells. This approach was later extended to the production of porous 3D PS-b-PEO scaffolds [100]. In this case, PEO nanodomains present throughout the highly porous 3D scaffolds were functionalized with RGD peptides. The 3D presentation of the RGD peptides, correlated directly to the nanodomain structures formed in the original BCP scaffold, was controlled by adjusting the molecular weight of the PS-b-PEO copolymer.

The effect of a chemical cue on cell behaviors due to the variations in the hydrophobic and hydrophilic patches on BCPs has been examined as well [93]. The BCP surface used for this investigation consisted of PMPC and poly(3-methacryloyloxy propyltris(trimethylsilyloxy) silane) (PMPTSSi). Different arrangements of nanoscopic polymer patches were obtained by BCP's phase reversal processes between the two blocks. The BCP surfaces were then used to produce cell-adhesive nanopatterns of FN molecules on the PMPTSSi patches for subsequent L929 cell adsorption. The BCP micellar geometry of the hydrophilic PMPC core and hydrophobic PMPTSSi matrix led to more adsorption of L929 cells than the opposite block arrangements for the core and matrix components [93].

3.4. Cells

Cell interaction with a material surface can steer various biological processes by playing an essential role in the regulation of cell viability, proliferation, and differentiation. Cell adhesion to the surfaces of artificial hearts and hollow dialysis fibers can cause undesirable outcomes such as platelet adhesion and thrombosis [174]. In contrast, promoting cell attachment to artificial scaffolds is critical in cell-based bioarrays and biosensors as well as in tissue engineering and regenerative medicine [190,191]. In some applications, both properties may be simultaneously needed. For instance, materials capable of promoting stem cells and, at the same time, inhibiting cancer cells on platform surfaces are desirable in bone regeneration after injury or pathology. As such, controlling polymer surfaces to facilitate or resist cell interactions to meet specific demands is an important consideration in devising biointerfaces [1]. There exists a great deal of fundamental knowledge and design principles to control cell–material interactions. For instance, it is generally understood that water wettability of a polymer surface is one of the key factors to determine cell behaviors. Protein adsorption, required for subsequent cell activities on the material surface, is known to be largely controlled by the water wettability of the material. However, most of the current knowledge set is based on cell interactions with macroscopic surfaces which may not be adequately carried over to explain cells interfacing nanoscopic surface features.

Cell behaviors on nanoscopic BCP patches with different hydrophobicity and hydrophilicity can drastically differ from what can be deduced by the average wettability values of macroscopic, homopolymer counterparts. Predicting cell behaviors as a function of a simple and single, structural or chemical parameter becomes difficult for BCP surfaces. For example, the lack of cell adhesion and cell proliferation behaviors is widely reported on the homogeneously prepared PMPC surface [81,192]. Yet, entirely different cell adhesion profiles were observed from a heterogeneously prepared PMPC polymer surface. A triblock copolymer platform, composed of hydrophilic PMPC and hydrophobic PDMS as A and B segments of the ABA-type BCP, was fabricated to present nanodomains of vertically arranged PDMS cylinders embedded in a PMPC matrix [81]. It was revealed that many L929 fibroblast cells adhered to the heterogeneously prepared, hydrophilic polymer surface with a water contact angle of less than 20°, even though the hydrophilic monomer composition in the heterogeneously prepared triblock platform was only around 45%. These results suggested that the segregated hydrophobic domains on the BCP platform should be considered for designing polymer-based biomaterials.

Although relatively fewer research attempts have been made for modulating cell behaviors specifically by BCP nanopatterns, there has been a growing interest in exploiting the unique advantages of BCPs to the development of cell-based bioarrays and biomaterials. BCPs can be readily designed to produce nanopatterned surfaces that are stable at physiological conditions, enough to sustain human and other cell viability. BCPs offer adjustable nanomorphology, versatile surface chemistry, and even a possibility for the facile development of non-cytotoxic supports from a plethora of available polymer materials. In addition, the different polymer chemistries, topographical features, and mechanical properties of BCPs can present a unique opportunity for their use as triggers or modifiers of biochemical and biophysical cues that are important to modulating cell behaviors. These aspects are highly beneficial in designing biocompatible materials for implant devices and tissue engineering applications. However, the typical size scale associated with cells, unlike those of individual proteins, are tens of micrometers or larger. As such, the size incompatibility between the characteristic dimensions of cells versus BCP nanopatterns can seemingly pose difficulties in correlating cell behaviors with any nanoscopic variations in the topological or chemical features of the underlying BCPs. Regardless, various research efforts have shown a promising sign that BCPs can be used to efficiently guide and modulate cell behaviors.

BCP Thin Films for Cells. Studies have reported that different micro- and nano-environments of BCPs can significantly change cellular behaviors by affecting the chemical, topographical, and mechanical nature of cell interactions. PS-b-P2VP nanopatterns were used as-formed without cell adhesion ligands. The effect of various solvent-annealed nanotopographies of the BCP on bactericidal efficiency as well as on cytotoxicity to mammalian cells were subsequently examined [109]. The morphology and viability of *Escherichia coli* (*E. coli*) as well as *Staphylococcus aureus* (*S. aureus*) cells were tested on various PS-b-P2VP templates whose results are shown in Figure 10. It was reported that the cylindrical templates with both the PS and P2VP blocks exposed to the surface exhibited a stronger bactericidal effect than the micellar templates with only PS exposed at the surface. The study also confirmed that the BCP nanopatterns were nontoxic to mammalian cells, making them an efficient platform to resist bacteria.

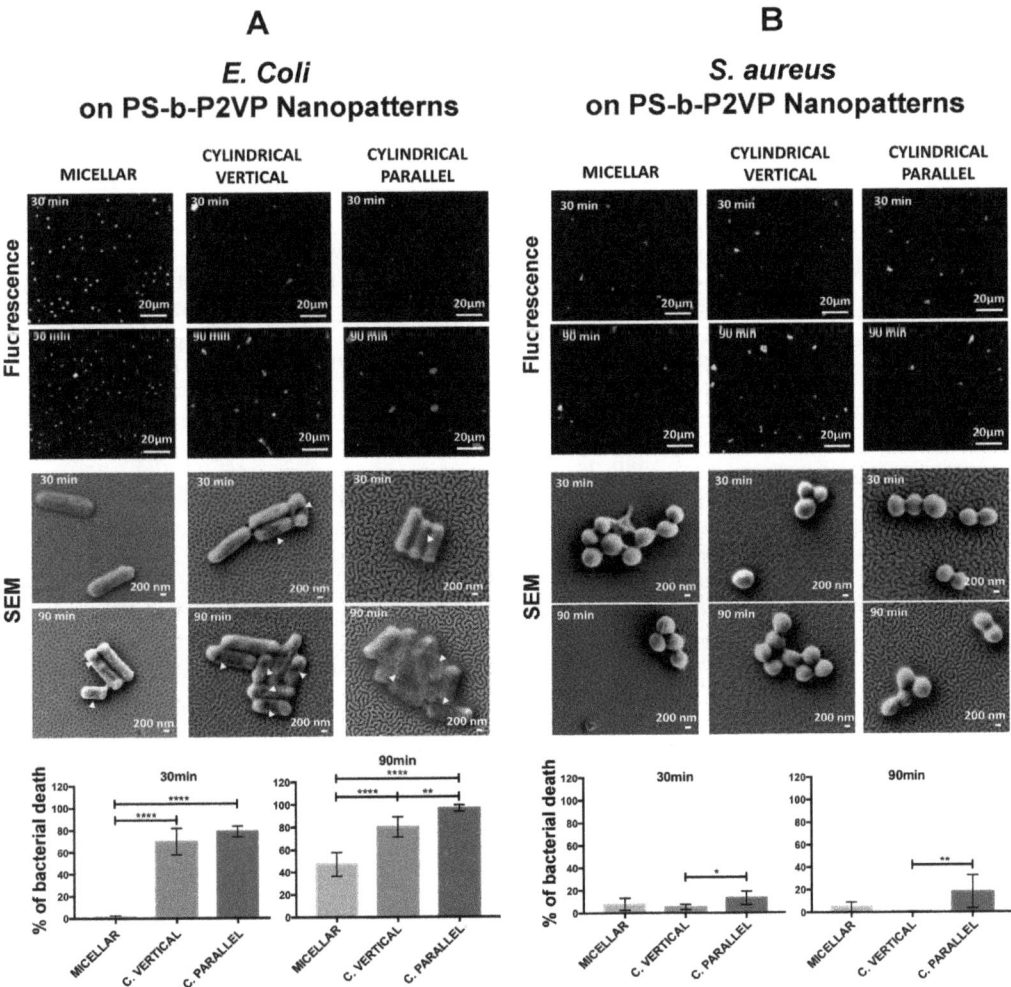

Figure 10. Various nanopatterned surfaces of PS-b-P2VP were used to assess the morphology and viability of (**A**) *E. coli* and (**B**) *S. aureus* cells. They included micellar (~40 nm in thickness, hexagonal micelles of 118 ± 27 nm in diameter), cylindrical vertical (~20 nm in thickness, PS domain of 61 ± 15 nm in width and P2VP domain of 116 ± 34 nm in width), and cylindrical parallel (~40 nm in thickness, PS domain of 62 ± 12 nm in width and P2VP domain of 70 ± 10 nm in width). (**A**) The bacterial cell viability of *E. coli* on the three PS-b-P2VP surfaces was analyzed by SYTO9 (green)/propidium iodide (red) method. SYTO9 was used to visualize those cells with intact and damaged membranes, while propidium iodide targeted only those cells with compromised membranes. When both were present, propidium iodide with a stronger affinity for the cell DNA displaced SYTO9 which, in turn, led to a decrease in green fluorescence inside the cells. White arrows mark damaged bacterial cell walls. Statistical differences: ** $p < 0.01$, **** $p < 0.0001$. (**B**) The fluorescence panels display the cell viability of *S. aureus* after live(green)/dead(red) assays, whereas the SEM panels show the cell morphology on the different PS-b-P2VP nanopatterns. Statistical differences: * $p < 0.05$, ** $p < 0.01$. Adapted with permission from Ref. [109] Copyright (2020) Elsevier.

In a different work, self-assembled nanopatterns of PS-b-P2VP with varying molecular weights and solvent vapor treatments were used to investigate cell morphology and cell adhesion related to bone healing [110]. Distinct cell responses were observed from two osteo-related cell types depending on the topography and chemistry of the BCP nanopatterns. Micellar nanopatterns assembled from a high molecular weight PS-b-P2VP (PS_{320}-b-$P2VP_{398}$) promoted the adhesion and spreading of human bone marrow mesenchymal stem cells (BMMSC), whereas the same surface exhibited an opposite effect on osteosarcoma cell line (SaOS-2). In another study, various nanotemplates of heights less than 10 nm were obtained from PS-b-P2VP as well as PS-b-P4VP, and employed for dermal fibroblasts and mesenchymal precursor cells [111]. Different rates of cell adhesion and proliferation were observed on the BCP surfaces, despite the use of BCP substrates with similar surface energies. These differences in cellular response due to the BCP surface-induced nanopatterns are displayed in Figure 11A. It was found that the fibroblasts and mesenchymal precursor cells preferred a BCP template with its nanostructures consisting of a larger nanodomain size and spacing, where the surface of 6 nm in width and 200 nm in spacing was favored over that of 3 nm in width and 160 nm in spacing.

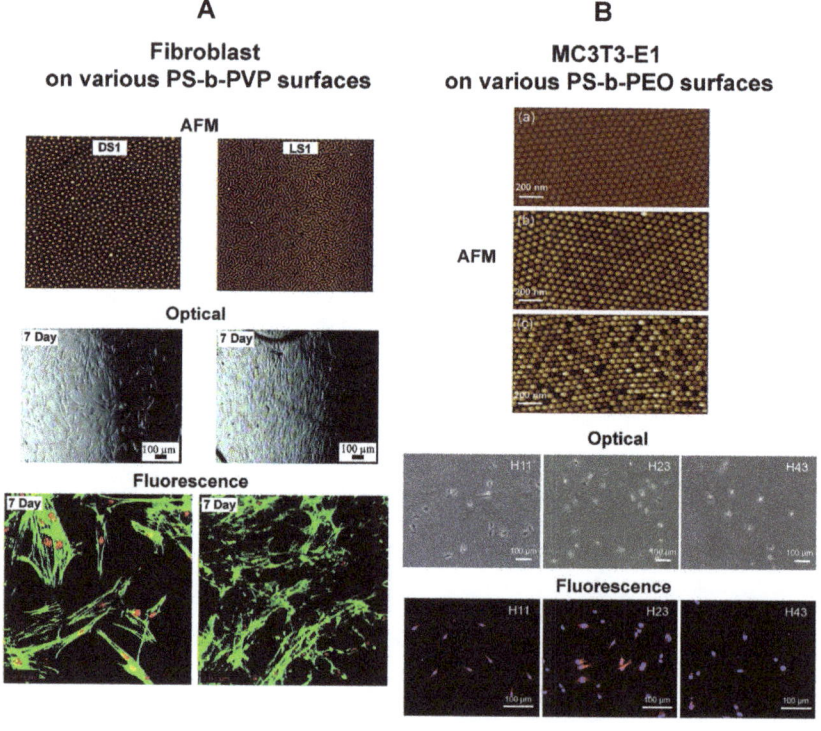

Figure 11. (A) Fibroblast cells cultured on two different PS-b-PVP templates of DS1 and LS1 were characterized at day 7 by optical microscopy and confocal fluorescence microscopy. The blank BCP templates before the cell culture are displayed in the AFM images (10×10 μm^2 in size). The DS1 template was prepared from PS_{1350}-b-$P2VP_{400}$ to exhibit dot-like nanopatterns of average 200 ± 10 nm in diameter, whereas the LS1 template from PS_{610}-b-$P4VP_{130}$ presented lamellar nanopatterns with an average width of 160 ± 7 nm. The confocal fluorescence images show cells stained with phalloidin (actin filaments, green) and propidium iodide (nucleus, red). When compared to fibroblast cells on DS1, earlier cell adhesion and enhanced proliferation were observed from the cells on the LS1 template. Adapted with permission from Ref. [111] Copyright (2007) American Chemical Society. (B) MC3T3-E1 cells cultured on various PS-b-PEO/DBSA templates were characterized by optical microscopy and confocal fluorescence microscopy. The AFM panels display PS-b-PEO templates with

PS nanopost heights of (**a**) 11 nm, (**b**) 23 nm, and (**c**) 43 nm. Bright areas correspond to PS domains and dark areas correspond to PEO domains. The average diameter of the PS nanoposts and the average center-to-center distance between PS nanoposts were kept constant as 54 nm and 71 nm, respectively, for all templates. The optical and confocal fluorescence images were captured 6 h post-seeding. The three PS-b-PEO/DBSA templates of different PS heights are marked as H11, H23, and H43. The red and blue contrasts in the fluorescence panels are due to vinculin and DAPI, respectively. Cells seeded onto a PS-b-PEO/DBSA surface with 23 nm-high nanoposts showed a much more stretched morphology. Cell growth rate and proliferation increased as the PS nanopost height increased from 11 nm to 23 nm, and then decreased with a further increase in the nanopost height. Adapted with permission from Ref. [112] Copyright (2014) Elsevier.

Cell polarization was also examined for MC3T3 osteoblasts on PS-b-P2VP surfaces [105]. The BCP micellar nanolithography used in this study enabled the formation of a quasi-hexagonal, 1–15 nm AuNP array with a tunable NP spacing of 15–250 nm. The AuNPs were biofunctionalized with c(-RGDfK-) in such a way that subsequent binding of integrin molecules to AuNPs followed the stoichiometry of one integrin to one particle. The gradient polarization ratio (GPR) of adherent cells was then analyzed in terms of the axial ratio of the most extended cell width along the direction parallel over perpendicular to the gradient direction on the BCP template. It was found that the cell morphology changed from a radial to elongated shape when the ligand patch spacing was changed from 50 nm to 80 nm. The strongest polarization of cell bodies occurred when the patch spacing ranged between 60 and 70 nm. The study also pointed out that the cells were highly sensitive to even a very small change in the spatial presentation of adhesion ligand patches and responded to a difference in patch spacing of ~1 nm [105].

In addition to the lateral nanoscale dimensions, vertical dimensions of BCP nanostructures that are important for cell behaviors were also identified. The role of nanoscale topology in cell adhesion and differentiation was examined on PS-b-PEO/dodecylbenzenesulfonic acid (PS-b-PEO/DBSA) [112]. The height of self-assembled PS nanoposts that functioned as cell-adhesion domains was varied from 11–43 nm, while keeping both the size and spacing of the nanoposts constant as 54 nm and 71 nm, respectively. The adhesion and growth of mouse preosteoblasts (MC3T3-E1) on these substrates were reported to reach the maximum when the nanopost was 23 nm in height. As for the cell differentiation of MC3T3-E1, the gene expression levels of core-binding factor α1 (Cbfa1) and osteocalcin (OCN) were significantly higher on the 23 nm-tall nanopost surface than on other surfaces. Such changes in cell behaviors that were triggered by the BCP nanotemplate height are summarized in Figure 11B.

The effect of the mechanical properties of an underlying surface on cell behaviors has been scrutinized as well. PAAm/bisAAm-b-PAA samples of various cross-linking densities were prepared to exhibit different mechanical properties [101]. The PAA segment of the BCP was then linked to an RGD peptide. Variation of mechanical properties of the initially polymerized PAAm block was achieved by adjusting the concentration of the cross-linker, bisAAm, while maintaining a constant concentration of AAm. The elastic moduli of the resulting sample surfaces ranged between 600 Pa and 3800 Pa. In subsequent cell studies, changes in cell density and spreading morphology were observed both for NIH 3T3 fibroblast and PaTu 8988t pancreatic tumor cells. The BCP architect of stiffer tethers led to more pronounced cell attachment compared to that of soft un-crosslinked tethers. The study concluded that the rigidity of the underlying BCP significantly influenced the cytoskeleton organization and focal adhesion formation.

Furthermore, aligned nanofeatures as well as triblock copolymer surfaces have been demonstrated for modulating cell behaviors. A gradient fibrous scaffold based on a triblock copolymer of polystyrene-block-poly(ethylene-co-butylene)-block-polystyrene (SEBS) was used to controllably guide the adhesion, spreading, and migration direction of endothelial cells (ECs) [113]. In this work, aligned electrospun fibers of SEBS were treated by selective solvent vapor annealing to produce micrometer and nanometer scale roughness on the surface. The resulting scaffolds texturized with graded fibrous structures were able to function as region-specific, topological guidance for the migration, adhesion, and spreading of the ECs. Nanofibrous micelles have also been prepared from BCPs to mimic the filamentous structure of native extracellular matrix (ECM) and subsequently used to regulate cellular response in tissue engineering. To this end, 2D-organized filomicelles of 46 nm in width, 200 μm in length, and 3–13 GPa in Young's modulus were produced from the BCP complex based on PS-b-PEO [94]. A combination of fabrication procedures such as out-of-equilibrium nanopattern assembly and soft lithography was employed in order to prepare micellar nanostructures that mimic those of collagen fibrils in native ECM. The fibrous nanostructures of PS-b-PEO were then aligned and immobilized onto a glass substrate and subsequently used for culturing NIH-3T3 fibroblasts. The study showed that the degree of cell alignment increased with the area density of micelles, demonstrating the BCP micelles' ability to topologically regulate the cellular behaviors on the surface. The area, density, and orientation of the ultra-long PS-b-PEO filomicelles played important roles in modulating the extent of cellular alignment and directionality of the fibroblasts.

4. Implications of BCP Nanobiotechnology in Biosensing and Biomaterials

4.1. Implications in Solid-State Protein Arrays

Solid-state arrays such as protein chips and protein microarrays are widely employed in genomics, proteomics, drug discovery, and clinical diagnostics [4,193–206], as depicted in Figure 12. These microarrays, typically prepared on a microwell or a slide, permit a high degree of multiplexed protein detection from a broad range of sample types including plasma, serum, tissues, and biofluids [194,195,201–203,207,208]. One critical aspect of the protein microarray technology is to precisely control the number and spatial distance of biomolecules that are linked to the array surface during its manufacturing stage [197,198,201,209–213]. The surface modification steps to pre-link biomolecules and print proteins on the array surface are necessary to provide specificity in subsequent biorecognition processes that occur between the pre-configured receptor molecules and target analytes. Another critical aspect of microarray technology is to provide stability and functionality of those protein molecules on the array surface. High and long-lasting bioactivity of the surface-bound molecules is a pre-requisite to ensuring successful bioassays and biodetection. The spot size and density in protein arrays are yet another important aspect of microarray technology. The protein spot size and density in conventional protein arrays are typically in the micrometer range. Shrinking these protein patterns down to the nanometer scale will be beneficial to the creation of miniaturized devices and sensors that are capable of low-volume, low-cost, and high-throughput assays. Reduced dimensions in biodevices will also permit minimally invasive detection.

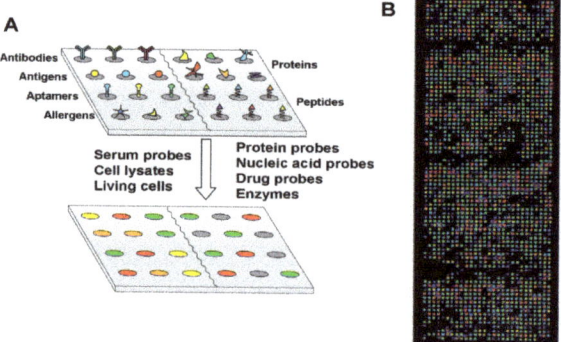

Figure 12. Examples of analytical protein microarrays are displayed. (**A**) Different types of ligands such as antibodies, antigens, DNA or RNA aptamers, carbohydrates or small molecules that are printed onto the surface of the protein array ensure high affinity and specificity to target analytes of

interest. The protein chip can be subsequently used for monitoring protein expression levels, protein profiling, and clinical diagnostics. (**B**) Protein microarrays were utilized to carry out multiplexed detection of antibodies in patient sera to tumor antigens. Over 1700 candidate tumor antigens were expressed and captured in a microarray format, and protein expression was detected using anti-glutathione S-transferase (GST) antibody. The arrays were then probed with sera from a healthy individual, and patients with melanoma, breast, and ovarian cancer. Adapted with permission from Ref. [214] Copyright (2008) American Chemical Society.

BCP-based schemes can represent a highly versatile and effective approach to rapidly prepare protein patterns into nanometer-sized spots and spacings. The approach relies solely on the self-assembly process of the polymers and that of the proteins. In addition, the chemical compositions of BCPs can be conveniently varied to offer multiple chemical functionalities as needed. Moreover, since the BCP strategy involves a thin film-based approach to produce protein nanopatterns, the resulting constructs can be easily integrated into a flexible sensor. Therefore, the BCP-based methodology is conducive to the recent trend in designing bioplatforms, i.e., miniaturization combined with chemical variety and mechanical flexibility. The advantages of the BCP-guided strategy present great potential for a low-cost, low-reagent volume, multiplexed, and high-throughput bioplatform whose architecture can be small, flexible, and portable. Overall, BCPs can greatly benefit the current field of protein micro-/nano-array technology.

4.2. Implications in Quantitative Bioanalyte Detection

The various research endeavors discussed earlier have begun to reveal intriguing behaviors of proteins and cells unique to nanoscale BCP interfaces. The fundamental knowledge obtained, in turn, can serve as guiding principles much needed for the design and application of next-generation nanobioarrays. One of the key findings directly influences the crucial aspect of reliably performing quantitative protein detection. In the BCP-guided assembly of proteins and cell adhesive molecules, it was possible to precisely control the number, surface density, and surface coverage of biomolecules [75,79,85,103,105,106,140,156,158]. The fact that such control in bioquantification can be attained through self-assembly by simply tuning the size of the underlying BCP nanodomain with respect to that of the biomolecule of interest is highly advantageous [37,75,85,140,156]. Another important discovery pertains to those systems whose length scales of the BCP surface features are commensurate with the characteristic dimensions of individual biomolecules [77,85,151,153,156,158]. In these cases, even a relatively small, topological and chemical change of the underlying BCP surface can result in strikingly different behaviors from biomolecules. This effect applies not only to those biomolecules directly interacting with BCP templates but also to those biomolecules indirectly coupled to the templates via prior-stage proteins or other bioligands.

The drastic surface-induced changes can be best evidenced in research efforts that systematically compared the interaction behaviors of the same proteins on BCP nanodomains relative to homopolymers, random copolymers, and polymer blends [35,76–78,85,87,156,158]. Salient features such as high protein density and tight 2D-packing represent those behaviors distinctively observed from nanoscale BCP platforms. When comparing the number of protein molecules per given polymer surface footprint, the largest number of proteins was found on nanoscale BCP platforms whose feature sizes were well-matched to that of an individual protein [77,85,151,156]. It was also found that the enhanced protein amount on a BCP surface was not a mere average of those amounts measured on its homopolymer counterparts. In addition, protein molecules on BCPs exhibited a well-ordered, close-packing behavior that resembled an orderly arrangement of atoms in a 2D inorganic crystal. In contrast, the spatial distribution of the same protein molecules did not show any spatial order on the homopolymer counterparts locally or globally.

BCPs' capacities summarized above can combinedly offer controllability and predictability in fabricating protein nanoarrays. The exact number, density, coverage, and mass of surface-bound protein molecules intended for a particular biodetection can be attained by varying the BCPs' topological cues (lateral and vertical dimensions, periodicity, and orientation of the nanodomains) with respect to the protein dimensions. They can also be attained by changing the BCPs' chemical cues (chemical composition, chemical interface, hydrophobicity, and surface charge) for modulating preferential protein–polymer segment interactions. When effectively controlled with these BCP surface-driven factors, the resulting bioarrays will offer a well-defined number of protein molecules organized into nanoscopic spots with a nanoscale periodicity which, in turn, can permit quantitative detection of bioanalytes in a highly reliable manner.

4.3. Implications in Stable Biosensors with High Functionality

The stability and activity of biomolecules are both extremely important criteria to consider in bioarray and biosensor applications. Even with a well-controlled number of biomolecules per spot, each of those immobilized biomolecules should be stable and active for the successful quantification of bioanalyte molecules and comparison of detection signals obtained from different spots in a bioarray. As discussed earlier, research efforts have confirmed that high biological stability and functionality are maintained for many different biomolecules assembled on BCP surfaces. Examples of those biomolecules included enzymes, antibodies, serum proteins, peptides, and cell adhesive proteins [76,78,79,85,87,103,105,106,140,158,215]. In comparison to macroscale homopolymer surfaces, biomolecules on nanoscale BCP surfaces are revealed to be much more stable and functional. The higher protein stability of BCPs is attributed to several factors such as the presence of high interfacial density, amphiphilicity, and protein–nanodomain size matching [35,85]. Relative to chemically uniform homopolymers, BCP surfaces can be fabricated to exhibit a high density of chemical interfaces between the nanoscopic domains of chemically distinctive polymer segments. As polymer segments in BCPs show different chemical properties of hydrophilicity/hydrophobicity and surface charge, the outer region of a nanodomain near the chemical interface presents a richer chemical environment than the center region of a nanodomain. As for proteins, their exterior surfaces are quite complex in chemical nature as well. The surface of a protein contains a large number of amino acid moieties with varying charges, polarity, and hydrophilicity/hydrophobicity. Hence, the richer the chemical environments are on a platform surface, the stabler the proteins are on the platform. BCP surfaces satisfy this requirement for protein stability.

In addition, the closer the BCP feature size is to that of a protein, the stabler the protein is on the platform. This is because various amino acid moieties on the exterior surface of the protein can be better stabilized by the amphiphilic environments of the BCP's interfacial regions [75,77,151,156,157,216,217]. Figure 13 displays synchrotron radiation-based, X-ray photoemission electron microscopy (XPEEM) data which maps the spatial distribution of HSA and the two polymer components of PS and PMMA on a PS/PMMA blend film. The densest layer of HSA was located at the interfaces between PS and PMMA. Similar localization behaviors were also observed from HSA adsorbed on a blend film of PS and polylactide (PLA), where the signals for proteins were highest along the interfacial lines between PS and PLA [216]. This can be also explained by the fact that the nanodomain regions near the chemical interface can stabilize a greater fraction of amino acid residues of the inherently amphiphile protein molecules, thus promoting protein binding at the interfacial as supposed to central sites of the nanodomains. The importance of the amphiphilicity provided by the underlying BCP surfaces is also emphasized in a cell study [81]. Stable cell adhesion was generally promoted on a BCP surface with its hydrophilic domains periodically spaced by hydrophobic domains, whereas a control surface consisting of only the hydrophilic polymer domain was repellant to cell adhesion. Therefore, in addition to the spatial nanopatterning ability to create self-assembled biomolecules, BCPs can provide the structural and chemical environments that are necessary to ensure the stability and functionality of biomolecules upon immobilization to bioarray and biosensor surfaces.

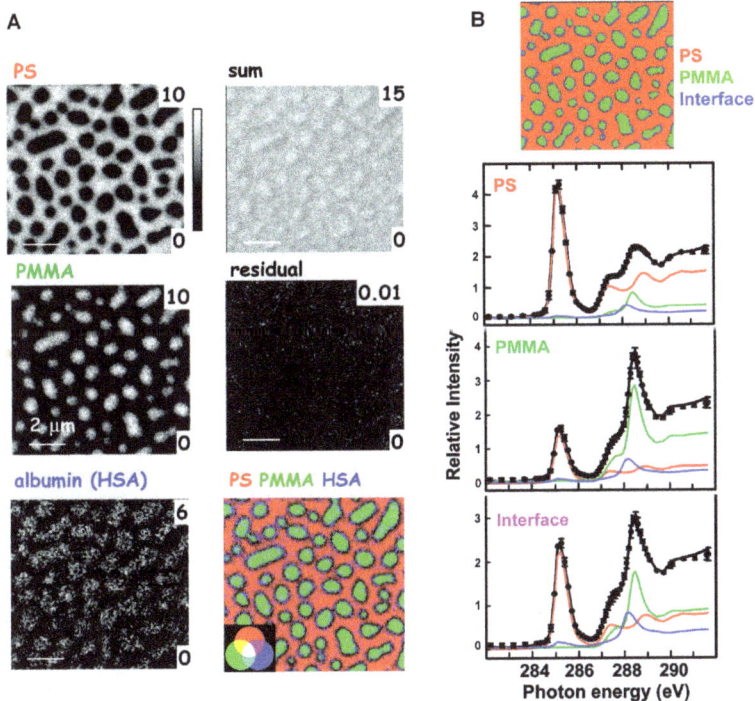

Figure 13. XPEEM component maps are displayed for HSA adsorbed on a PS/PMMA blend film. (**A**) The PS, PMMA, and HSA component maps show the location of each component on the film. The numbers in the upper and lower right of each component map are the minimum and maximum thicknesses in nm for the gray scales. The gray scale for the residual fit is the deviation of the fit and the measured signal, averaged over all photon energies. A composite map is also provided in the last panel where PS, PMMA, and HSA are shown in red, green, and blue, respectively. (**B**) The composite map and the plots for each component are obtained from HSA deposited on the PS/PMMA film using five different pH conditions. PS, PMMA, and HSA signals are shown in red, green, and blue, respectively. All data were derived from C 1s image sequences. Reproduced with permission from Ref. [217] Copyright (2008) American Chemical Society.

4.4. Implications in Tuning Protein Resistance

Being able to tune the protein resistance of a polymer surface is extremely important for the development of biomaterials and biomedical devices. The conventional way to attain low or high protein resistance is by changing the chemical nature of a polymer [218]. The main strategies to attain protein resistance depend on improving surface hydrophilicity through a chemical modification process or incorporation of surface modifying additives. Such surface hydrophilization agents can be incorporated into polymer surfaces either by physisorption, hydrogel network formation, surface grafting, layer-by-layer (LbL) assembly, or additive blending with base polymers [218–220]. In some cases, both the anti-biofouling capacity and protein resistance of BCPs are increased by chemically modifying the BCP side chains. For example, BCPs containing fluoroalkyl side chains were produced to form protein-resistant surfaces of polystyrene-block-poly(ethoxylated fluoroalkyl acrylate) (PS-b-PAA–AMP) and poly(hydroxyethylacrylamide)-block-poly(1H,1H-pentafluoropropyl methacrylate) (PHEAA-b-PFMA) [221,222].

It is worthwhile to point out a highly promising but yet-to-be-explored pathway that is different from the conventional routes discussed above. Insights from the aforementioned research endeavors involving BCPs suggest that protein resistance can be effectively tuned by the size scale of polymer surface features, without having to modify their chemical compositions. For instance, the difference in IgG binding behavior to PMMA is intriguing when it is compared between the cases of PMMA homopolymer versus PS-b-PMMA surfaces [156]. As the separation distance between the chemical interfaces of the polymer segments approaches several tens of nm in the BCP template from infinity in the case of PMMA homopolymer, no IgG molecules bind to the PMMA areas on the nanoscale BCP template. This is drastically different than the behavior of the same protein on PMMA homopolymer, where IgG molecules readily bind to PMMA. This outcome suggests that the nature of protein fouling and antifouling behaviors to given polymers can be dramatically changed when a polymer surface of the same chemical composition becomes nanosized and surrounded by chemical interfaces. It further indicates that the tendency of a given polymer to act as a protein-attracting or -repelling platform may be altered by simply tuning the size scale of the polymer interface down to that of an individual protein, instead of changing the chemical composition of the polymer.

4.5. Demonstration of Biosensors

Several examples are found in the literature in which the BCP thin film-based biomolecules were utilized in optical and electrochemical biosensors with specific sensor characteristics such as the detection limit and dynamic range [114–117]. Most of these studies involve a conventional material surface as a base platform onto which BCP nanostructures were applied for modification with biomolecules. For example, an approach to use AuNPs in conjunction with a BCP has been successfully employed to create a highly sensitive, BCP-templated, optical fiber for DNA detection. Figure 14 displays the overall approach for the fiber optic sensor using AuNPs/BCP. A conventional optical fiber surface was first treated with PS-b-P4VP whose nanopatterns were then used for linking AuNPs to the P4VP nanodomains [114]. Single-stranded DNA (ssDNA) hybridization reactions for *rop* B gene were carried out by using the BCP-templated LSPR sensor, whose surface contained capture ssDNA tethered to AuNPs as signal amplification tags. The employment of AuNPs enabled an optical phenomenon known as localized surface plasmon resonance (LSPR) which, in turn, led to high sensitivity in the detection of *rop* B gene related to Rifampicin-resistant tuberculosis. The study reported a detection limit of 67 pM and a dynamic range of 10^{-10}–10^{-6} M for the *rop* B detection.

BCP thin film-based efforts have been made to develop electrochemical sensors as well. The surface of a conventional Pt electrode was first modified with a film of PS-b-P4VP micelles. The BCP-modified electrode surface was subsequently treated to immobilize glucose oxidase (GOx) for the detection of glucose [115]. The study demonstrated that the BCP surface provided excellent stability for the enzyme molecules. The enzymatic activity and selectivity were maintained on the BCP surface for the glucose detection, resulting in a detection limit of 0.05 µM and a linear range of 10–4500 µM. Similarly, a BCP thin film of poly(n-butylmethacrylate)-block-poly(N,N-dimethylaminoethyl methacrylate) (PnBMA-b-PDMAEMA) micelles was adsorbed to the surface of conductive materials of graphite and Au. Then, choline oxidase (ChO) was immobilized onto the BCP template to examine the enzymatic activity towards choline [116]. The resulting amperometric choline sensor exhibited a detection limit of 30 nM and a linear range of 30 nM–100 µM. In a different study, a BCP thin film of PS-b-P4VP coupled with CuO nanodots was developed into an electrochemical biosensor and subsequently used for sensitive and selective determination of dopamine (DA) [117]. In this work, the BCP nanotemplate was prepared atop an ITO substrate that served as a working electrode. A porous nanotemplate was then generated via solvent vapor annealing of the BCP thin film, after which a copper nitrate solution was spun-coated onto the porous nanostructure. An ensuing treatment of UV/O_3 yielded CuO nanodots on the film. Figure 15A displays the overall approach used to produce a BCP-based DA sensor with CuO nanodots as an active DA-sensing element. The voltametric behavior of the oxidation reaction of DA on the sensor was examined by using cyclic voltammetry and differential pulse voltammetry. The current signal of the sensor was linear within a DA concentration range of 0.12–56.87 µM, with a sensitivity of 326.91 µA mM^{-1}cm^{-2} and a detection limit of 0.03 µM. Figure 15B,C show the analytical characteristics of the DA sensor. The work demonstrated that a BCP-based approach can be applied to create a robust and reliable DA sensor that has a critical importance in neurological disorders.

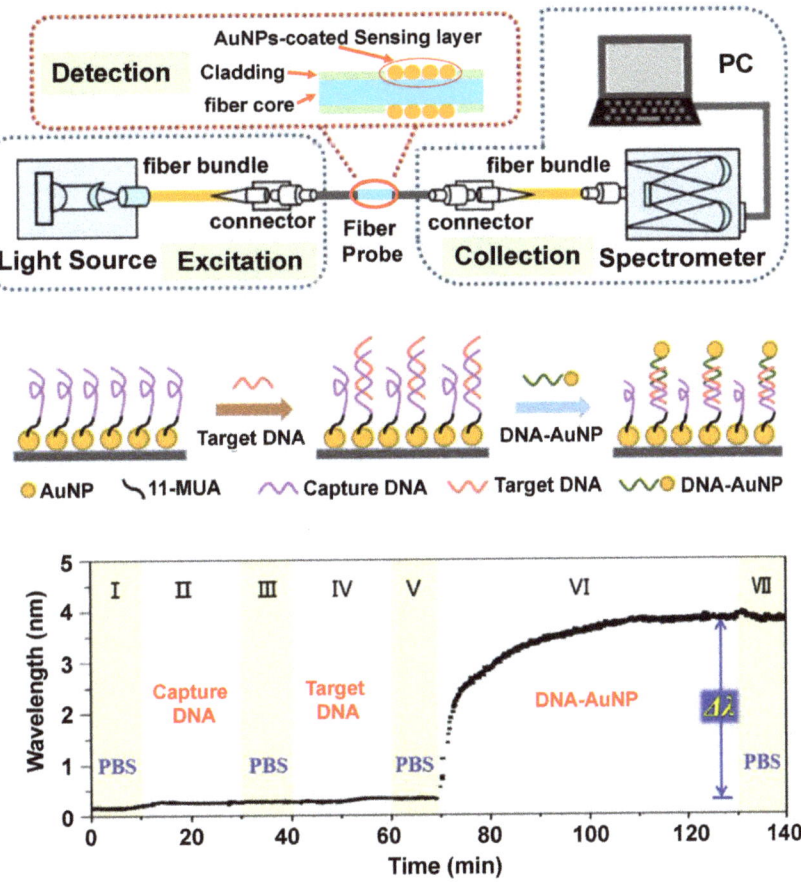

Figure 14. The schematics display an AuNP-based, fiber-optic LSPR sensor using a BCP-templating technique that was employed for the detection of ssDNA hybridization. The plot of wavelength versus time shows the sensor responses for different DNA fragments (capture DNA, target DNA, and DNA-AuNP) as well as for phosphate-buffered solution (PBS) during the hybridization reactions. Adapted with permission from Ref. [114] Copyright (2021) MDPI.

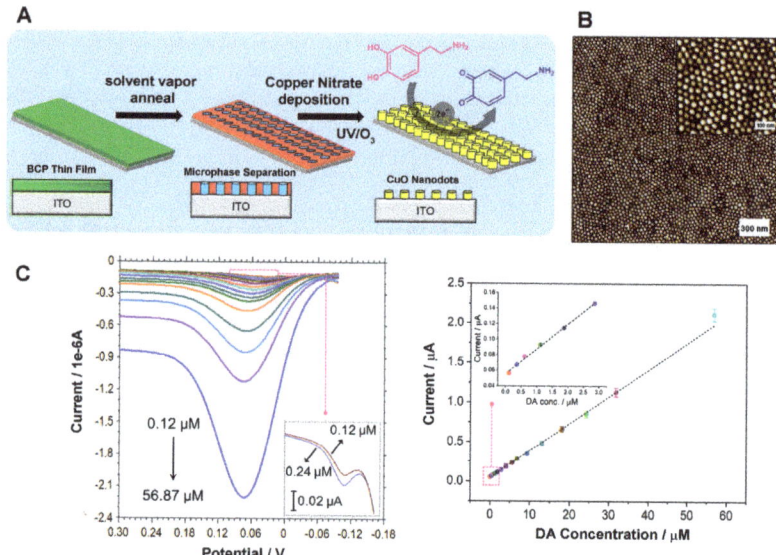

Figure 15. (**A**) The schematic shows the PS-b-P4VP thin film-based sensor fabrication process for developing CuO nanodots on an ITO substrate as well as the mechanism for the sensor operation in detecting DA. (**B**) The AFM image displays CuO nanodots on a Si substrate produced by using the PS-b-P4VP thin film and copper nitrate infiltration process. (**C**) The plots in the left panel correspond to the differential pulse voltammetry (DPV) data of the sensor in the presence of various DA concentrations in PBS pH 7.4. The calibration graph of the sensor for the determination of DA is provided in the right panel. Adapted with permission from Ref. [117] Copyright (2019) American Chemical Society.

Overall, biosensing applications that specifically exploit biomolecules on BCP nanopatterns are currently limited. However, the versatility and flexibility of BCPs in meeting the growing demand for miniaturized and flexible sensor arrays can open additional pathways for practical applications. The likelihood for the development of next-generation BCP-based biosensors and biomaterials is even greater when considering a plethora of biomolecules that can be easily self-assembled and guided by BCP nanopatterns. New types of BCP-based biointerfaces such as those enabling stimuli-responsive and time-programmed release of biomolecules may soon be realized as the research field becomes more mature. Continuous efforts in this direction will ultimately provide BCP-based, preventive and therapeutic measures in nanomedicine.

5. Outlook and Conclusions

The pioneering research endeavors on BCP-biomolecule nanoassembly and nanoconstructs discussed in this paper present new avenues for building biosensors, biodevices, and biomaterials with exquisite control at the nanoscale level. Many inspiring outcomes have already begun to provide definitive experimental evidence as well as computer simulation predictions on the mechanisms and kinetics of biomolecule–nanoscale surface interactions at the single-molecule level. One of the key messages from previous studies is that the behaviors of biomolecules on nanoscale, chemically rich surfaces are drastically different from those observed on bulk or macroscopic surfaces with no chemical variations. Another important lesson from the earlier efforts is that the interaction behaviors of biomolecules become even more distinct on surfaces that feature structural and chemical variations at the length scale commensurate with single biomolecule dimensions.

Future research engagements are still highly warranted in order to address the fundamental questions that have not yet been answered and further guide the current capacity to rationally design custom-tailored biointerfaces tuned at the nanoscale. Compared to what is currently known for single-component biomolecule behaviors on BCP surfaces, interfacial interactions involving multicomponent biomolecules are far less understood. Much of the dynamic, time-dependent interaction behaviors of biomolecules on BCP surfaces are still unknown. The structure–function relationships should be determined from a wider range of biomolecules on BCP surfaces beyond the present knowledge. Such efforts will provide answers to how structural and chemical variations on BCP surfaces can be used to ultimately modulate the function of the surface-bound biomolecules. Relative to those systems of proteins, cell adhesive molecules and cells, little is understood about the interaction behaviors of nucleic acids such as DNAs on BCP nanopatterns. Unlike cell adhesion, less is known for cell alignment, migration, and differentiation on BCP surfaces. It largely remains unclear the exact mechanisms through which cells sense a very small variation in the topological, mechanical, and chemical on BCP surfaces other than those cases where local directional cues are varied by the density of adhesion molecules. Similarly, how the nanoscale structures of proteins on BCP surfaces affect cell behaviors is unclear.

Looking ahead, plenty of exciting research opportunities await to advance the research field of BCP nanobiotechnology by capitalizing on the fundamental groundwork discussed in this paper. Future research efforts are needed to explicitly demonstrate biomedical applications that are beyond those limited examples currently available at the proof-of-concept level. Future developments in this direction should be geared towards significantly improving their practical applications. For this goal, additional research endeavors are warranted for the development of novel BCP–biomolecule nanoconstructs that are small and noninvasive with built-in chemical and biological functionalities needed in biosensing and biomaterials. In addition, the practical utilities of BCP–biomolecule nanoarrays as miniaturized, high-throughput, and flexible biosensing platforms are still to be proven. Incorporations of BCP–biomolecule nanoconstructs into field-ready, medical implants and medical devices are also to be demonstrated. For these applications, new types of BCP-based biointerfaces are to be designed for controlling protein- and antimicrobial-resistance as well as for time-dependent loading/unloading of cargo molecules into/from the biointerfaces. For cell-based applications, nanotextured BCP biomaterials are yet to be developed and demonstrated for selectively propagating a particular cell type and further optimizing specific cell adhesion and proliferation.

In summary, nanoscale-controlled surface organization of functional biomolecules guided by BCP thin films presents a new paradigm in nanobiotechnology. BCP nanobiotechnology can enable the future development of nanobiosensors and nanobiomaterials with superb spatial and functional control whose fabrication processes are solely driven by the self-assembly of the component elements. New intriguing characteristics of biomolecule interactions have begun to emerge from the previous research efforts highlighted in this paper. These research endeavors have laid the important groundwork for advancing BCP nanobiotechnology. Future research directions should focus on narrowing the current knowledge gap in nanoscale versus macroscale as well as ensemble versus single biomolecule behaviors on polymer surfaces and, also, on promoting the practical utility of BCP–biomolecule nanoconstructs in biosensing, biomaterials, and medical devices.

Author Contributions: Writing—original draft preparation, M.R.C.S. and J.-i.H.; writing—review and editing, M.R.C.S., D.H.C. and J.-i.H.; funding acquisition, J.-i.H. All authors have read and agreed to the published version of the manuscript.

Funding: The authors acknowledge financial support on this work by the National Science Foundation (Award No. CHE1903857) from the Macromolecular, Supramolecular and Nanochemistry Program under the Division of Chemistry.

Institutional Review Board Statement: Not applicable.

Data Availability Statement: Not applicable.

Conflicts of Interest: The authors declare no conflicts of interest.

References

1. Gupta, N.; Renugopalakrishnan, V.; Liepmann, D.; Paulmurugan, R.; Malhotra, B.D. Cell-Based Biosensors: Recent Tends, Challenges and Future Perspectives. *Biosens. Bioelectron.* **2019**, *141*, 111435. [CrossRef] [PubMed]
2. Fruncillo, S.; Su, X.; Liu, H.; Wong, L.S. Lithographic Processes for the Scalable Fabrication of Micro- and Nanostructures for Biochips and Biosensors. *ACS Sens.* **2021**, *6*, 2002–2024. [CrossRef]
3. Jin, X.; Li, G.; Xu, T.; Su, L.; Yan, D.; Zhang, X. Fully Integrated Flexible Biosensor for Wearable Continuous Glucose Monitoring. *Biosens. Bioelectron.* **2022**, *196*, 113760. [CrossRef]
4. Seo, Y.; Jeong, S.; Lee, J.; Choi, H.S.; Kim, J.; Lee, H. Innovations in Biomedical Nanoengineering: Nanowell Array Biosensor. *Nano Converg.* **2018**, *5*, 9. [CrossRef] [PubMed]
5. Wu, W.; Wang, L.; Yang, Y.; Du, W.; Ji, W.; Fang, Z.; Hou, X.; Wu, Q.; Zhang, C.; Li, L. Optical Flexible Biosensors: From Detection Principles to Biomedical Applications. *Biosens. Bioelectron.* **2022**, *210*, 114328. [CrossRef] [PubMed]
6. Mahzabeen, F.; Vermesh, O.; Levi, J.; Tan, M.; Alam, I.S.; Chan, C.T.; Gambhir, S.S.; Harris, J.S. Real-Time Point-of-Care Total Protein Measurement with a Miniaturized Optoelectronic Biosensor and Fast Fluorescence-Based Assay. *Biosens. Bioelectron.* **2021**, *180*, 112823. [CrossRef]
7. Wang, Y.; Yang, B.; Hua, Z.; Zhang, J.; Guo, P.; Hao, D.; Gao, Y.; Huang, J. Recent Advancements in Flexible and Wearable Sensors for Biomedical and Healthcare Applications. *J. Phys. D Appl. Phys.* **2022**, *55*, 134001. [CrossRef]
8. Soleymani, L.; Li, F. Mechanistic Challenges and Advantages of Biosensor Miniaturization into the Nanoscale. *ACS Sens.* **2017**, *2*, 458–467. [CrossRef]
9. Xu, M.; Obodo, D.; Yadavalli, V.K. The Design, Fabrication, and Applications of Flexible Biosensing Devices. *Biosens. Bioelectron.* **2019**, *124–125*, 96–114. [CrossRef]
10. Bhattarai, P.; Hameed, S. Basics of Biosensors and Nanobiosensors. In *Nanobiosensors: From Design to Applications*; Wiley: Hoboken, NJ, USA, 2020; pp. 1–22.
11. Mahardika, I.H.; Naorungroj, S.; Khamcharoen, W.; Kin, S.; Rodthongkum, N.; Chailapakul, O.; Shin, K. Point-of-Care Testing (POCT) Devices for DNA Detection: A Comprehensive Review. *Adv. NanoBiomed Res.* **2023**, *3*, 2300058. [CrossRef]
12. Derkus, B. Applying the Miniaturization Technologies for Biosensor Design. *Biosens. Bioelectron.* **2016**, *79*, 901–913. [CrossRef]
13. Stephens, A.D.; Song, Y.; McClellan, B.L.; Su, S.-H.; Xu, S.; Chen, K.; Castro, M.G.; Singer, B.H.; Kurabayashi, K. Miniaturized Microarray-Format Digital ELISA Enabled by Lithographic Protein Patterning. *Biosens. Bioelectron.* **2023**, *237*, 115536. [CrossRef] [PubMed]
14. Sathish, S.; Ishizu, N.; Shen, A.Q. Air Plasma-Enhanced Covalent Functionalization of Poly(methyl methacrylate): High-Throughput Protein Immobilization for Miniaturized Bioassays. *ACS Appl. Mater. Interfaces* **2019**, *11*, 46350–46360. [CrossRef] [PubMed]
15. Windmiller, J.R.; Wang, J. Wearable Electrochemical Sensors and Biosensors: A Review. *Electroanalysis* **2013**, *25*, 29–46. [CrossRef]
16. Wang, T.; Yang, H.; Qi, D.; Liu, Z.; Cai, P.; Zhang, H.; Chen, X. Mechano-Based Transductive Sensing for Wearable Healthcare. *Small* **2018**, *14*, 1702933. [CrossRef]
17. Handrea-Dragan, I.M.; Botiz, I.; Tatar, A.-S.; Boca, S. Patterning at the Micro/Nano-Scale: Polymeric Scaffolds for Medical Diagnostic and Cell-Surface Interaction Applications. *Colloids Surf. B Biointerfaces* **2022**, *218*, 112730. [CrossRef]
18. Chiu, D.T.; Jeon, N.L.; Huang, S.; Kane, R.S.; Wargo, C.J.; Choi, I.S.; Ingber, D.E.; Whitesides, G.M. Patterned Deposition of Cells and Proteins onto Surfaces by Using Three-Dimensional Microfluidic Systems. *Proc. Natl. Acad. Sci. USA* **2000**, *97*, 2408–2413. [CrossRef] [PubMed]
19. Kane, R.S.; Takayama, S.; Ostuni, E.; Ingber, D.E.; Whitesides, G.M. Patterning Proteins and Cells Using Soft Lithography. *Biomaterials* **1999**, *20*, 2363–2376. [CrossRef] [PubMed]
20. Marchesan, S.; Easton, C.D.; Styan, K.E.; Leech, P.; Gengenbach, T.R.; Forsythe, J.S.; Hartley, P.G. SU-8 Photolithography on Reactive Plasma Thin-films: Coated Microwells for Peptide Display. *Colloids Surf. B Biointerfaces* **2013**, *108*, 313–321. [CrossRef]
21. Alvarado, R.E.; Nguyen, H.T.; Pepin-Donat, B.; Lombard, C.; Roupioz, Y.; Leroy, L. Optically Assisted Surface Functionalization for Protein Arraying in Aqueous Media. *Langmuir* **2017**, *33*, 10511–10516. [CrossRef]
22. Kargl, R.; Mohan, T.; Köstler, S.; Spirk, S.; Doliška, A.; Stana-Kleinschek, K.; Ribitsch, V. Functional Patterning of Biopolymer Thin Films Using Enzymes and Lithographic Methods. *Adv. Funct. Mater.* **2013**, *23*, 308–315. [CrossRef]
23. Guyomard-Lack, A.; Delorme, N.; Moreau, C.; Bardeau, J.-F.; Cathala, B. Site-Selective Surface Modification Using Enzymatic Soft Lithography. *Langmuir* **2011**, *27*, 7629–7634. [CrossRef]
24. Gdor, E.; Shemesh, S.; Magdassi, S.; Mandler, D. Multienzyme Inkjet Printed 2D Arrays. *ACS Appl. Mater. Interfaces* **2015**, *7*, 17985–17992. [CrossRef] [PubMed]
25. Li, X.; Liu, B.; Pei, B.; Chen, J.; Zhou, D.; Peng, J.; Zhang, X.; Jia, W.; Xu, T. Inkjet Bioprinting of Biomaterials. *Chem. Rev.* **2020**, *120*, 10793–10833. [CrossRef] [PubMed]
26. Kim, S.; Marelli, B.; Brenckle, M.A.; Mitropoulos, A.N.; Gil, E.-S.; Tsioris, K.; Tao, H.; Kaplan, D.L.; Omenetto, F.G. All-water-based Electron-beam Lithography using Silk as a Resist. *Nat. Nanotech.* **2014**, *9*, 306–310. [CrossRef] [PubMed]
27. Bat, E.; Lee, J.; Lau, U.Y.; Maynard, H.D. Trehalose Glycopolymer Resists Allow Direct Writing of Protein Patterns by Electron-beam Lithography. *Nat. Commun.* **2015**, *6*, 6654. [CrossRef] [PubMed]

28. Mancini, R.J.; Paluck, S.J.; Bat, E.; Maynard, H.D. Encapsulated Hydrogels by E-beam Lithography and Their Use in Enzyme Cascade Reactions. *Langmuir* **2016**, *32*, 4043–4051. [CrossRef] [PubMed]
29. Liu, X.; Kumar, M.; Calo', A.; Albisetti, E.; Zheng, X.; Manning, K.B.; Elacqua, E.; Weck, M.; Ulijn, R.V.; Riedo, E. High-Throughput Protein Nanopatterning. *Faraday Discuss.* **2019**, *219*, 33–43. [CrossRef]
30. Chai, J.; Wong, L.S.; Giam, L.; Mirkin, C.A. Single-Molecule Protein Arrays Enabled by Scanning Probe Block Copolymer Lithography. *Proc. Natl. Acad. Sci. USA* **2011**, *108*, 19521–19525. [CrossRef]
31. Xie, Z.; Chen, C.; Zhou, X.; Gao, T.; Liu, D.; Miao, Q.; Zheng, Z. Massively Parallel Patterning of Complex 2D and 3D Functional Polymer Brushes by Polymer Pen Lithography. *ACS Appl. Mater. Interfaces* **2014**, *6*, 11955–11964. [CrossRef]
32. Merino, S.; Retolaza, A.; Trabadelo, V.; Cruz, A.; Heredia, P.; Aldunćin, J.A.; Mecerreyes, D.; Fernández-Cuesta, I.; Borrisé, X.; Pérez-Murano, F. Protein Patterning on the Micro- and Nanoscale by Thermal Nanoimprint Lithography on a New Functionalized Copolymer. *J. Vac. Sci. Technol. B* **2009**, *27*, 2439–2443. [CrossRef]
33. Fontelo, R.; Reis, R.L.; Novoa-Carballal, R.; Pashkuleva, I. Preparation, Properties, and Bioapplications of Block Copolymer Nanopatterns. *Adv. Healthc. Mater.* **2023**, *13*, 2301810. [CrossRef]
34. Gu, X.; Gunkel, I.; Russell, T.P. Pattern Transfer Using Block Copolymers. *Philos. Trans. R. Soc. A* **2013**, *371*, 20120306. [CrossRef]
35. Cho, D.H.; Xie, T.; Truong, J.; Stoner, A.C.; Hahm, J. Recent Advances Towards Single Biomolecule Level Understanding of Protein Adsorption Phenomena Unique to Nanoscale Polymer Surfaces with Chemical Variations. *Nano Res.* **2020**, *13*, 1295–1317. [CrossRef]
36. Hahm, J. Fundamentals of Nanoscale Polymer-Protein Interactions and Potential Contributions to Solid-state Nanobioarrays. *Langmuir* **2014**, *30*, 9891–9904. [CrossRef]
37. Cho, D.H.; Hahm, J.-I. Protein–Polymer Interaction Characteristics Unique to Nanoscale Interfaces: A Perspective on Recent Insights. *J. Phys. Chem. B* **2021**, *125*, 6040–6057. [CrossRef] [PubMed]
38. Biswas, A.; Bayer, I.S.; Biris, A.S.; Wang, T.; Dervishi, E.; Faupel, F. Advances in Top–down and Bottom–up Surface Nanofabrication: Techniques, Applications & Future Prospects. *Adv. Colloid Interface Sci.* **2012**, *170*, 2–27. [PubMed]
39. Yorulmaz Avsar, S.; Kyropoulou, M.; Di Leone, S.; Schoenenberger, C.-A.; Meier, W.P.; Palivan, C.G. Biomolecules Turn Self-Assembling Amphiphilic Block Co-Polymer Platforms into Biomimetic Interfaces. *Front. Chem.* **2019**, *6*, 645. [CrossRef]
40. Bates, F.S.; Fredrickson, G.H. Block Copolymers—Designer Soft Materials. *Phys. Today* **1999**, *52*, 32–38. [CrossRef]
41. Shull, K.R. Mean-Field Theory of Block Copolymers: Bulk Melts, Surfaces, and Thin Films. *Macromolecules* **1992**, *25*, 2122–2133. [CrossRef]
42. Cummins, C.; Lundy, R.; Walsh, J.J.; Ponsinet, V.; Fleury, G.; Morris, M.A. Enabling Future Nanomanufacturing Through Block Copolymer Self-Assembly: A Review. *Nano Today* **2020**, *35*, 100936. [CrossRef]
43. Feng, H.; Lu, X.; Wang, W.; Kang, N.-G.; Mays, J.W. Block Copolymers: Synthesis, Self-Assembly, and Applications. *Polymers* **2017**, *9*, 494. [CrossRef] [PubMed]
44. Li, W.; Liu, M.; Qiu, F.; Shi, A.-C. Phase Diagram of Diblock Copolymers Confined in Thin Films. *J. Phys. Chem. B* **2013**, *117*, 5280–5288. [CrossRef] [PubMed]
45. Tseng, Y.-C.; Darling, S.B. Block Copolymer Nanostructures for Technology. *Polymers* **2010**, *2*, 470–489. [CrossRef]
46. Segalman, R.A.; McCulloch, B.; Kirmayer, S.; Urban, J.J. Block Copolymers for Organic Optoelectronics. *Macromolecules* **2009**, *42*, 9205–9216. [CrossRef]
47. Darling, S.B. Directing the Self-assembly of Block Copolymers. *Prog. Polym. Sci.* **2007**, *32*, 1152–1204. [CrossRef]
48. Wang, R.-Y.; Park, M.J. Self-Assembly of Block Copolymers with Tailored Functionality: From the Perspective of Intermolecular Interactions. *Annu. Rev. Mater. Res.* **2020**, *50*, 521–549. [CrossRef]
49. Epps, T.H.; O'Reilly, R.K. Block Copolymers: Controlling Nanostructure to Generate Functional Materials–Synthesis, Characterization, and Engineering. *Chem. Sci.* **2016**, *7*, 1674–1689. [CrossRef] [PubMed]
50. Schricker, S.R.; Palacio, M.L.B.; Bhushan, B. Designing Nanostructured Block Copolymer Surfaces to Control Protein Adhesion. *Philos. Trans. R. Soc. A* **2012**, *370*, 2348–2380. [CrossRef]
51. Keddie, D.J. A Guide to the Synthesis of Block Copolymers Using Reversible-Addition Fragmentation Chain Transfer (RAFT) Polymerization. *Chem. Soc. Rev.* **2014**, *43*, 496–505. [CrossRef]
52. Dworakowska, S.; Lorandi, F.; Gorczyński, A.; Matyjaszewski, K. Toward Green Atom Transfer Radical Polymerization: Current Status and Future Challenges. *Adv. Sci.* **2022**, *9*, 2106076. [CrossRef] [PubMed]
53. Wang, R.; Wei, Q.; Sheng, W.; Yu, B.; Zhou, F.; Li, B. Driving Polymer Brushes from Synthesis to Functioning. *Angew. Chem. Int. Ed.* **2023**, *62*, e202219312. [CrossRef] [PubMed]
54. Dau, H.; Jones, G.R.; Tsogtgerel, E.; Nguyen, D.; Keyes, A.; Liu, Y.-S.; Rauf, H.; Ordonez, E.; Puchelle, V.; Basbug Alhan, H.; et al. Linear Block Copolymer Synthesis. *Chem. Rev.* **2022**, *122*, 14471–14553. [CrossRef] [PubMed]
55. Angelopoulou, P.P.; Moutsios, I.; Manesi, G.-M.; Ivanov, D.A.; Sakellariou, G.; Avgeropoulos, A. Designing High χ Copolymer Materials for Nanotechnology Applications: A Systematic Bulk vs. Thin Films Approach. *Prog. Polym. Sci.* **2022**, *135*, 101625. [CrossRef]
56. Park, S.J.; Bates, F.S.; Dorfman, K.D. Complex Phase Behavior in Binary Blends of AB Diblock Copolymer and ABC Triblock Terpolymer. *Macromolecules* **2023**, *56*, 1278–1288. [CrossRef]
57. Hrubý, M.; Filippov, S.K.; Štěpánek, P. Biomedical Application of Block Copolymers. In *Macromolecular Self-Assembly*; Billon, L., Borisov, O., Eds.; John Wiley & Sons, Inc.: Hoboken, NJ, USA, 2016; pp. 231–250.

58. Khandpur, A.K.; Foerster, S.; Bates, F.S.; Hamley, I.W.; Ryan, A.J.; Bras, W.; Almdal, K.; Mortensen, K. Polyisoprene-Polystyrene Diblock Copolymer Phase Diagram Near the Order-Disorder Transition. *Macromolecules* **1995**, *28*, 8796–8806. [CrossRef]
59. Samaddar, P.; Deep, A.; Kim, K.-H. An Engineering Insight into Block Copolymer Self-Assembly: Contemporary Application from Biomedical Research to Nanotechnology. *Chem. Eng. J.* **2018**, *342*, 71–89. [CrossRef]
60. Tang, P.; Qiu, F.; Zhang, H.; Yang, Y. Morphology and Phase Diagram of Complex Block Copolymers: ABC Linear Triblock Copolymers. *Phys. Rev. E* **2004**, *69*, 031803. [CrossRef] [PubMed]
61. Bates, F.S.; Fredrickson, G.H. Block Copolymer Thermodynamics-Theory and Experiment. *Annu. Rev. Phys. Chem.* **1990**, *41*, 525–557. [CrossRef]
62. Mastroianni, S.E.; Epps, T.H. Interfacial Manipulations: Controlling Nanoscale Assembly in Bulk, Thin film, and Solution Block Copolymer Systems. *Langmuir* **2013**, *29*, 3864–3878. [CrossRef]
63. Matsen, M.W.; Bates, F.S. Unifying Weak- and Strong-segregation Block Copolymer Theories. *Macromolecules* **1996**, *29*, 1091–1098. [CrossRef]
64. Matsen, M.W.; Schick, M. Microphase Separation in Starblock Copolymer Melts. *Macromolecules* **1994**, *27*, 6761–6767. [CrossRef]
65. Park, J.; Winey, K.I. Double Gyroid Morphologies in Precise Ion-Containing Multiblock Copolymers Synthesized via Step-Growth Polymerization. *JACS Au* **2022**, *2*, 1769–1780. [CrossRef] [PubMed]
66. Park, C.; Yoon, J.; Thomas, E.L. Enabling Nanotechnology with Self Assembled Block Copolymer Patterns. *Polymer* **2003**, *44*, 6725–6760. [CrossRef]
67. Lohmüller, T.; Aydin, D.; Schwieder, M.; Morhard, C.; Louban, I.; Pacholski, C.; Spatz, J.P. Nanopatterning by Block Copolymer Micelle Nanolithography and Bioinspired Applications. *Biointerphases* **2011**, *6*, MR1–MR12. [CrossRef] [PubMed]
68. Ranasinghe, D.R.; Doerk, G.; Aryal, B.R.; Pang, C.; Davis, R.C.; Harb, J.N.; Woolley, A.T. Block Copolymer Self-Assembly to Pattern Gold Nanodots for Site-Specific Placement of DNA Origami and Attachment of Nanomaterials. *Nanoscale* **2023**, *15*, 2188–2196. [CrossRef]
69. Singh, A.N.; Thakre, R.D.; More, J.C.; Sharma, P.K.; Agrawal, Y.K. Block Copolymer Nanostructures and Their Applications: A Review. *Polym. Plast. Technol. Eng.* **2015**, *54*, 1077–1095. [CrossRef]
70. Kim, B.H.; Kim, J.Y.; Kim, S.O. Directed Self-Assembly of Block Copolymers for Universal Nanopatterning. *Soft Matter* **2013**, *9*, 2780–2786. [CrossRef]
71. Yang, G.G.; Choi, H.J.; Han, K.H.; Kim, J.H.; Lee, C.W.; Jung, E.I.; Jin, H.M.; Kim, S.O. Block Copolymer Nanopatterning for Nonsemiconductor Device Applications. *ACS Appl. Mater. Interfaces* **2022**, *14*, 12011–12037. [CrossRef]
72. Killops, K.L.; Gupta, N.; Dimitriou, M.D.; Lynd, N.A.; Jung, H.; Tran, H.; Bang, J.; Campos, L.M. Nanopatterning Biomolecules by Block Copolymer Self-Assembly. *ACS Macro Lett.* **2012**, *1*, 758–763. [CrossRef]
73. Oleske, K.W.; Barteau, K.P.; Turker, M.Z.; Beaucage, P.A.; Estroff, L.A.; Wiesner, U. Block Copolymer Directed Nanostructured Surfaces as Templates for Confined Surface Reactions. *Macromolecules* **2017**, *50*, 542–549. [CrossRef]
74. Hu, H.; Gopinadhan, M.; Osuji, C.O. Directed Self-Assembly of Block Copolymers: A Tutorial Review of Strategies for Enabling Nanotechnology with Soft Matter. *Soft Matter* **2014**, *10*, 3867–3889. [CrossRef] [PubMed]
75. Kumar, N.; Hahm, J. Nanoscale Protein Patterning using Self-Assembled Diblock Copolymers. *Langmuir* **2005**, *21*, 6652–6655. [CrossRef] [PubMed]
76. Cho, D.H.; Xie, T.; Mulcahey, P.J.; Kelleher, N.P.; Hahm, J.-I. Distinctive Adsorption Mechanism and Kinetics of Immunoglobulin G on a Nanoscale Polymer Surface. *Langmuir* **2022**, *38*, 1458–1470. [CrossRef] [PubMed]
77. Lau, K.H.A.; Bang, J.; Hawker, C.J.; Kim, D.H.; Knoll, W. Modulation of Protein–Surface Interactions on Nanopatterned Polymer Films. *Biomacromolecules* **2009**, *10*, 1061–1066. [CrossRef] [PubMed]
78. Xie, T.; Chattoraj, J.; Mulcahey, P.J.; Kelleher, N.P.; Del Gado, E.; Hahm, J.-I. Revealing the Principal Attributes of Protein Adsorption on Block Copolymer Surfaces with Direct Experimental Evidence at the Single Protein Level. *Nanoscale* **2018**, *10*, 9063–9076. [CrossRef] [PubMed]
79. Kumar, N.; Parajuli, O.; Dorfman, A.; Kipp, D.; Hahm, J. Activity Study of Self-Assembled Proteins on Nanoscale Diblock Copolymer Templates. *Langmuir* **2007**, *23*, 7416–7422. [CrossRef] [PubMed]
80. Tran, H.; Ronaldson, K.; Bailey, N.A.; Lynd, N.A.; Killops, K.L.; Vunjak-Novakovic, G.; Campos, L.M. Hierarchically Ordered Nanopatterns for Spatial Control of Biomolecules. *ACS Nano* **2014**, *8*, 11846–11853. [CrossRef] [PubMed]
81. Seo, J.-H.; Matsuno, R.; Takai, M.; Ishihara, K. Cell Adhesion on Phase-separated Surface of Block Copolymer Composed of Poly(2-methacryloyloxyethyl phosphorylcholine) and Poly(dimethylsiloxane). *Biomaterials* **2009**, *30*, 5330–5340. [CrossRef]
82. Stel, B.; Gunkel, I.; Gu, X.; Russell, T.P.; De Yoreo, J.; Lingenfelder, M. Contrasting Chemistry of Block Copolymer Films Controls the Dynamics of Protein Self-Assembly at the Nanoscale. *ACS Nano* **2019**, *13*, 4018–4027. [CrossRef]
83. Akkineni, S.; Doerk, G.S.; Shi, C.; Jin, B.; Zhang, S.; Habelitz, S.; De Yoreo, J.J. Biomimetic Mineral Synthesis by Nanopatterned Supramolecular-Block Copolymer Templates. *Nano Lett.* **2023**, *23*, 4290–4297. [CrossRef] [PubMed]
84. Song, S.; Ravensbergen, K.; Alabanza, A.; Soldin, D.; Hahm, J.-I. Distinct Adsorption Configurations and Self-Assembly Characteristics of Fibrinogen on Chemically Uniform and Alternating Surfaces including Block Copolymer Nanodomains. *ACS Nano* **2014**, *8*, 5257–5269. [CrossRef]
85. Xie, T.; Vora, A.; Mulcahey, P.J.; Nanescu, S.E.; Singh, M.; Choi, D.S.; Huang, J.K.; Liu, C.-C.; Sanders, D.P.; Hahm, J. Surface Assembly Configurations and Packing Preferences of Fibrinogen Mediated by the Periodicity and Alignment Control of Block Copolymer Nanodomains. *ACS Nano* **2016**, *10*, 7705–7720. [CrossRef]

86. Matsusaki, M.; Omichi, M.; Kadowaki, K.; Kim, B.H.; Kim, S.O.; Maruyama, I.; Akashi, M. Protein Nanoarrays on a Highly-Oriented Lamellar Surface. *Chem. Commun.* **2010**, *46*, 1911–1913. [CrossRef]
87. Song, S.; Xie, T.; Ravensbergen, K.; Hahm, J.-I. Ascertaining Effects of Nanoscale Polymeric Interfaces on Competitive Protein Adsorption at the Individual Protein Level. *Nanoscale* **2016**, *8*, 3496–3509. [CrossRef] [PubMed]
88. Shen, L.; Zhu, J.; Liang, H. Heterogeneous Patterns on Block Copolymer Thin Film via Solvent Annealing: Effect on Protein Adsorption. *J. Chem. Phys.* **2015**, *142*, 101908. [CrossRef]
89. Shen, L.; Garland, A.; Wang, Y.; Li, Z.; Bielawski, C.W.; Guo, A.; Zhu, X.-Y. Two Dimensional Nanoarrays of Individual Protein Molecules. *Small* **2012**, *8*, 3169–3174. [CrossRef] [PubMed]
90. Palacio, M.L.B.; Schricker, S.R.; Bhushan, B. Block Copolymer Arrangement and Composition Effects on Protein Conformation Using Atomic Force Microscope-Based Antigen–Antibody Adhesion. *J. Biomed. Mater. Res.* **2012**, *100A*, 978–988. [CrossRef]
91. George, P.A.; Doran, M.R.; Croll, T.I.; Munro, T.P.; Cooper-White, J.J. Nanoscale Presentation of Cell Adhesive Molecules via Block Copolymer Self-Assembly. *Biomaterials* **2009**, *30*, 4732–4737. [CrossRef]
92. Liu, D.; Che Abdullah, C.A.; Sear, R.P.; Keddie, J.L. Cell Adhesion on Nanopatterned Fibronectin Substrates. *Soft Matter* **2010**, *6*, 5408–5416. [CrossRef]
93. Hiraguchi, Y.; Nagahashi, K.; Shibayama, T.; Hayashi, T.; Yano, T.-A.; Kushiro, K.; Takai, M. Effect of the Distribution of Adsorbed Proteins on Cellular Adhesion Behaviors Using Surfaces of Nanoscale Phase-Reversed Amphiphilic Block Copolymers. *Acta Biomater.* **2014**, *10*, 2988–2995. [CrossRef] [PubMed]
94. Zhang, K.; Arranja, A.; Chen, H.; Mytnyk, S.; Wang, Y.; Oldenhof, S.; van Esch, J.H.; Mendes, E. A Nano-Fibrous Platform of Copolymer Patterned Surfaces for Controlled Cell Alignment. *RSC Adv.* **2018**, *8*, 21777–21785. [CrossRef] [PubMed]
95. Liu, D.; Wang, T.; Keddie, J.L. Protein Nanopatterning on Self-Organized Poly(styrene-b-isoprene) Thin Film Templates. *Langmuir* **2009**, *25*, 4526–4534. [CrossRef] [PubMed]
96. Osypova, A.; Magnin, D.; Sibret, P.; Aqil, A.; Jérôme, C.; Dupont-Gillain, C.; Pradier, C.M.; Demoustier-Champagne, S.; Landoulsi, J. Dual Stimuli-Responsive Coating Designed Through Layer-By-Layer Assembly of PAA-b-PNIPAM Block Copolymers for the Control of Protein Adsorption. *Soft Matter* **2015**, *11*, 8154–8164. [CrossRef] [PubMed]
97. Reynhout, I.C.; Delaittre, G.; Kim, H.-C.; Nolte, R.J.M.; Cornelissen, J.J.L.M. Nanoscale Organization of Proteins via Block Copolymer Lithography and Non-Covalent Bioconjugation. *J. Mater. Chem. B* **2013**, *1*, 3026–3030. [CrossRef] [PubMed]
98. Presley, A.D.; Chang, J.J.; Xu, T. Directed Co-Assembly of Heme Proteins with Amphiphilic Block Copolymers Toward Functional Biomolecular Materials. *Soft Matter* **2011**, *7*, 172–179. [CrossRef]
99. Horrocks, M.S.; Kollmetz, T.; O'Reilly, P.; Nowak, D.; Malmström, J. Quantitative Analysis of Biomolecule Release from Polystyrene-Block-Polyethylene Oxide Thin Films. *Soft Matter* **2022**, *18*, 4513–4526. [CrossRef] [PubMed]
100. George, P.A.; Quinn, K.; Cooper-White, J.J. Hierarchical Scaffolds via Combined Macro- and Micro-phase Separation. *Biomaterials* **2010**, *31*, 641–647. [CrossRef] [PubMed]
101. Lilge, I.; Schönherr, H. Block Copolymer Brushes for Completely Decoupled Control of Determinants of Cell–Surface Interactions. *Angew. Chem. Int. Ed.* **2016**, *55*, 13114–13117. [CrossRef]
102. Malmström, J.; Wason, A.; Roache, F.; Yewdall, N.A.; Radjainia, M.; Wei, S.; Higgins, M.J.; Williams, D.E.; Gerrard, J.A.; Travas-Sejdic, J. Protein Nanorings Organized by Poly(styrene-block-ethylene oxide) Self-Assembled Thin Films. *Nanoscale* **2015**, *7*, 19940–19948. [CrossRef]
103. Parajuli, O.; Gupta, A.; Kumar, N.; Hahm, J. Evaluation of Enzymatic Activity on Nanoscale PS-b-PMMA Diblock Copolymer Domains. *J. Phys. Chem. B* **2007**, *111*, 14022–14027. [CrossRef] [PubMed]
104. Auriemma, F.; De Rosa, C.; Malafronte, A.; Di Girolamo, R.; Santillo, C.; Gerelli, Y.; Fragneto, G.; Barker, R.; Pavone, V.; Maglio, O.; et al. Nano-in-Nano Approach for Enzyme Immobilization Based on Block Copolymers. *ACS Appl. Mater. Interfaces* **2017**, *9*, 29318–29327. [CrossRef] [PubMed]
105. Arnold, M.; Hirschfeld-Warneken, V.C.; Lohmüller, T.; Heil, P.; Blümmel, J.; Cavalcanti-Adam, E.A.; López-García, M.; Walther, P.; Kessler, H.; Geiger, B.; et al. Induction of Cell Polarization and Migration by a Gradient of Nanoscale Variations in Adhesive Ligand Spacing. *Nano Lett.* **2008**, *8*, 2063–2069. [CrossRef] [PubMed]
106. Arnold, M.; Cavalcanti-Adam, E.A.; Glass, R.; Blümmel, J.; Eck, W.; Kantlehner, M.; Kessler, H.; Spatz, J.P. Activation of Integrin Function by Nanopatterned Adhesive Interfaces. *Chem. Phys. Chem.* **2004**, *5*, 383–388. [CrossRef] [PubMed]
107. Bandyopadhyay, K.; Tan, E.; Ho, L.; Bundick, S.; Baker, S.M.; Niemz, A. Deposition of DNA-Functionalized Gold Nanospheres into Nanoporous Surfaces. *Langmuir* **2006**, *22*, 4978–4984. [CrossRef] [PubMed]
108. Pearson, A.C.; Pound, E.; Woolley, A.T.; Linford, M.R.; Harb, J.N.; Davis, R.C. Chemical Alignment of DNA Origami to Block Copolymer Patterned Arrays of 5 nm Gold Nanoparticles. *Nano Lett.* **2011**, *11*, 1981–1987. [CrossRef]
109. Fontelo, R.; Soares da Costa, D.; Reis, R.L.; Novoa-Carballal, R.; Pashkuleva, I. Bactericidal Nanopatterns Generated by Block Copolymer Self-Assembly. *Acta Biomater.* **2020**, *112*, 174–181. [CrossRef] [PubMed]
110. Fontelo, R.; da Costa, D.S.; Reis, R.L.; Novoa-Carballal, R.; Pashkuleva, I. Block Copolymer Nanopatterns Affect Cell Spreading: Stem Versus Cancer Bone Cells. *Colloids Surf. B Biointerfaces* **2022**, *219*, 112774. [CrossRef] [PubMed]
111. Khor, H.L.; Kuan, Y.; Kukula, H.; Tamada, K.; Knoll, W.; Moeller, M.; Hutmacher, D.W. Response of Cells on Surface-Induced Nanopatterns: Fibroblasts and Mesenchymal Progenitor Cells. *Biomacromolecules* **2007**, *8*, 1530–1540. [CrossRef]

112. Jeong, E.J.; Lee, J.W.; Kwark, Y.-J.; Kim, S.H.; Lee, K.Y. The Height of Cell-Adhesive Nanoposts Generated by Block Copolymer/Surfactant Complex Systems Influences the Preosteoblast Phenotype. *Colloids Surf. B Biointerfaces* **2014**, *123*, 679–684. [CrossRef]
113. Chen, L.; Yu, Q.; Jia, Y.; Xu, M.; Wang, Y.; Wang, J.; Wen, T.; Wang, L. Micro-and-Nanometer Topological Gradient of Block Copolymer Fibrous Scaffolds Towards Region-Specific Cell Regulation. *J. Colloid Interface Sci.* **2022**, *606*, 248–260. [CrossRef] [PubMed]
114. Lu, M.; Peng, W.; Lin, M.; Wang, F.; Zhang, Y. Gold Nanoparticle-Enhanced Detection of DNA Hybridization by a Block Copolymer-Templating Fiber-Optic Localized Surface Plasmon Resonance Biosensor. *Nanomaterials* **2021**, *11*, 616. [CrossRef]
115. Guo, T.; Gao, J.; Qin, X.; Zhang, X.; Xue, H. A Novel Glucose Biosensor Based on Hierarchically Porous Block Copolymer Film. *Polymers* **2018**, *10*, 723. [CrossRef] [PubMed]
116. Sigolaeva, L.V.; Günther, U.; Pergushov, D.V.; Gladyr, S.Y.; Kurochkin, I.N.; Schacher, F.H. Sequential pH-Dependent Adsorption of Ionic Amphiphilic Diblock Copolymer Micelles and Choline Oxidase Onto Conductive Substrates: Toward the Design of Biosensors. *Macromol. Biosci.* **2014**, *14*, 1039–1051. [CrossRef] [PubMed]
117. Bas, S.Z.; Cummins, C.; Selkirk, A.; Borah, D.; Ozmen, M.; Morris, M.A. A Novel Electrochemical Sensor Based on Metal Ion Infiltrated Block Copolymer Thin Films for Sensitive and Selective Determination of Dopamine. *ACS Appl. Nano Mater.* **2019**, *2*, 7311–7318. [CrossRef]
118. Li, S.; Jiang, Y.; Chen, J.Z.Y. Morphologies and Phase Diagrams of ABC Star Triblock Copolymers Confined in a Spherical Cavity. *Soft Matter* **2013**, *9*, 4843–4854. [CrossRef]
119. Sun, M.; Wang, P.; Qiu, F.; Tang, P.; Zhang, H.; Yang, Y. Morphology and Phase Diagram of ABC Linear Triblock Copolymers: Parallel Real-Space Self-Consistent-Field-Theory Simulation. *Phys. Rev. E* **2008**, *77*, 016701. [CrossRef] [PubMed]
120. Epps, T.H.; Cochran, E.W.; Hardy, C.M.; Bailey, T.S.; Waletzko, R.S.; Bates, F.S. Network Phases in ABC Triblock Copolymers. *Macromolecules* **2004**, *37*, 7085–7088. [CrossRef]
121. Tang, P.; Qiu, F.; Zhang, H.; Yang, Y. Morphology and Phase Diagram of Complex Block Copolymers: ABC Star Triblock Copolymers. *J. Phys. Chem. B* **2004**, *108*, 8434–8438. [CrossRef]
122. Tyler, C.A.; Qin, J.; Bates, F.S.; Morse, D.C. SCFT Study of Nonfrustrated ABC Triblock Copolymer Melts. *Macromolecules* **2007**, *40*, 4654–4668. [CrossRef]
123. Zheng, W.; Wang, Z.-G. Morphology of ABC Triblock Copolymers. *Macromolecules* **1995**, *28*, 7215–7223. [CrossRef]
124. Matsen, M.W. Effect of Architecture on the Phase Behavior of AB-Type Block Copolymer Melts. *Macromolecules* **2012**, *45*, 2161–2165. [CrossRef]
125. Lodge, T.P. Block Copolymers: Long-Term Growth with Added Value. *Macromolecules* **2020**, *53*, 2–4. [CrossRef]
126. Song, S.; Milchak, M.; Zhou, H.B.; Lee, T.; Hanscom, M.; Hahm, J.I. Nanoscale Protein Arrays of Rich Morphologies via Self-Assembly on Chemically Treated Diblock Copolymer Surfaces. *Nanotechnology* **2013**, *24*, 095601. [CrossRef] [PubMed]
127. Wang, Q.; Yan, Q.; Nealey, P.F.; de Pablo, J.J. Monte Carlo Simulations of Diblock Copolymer Thin Films Confined Between Two Homogeneous Surfaces. *J. Chem. Phys.* **2000**, *112*, 450–464. [CrossRef]
128. Morkved, T.L.; Lopes, W.A.; Hahm, J.; Sibener, S.J.; Jaeger, H.M. Silicon Nitride Membrane Substrates for the Investigation of Local Structure in Polymer Thin Films. *Polymer* **1998**, *39*, 3871–3875. [CrossRef]
129. Fasolka, M.J.; Mayes, A.M. Block Copolymer Thin Films: Physics and Applications. *Annu. Rev. Mater. Res.* **2001**, *31*, 323–355. [CrossRef]
130. Segalman, R.A. Patterning with Block Copolymer Thin Films. *Mater. Sci. Eng. R Rep.* **2005**, *48*, 191–226. [CrossRef]
131. Mai, Y.; Eisenberg, A. Self-Assembly of Block Copolymers. *Chem. Soc. Rev.* **2012**, *41*, 5969–5985. [CrossRef]
132. Tritschler, U.; Pearce, S.; Gwyther, J.; Whittell, G.R.; Manners, I. 50th Anniversary Perspective: Functional Nanoparticles from the Solution Self-Assembly of Block Copolymers. *Macromolecules* **2017**, *50*, 3439–3463. [CrossRef]
133. Karayianni, M.; Pispas, S. Block Copolymer Solution Self-Assembly: Recent Advances, Emerging Trends, and Applications. *J. Polym. Sci.* **2021**, *59*, 1874–1898. [CrossRef]
134. Hamley, I.W. Nanostructure Fabrication Using Block Copolymers. *Nanotechnol.* **2003**, *14*, R39. [CrossRef]
135. Alexandridis, P.; Lindman, B. *Amphiphilic Block Copolymers: Self-Assembly and Applications*; Elsevier: Amsterdam, The Netherlands, 2000.
136. Blanazs, A.; Armes, S.P.; Ryan, A.J. Self-Assembled Block Copolymer Aggregates: From Micelles to Vesicles and their Biological Applications. *Macromol. Rapid Commun.* **2009**, *30*, 267–277. [CrossRef] [PubMed]
137. Jain, S.; Bates, F.S. On the Origins of Morphological Complexity in Block Copolymer Surfactants. *Science* **2003**, *300*, 460–464. [CrossRef] [PubMed]
138. Gao, Z.; Eisenberg, A. A Model of Micellization for Block Copolymers in Solutions. *Macromolecules* **1993**, *26*, 7353–7360. [CrossRef]
139. Jin, C.; Olsen, B.C.; Luber, E.J.; Buriak, J.M. Nanopatterning via Solvent Vapor Annealing of Block Copolymer Thin Films. *Chem. Mater.* **2017**, *29*, 176–188. [CrossRef]
140. Kumar, N.; Parajuli, O.; Hahm, J. Two-Dimensionally Self-Arranged Protein Nanoarrays on Diblock Copolymer Templates. *J. Phys. Chem. B* **2007**, *111*, 4581–4587. [CrossRef] [PubMed]
141. Zhang, L.; Yu, K.; Eisenberg, A. Ion-Induced Morphological Changes in "Crew-Cut" Aggregates of Amphiphilic Block Copolymers. *Science* **1996**, *272*, 1777–1779. [CrossRef] [PubMed]

142. Zhang, L.; Eisenberg, A. Multiple Morphologies and Characteristics of "Crew-Cut" Micelle-like Aggregates of Polystyrene-b-poly(acrylic acid) Diblock Copolymers in Aqueous Solutions. *J. Am. Chem. Soc.* **1996**, *118*, 3168–3181. [CrossRef]
143. Bhargava, P.; Zheng, J.X.; Li, P.; Quirk, R.P.; Harris, F.W.; Cheng, S.Z.D. Self-Assembled Polystyrene-block-poly(ethylene oxide) Micelle Morphologies in Solution. *Macromolecules* **2006**, *39*, 4880–4888. [CrossRef]
144. Lin, Y.; Böker, A.; He, J.; Sill, K.; Xiang, H.; Abetz, C.; Li, X.; Wang, J.; Emrick, T.; Long, S.; et al. Self-Directed Self-assembly of Nanoparticle/Copolymer mixtures. *Nature* **2005**, *434*, 55–59. [CrossRef] [PubMed]
145. Lee, C.; Kim, S.H.; Russell, T.P. Controlling Orientation and Functionalization in Thin Films of Block Copolymers. *Macromol. Rapid Commun.* **2009**, *30*, 1674–1678. [CrossRef] [PubMed]
146. Song, S.; Milchak, M.; Zhou, H.; Lee, T.; Hanscom, M.; Hahm, J.-I. Elucidation of Novel Nanostructures by Time-Lapse Monitoring of Polystyrene-block-Polyvinylpyridine under Chemical Treatment. *Langmuir* **2012**, *28*, 8384–8391. [CrossRef] [PubMed]
147. Hannon, A.F.; Bai, W.; Alexander-Katz, A.; Ross, C.A. Simulation Methods for Solvent Vapor Annealing of Block Copolymer Thin Films. *Soft Matter* **2015**, *11*, 3794–3805. [CrossRef] [PubMed]
148. Lopes, W.A.; Jaeger, H.M. Hierarchical Self-assembly of Metal Nanostructures on Diblock Copolymer Scaffolds. *Nature* **2001**, *414*, 735–738. [CrossRef] [PubMed]
149. Darling, S.B.; Yufa, N.A.; Cisse, A.L.; Bader, S.D.; Sibener, S.J. Self-Organization of FePt Nanoparticles on Photochemically Modified Diblock Copolymer Templates. *Adv. Mater.* **2005**, *17*, 2446–2450. [CrossRef]
150. Brassat, K.; Lindner, J.K.N. Nanoscale Block Copolymer Self-Assembly and Microscale Polymer Film Dewetting: Progress in Understanding the Role of Interfacial Energies in the Formation of Hierarchical Nanostructures. *Adv. Mater. Interfaces* **2020**, *7*, 1901565. [CrossRef]
151. Lau, K.H.A.; Bang, J.; Kim, D.H.; Knoll, W. Self-Assembly of Protein Nanoarrays on Block Copolymer Templates. *Adv. Funct. Mater.* **2008**, *18*, 3148–3157. [CrossRef]
152. Malmström, J.; Travas-Sejdic, J. Block Copolymers for Protein Ordering. *J. Appl. Polym. Sci.* **2014**, *131*, 40360. [CrossRef]
153. Firkowska-Boden, I.; Zhang, X.; Jandt, K.D. Controlling Protein Adsorption Through Nanostructured Polymeric Surfaces. *Adv. Healthc. Mater.* **2018**, *7*, 1700995. [CrossRef]
154. Hahm, J. Polymeric Surface-Mediated, High-Density Nano-Assembly of Functional Protein Arrays. *J. Biomed. Nanotechnol.* **2011**, *7*, 731–742. [CrossRef]
155. Bhushan, B.; Schricker, S.R. A Review of Block Copolymer-Based Biomaterials That Control Protein and Cell Interactions. *J. Biomed. Mater. Res.* **2014**, *102*, 2467–2480. [CrossRef] [PubMed]
156. Kumar, N.; Parajuli, O.; Gupta, A.; Hahm, J. Elucidation of Protein Adsorption Behavior on Polymeric Surfaces: Towards High Density, High Payload, Protein Templates. *Langmuir* **2008**, *24*, 2688–2694. [CrossRef]
157. Leung, B.O.; Hitchcock, A.P.; Cornelius, R.M.; Brash, J.L.; Scholl, A.; Doran, A. Using X-PEEM to Study Biomaterials: Protein and Peptide Adsorption to a Polystyrene–Poly(methyl methacrylate)-b-Polyacrylic Acid Blend. *J. Electron. Spectrosc. Relat. Phenom.* **2012**, *185*, 406–416. [CrossRef]
158. Keller, T.F.; Schönfelder, J.; Reichert, J.; Tuccitto, N.; Licciardello, A.; Messina, G.M.L.; Marletta, G.; Jandt, K.D. How the Surface Nanostructure of Polyethylene Affects Protein Assembly and Orientation. *ACS Nano* **2011**, *5*, 3120–3131. [CrossRef] [PubMed]
159. Colman, R.W.; Hirsh, J.; Marder, V.J.; Clowes, A.W.; George, J.N. *Hemostasis and Thrombosis: Basic Principles and Clinical Practice*; Lippincott Williams & Wilkins: Philadelphia, PA, USA, 2001.
160. Zuyderhoff, E.M.; Dupont-Gillain, C.C. Nano-organized Collagen Layers Obtained by Adsorption on Phase-Separated Polymer Thin Films. *Langmuir* **2012**, *28*, 2007–2014. [CrossRef] [PubMed]
161. Shiohara, A.; Prieto-Simon, B.; Voelcker, N.H. Porous Polymeric Membranes: Fabrication Techniques and Biomedical Applications. *J. Mater. Chem. B* **2021**, *9*, 2129–2154. [CrossRef] [PubMed]
162. Puiggalí-Jou, A.; del Valle, L.J.; Alemán, C. Biomimetic Hybrid Membranes: Incorporation of Transport Proteins/Peptides into Polymer Supports. *Soft Matter* **2019**, *15*, 2722–2736. [CrossRef]
163. Droumaguet, B.L.; Grande, D. Diblock and Triblock Copolymers as Nanostructured Precursors to Functional Nanoporous Materials: From Design to Application. *ACS Appl. Mater. Interfaces* **2023**, *15*, 58023–58040. [CrossRef]
164. Thurn-Albrecht, T.; Schotter, J.; Kästle, G.A.; Emley, N.; Shibauchi, T.; Krusin-Elbaum, L.; Guarini, K.; Black, C.T.; Tuominen, M.T.; Russell, T.P. Ultrahigh-Density Nanowire Arrays Grown in Self-Assembled Diblock Copolymer Templates. *Science* **2000**, *290*, 2126–2129. [CrossRef]
165. Olayo-Valles, R.; Lund, M.S.; Leighton, C.; Hillmyer, M.A. Large Area Nanolithographic Templates by Selective Etching of Chemically Stained Block Copolymer Thin Films. *J. Mater. Chem.* **2004**, *14*, 2729–2731. [CrossRef]
166. Zalusky, A.S.; Olayo-Valles, R.; Wolf, J.H.; Hillmyer, M.A. Ordered Nanoporous Polymers from Polystyrene–Polylactide Block Copolymers. *J. Am. Chem. Soc.* **2002**, *124*, 12761–12773. [CrossRef] [PubMed]
167. Leiston-Belanger, J.M.; Russell, T.P.; Drockenmuller, E.; Hawker, C.J. A Thermal and Manufacturable Approach to Stabilized Diblock Copolymer Templates. *Macromolecules* **2005**, *38*, 7676–7683. [CrossRef]
168. Bamford, C.H.; Cooper, S.L.; Tsuruta, T. *The Vroman Effect*; VSP BV: Utrecht, The Netherlands, 1992.
169. Hirsh, S.L.; McKenzie, D.R.; Nosworthy, N.J.; Denman, J.A.; Sezerman, O.U.; Bilek, M.M.M. The Vroman Effect: Competitive Protein Exchange with Dynamic Multilayer Protein Aggregates. *Colloids Surf. B Biointerfaces* **2013**, *103*, 395–404. [CrossRef]
170. Jung, S.-Y.; Lim, S.-M.; Albertorio, F.; Kim, G.; Gurau, M.C.; Yang, R.D.; Holden, M.A.; Cremer, P.S. The Vroman Effect: A Molecular Level Description of Fibrinogen Displacement. *J. Am. Chem. Soc.* **2003**, *125*, 12782–12786. [CrossRef] [PubMed]

171. Krishnan, A.; Siedlecki, C.A.; Vogler, E.A. Mixology of Protein Solutions and the Vroman Effect. *Langmuir* **2004**, *20*, 5071–5078. [CrossRef] [PubMed]
172. Palacio, M.L.B.; Schricker, S.R.; Bhushan, B. Bioadhesion of Various Proteins on Random, Diblock and Triblock Copolymer Surfaces and the Effect of pH Conditions. *J. R. Soc. Interface* **2011**, *8*, 630–640. [CrossRef] [PubMed]
173. Slack, S.M.; Horbett, T.A. The Vroman Effect: A Critical Review. In *Proteins at Interfaces II: Fundamentals and Applications*; Horbett, T.A., Brash, J.L., Eds.; ACS: Washington, DC, USA, 1995; pp. 112–128.
174. Horbett, T.A. The Role of Adsorbed Proteins in Tissue Response to Biomaterials. In *Biomaterials Science-An Introduction to Materials in Medicine*, 2nd ed.; Ratner, B.D., Hoffman, A.S., Schoen, F.J., Lemons, J.E., Eds.; Elsevier Academic Press: New York, NY, USA, 2004; pp. 237–246.
175. Wan, J.; Thomas, M.S.; Guthrie, S.; Vullev, V.I. Surface-Bound Proteins with Preserved Functionality. *Ann. Biomed. Eng.* **2009**, *37*, 1190–1205. [CrossRef]
176. Cha, T.W.; Guo, A.; Zhu, X.-Y. Enzymatic Activity on a Chip: The Critical Role of Protein Orientation. *Proteomics* **2005**, *5*, 416–419. [CrossRef]
177. Nudelman, F.; Sommerdijk, N.A.J.M. Biomineralization as an Inspiration for Materials Chemistry. *Angew. Chem. Int. Ed.* **2012**, *51*, 6582–6596. [CrossRef]
178. Davis, S.A.; Dujardin, E.; Mann, S. Biomolecular Inorganic Materials Chemistry. *Curr. Opin. Solid State Mater. Sci.* **2003**, *7*, 273–281. [CrossRef]
179. Deng, X.; Hasan, A.; Elsharkawy, S.; Tejeda-Montes, E.; Tarakina, N.V.; Greco, G.; Nikulina, E.; Stormonth-Darling, J.M.; Convery, N.; Rodriguez-Cabello, J.C.; et al. Topographically Guided Hierarchical Mineralization. *Mater. Today Bio* **2021**, *11*, 100119. [CrossRef] [PubMed]
180. Cui, F.-Z.; Ge, J. New Observations of the Hierarchical Structure of Human Enamel, from Nanoscale to Microscale. *J. Tissue Eng. Regen. Med.* **2007**, *1*, 185–191. [CrossRef]
181. Beniash, E.; Stifler, C.A.; Sun, C.-Y.; Jung, G.S.; Qin, Z.; Buehler, M.J.; Gilbert, P.U.P.A. The Hidden Structure of Human Enamel. *Nat. Commun.* **2019**, *10*, 4383. [CrossRef] [PubMed]
182. Politakos, N. Block Copolymers in 3D/4D Printing: Advances and Applications as Biomaterials. *Polymers* **2023**, *15*, 322. [CrossRef] [PubMed]
183. Epple, M.; Ganesan, K.; Heumann, R.; Klesing, J.; Kovtun, A.; Neumann, S.; Sokolova, V. Application of Calcium Phosphate Nanoparticles in Biomedicine. *J. Mater. Chem.* **2010**, *20*, 18–23. [CrossRef]
184. Hou, X.; Zhang, L.; Zhou, Z.; Luo, X.; Wang, T.; Zhao, X.; Lu, B.; Chen, F.; Zheng, L. Calcium Phosphate-Based Biomaterials for Bone Repair. *J. Funct. Biomater.* **2022**, *13*, 187. [CrossRef] [PubMed]
185. Plummer, S.T.; Wang, Q.; Bohn, P.W.; Stockton, R.; Schwartz, M.A. Electrochemically Derived Gradients of the Extracellular Matrix Protein Fibronectin on Gold. *Langmuir* **2003**, *19*, 7528–7536. [CrossRef]
186. Kang, C.E.; Gemeinhart, E.J.; Gemeinhart, R.A. Cellular Alignment by Grafted Adhesion Peptide Surface Density Gradients. *J. Biomed. Mater. Res.* **2004**, *71A*, 403–411. [CrossRef]
187. Dertinger, S.K.W.; Jiang, X.; Li, Z.; Murthy, V.N.; Whitesides, G.M. Gradients of Substrate-bound Laminin Orient Axonal Specification of Neurons. *Proc. Natl. Acad. Sci. USA* **2002**, *99*, 12542–12547. [CrossRef]
188. Maheshwari, G.; Brown, G.; Lauffenburger, D.A.; Wells, A.; Griffith, L.G. Cell Adhesion and Motility Depend on Nanoscale RGD Clustering. *J. Cell Sci.* **2000**, *113*, 1677–1686. [CrossRef] [PubMed]
189. Geiger, B.; Bershadsky, A.; Pankov, R.; Yamada, K.M. Transmembrane Crosstalk between the Extracellular Matrix and the Cytoskeleton. *Nat. Rev. Mol. Cell Biol.* **2001**, *2*, 793–805. [CrossRef] [PubMed]
190. Chen, L.; Yan, C.; Zheng, Z. Functional Polymer Surfaces for Controlling Cell Behaviors. *Mater. Today* **2018**, *21*, 38–59. [CrossRef]
191. Gallo, E.; Rosa, E.; Diaferia, C.; Rossi, F.; Tesauro, D.; Accardo, A. Systematic Overview of Soft Materials as a Novel Frontier for MRI Contrast Agents. *RSC Adv.* **2020**, *10*, 27064–27080. [CrossRef] [PubMed]
192. Kojima, C.; Katayama, R.; Lien Nguyen, T.; Oki, Y.; Tsujimoto, A.; Yusa, S.-i.; Shiraishi, K.; Matsumoto, A. Different Antifouling Effects of Random and Block Copolymers Comprising 2-methacryloyloxyethyl Phosphorylcholine and Dodecyl Methacrylate. *Eur. Polym. J.* **2020**, *136*, 109932. [CrossRef]
193. Angenendt, P.; Glökler, J.; Konthur, Z.; Lehrach, H.; Cahill, D.J. 3D Protein Microarrays: Performing Multiplex Immunoassays on a Single Chip. *Anal. Chem.* **2003**, *75*, 4368–4372. [CrossRef] [PubMed]
194. Cahill, D.J. Protein and Antibody Arrays and Their Medical Applications. *J. Immunol. Methods* **2001**, *250*, 81–91. [CrossRef]
195. Gong, P.; Grainger, D.W. *Microarrays: Methods and Protocols*, 2nd ed.; Humana Press: Totowa, NJ, USA, 2007.
196. Kersten, B.; Wanker, E.E.; Hoheisel, J.D.; Angenendt, P. Multiplexed Approaches in Protein Microarray Technology. *Expert Rev. Proteom.* **2005**, *2*, 499–510. [CrossRef]
197. MacBeath, G. Protein Microarrays and Proteomics. *Nat. Genet.* **2002**, *32*, 526–532. [CrossRef]
198. MacBeath, G.; Schreiber, S.L. Printing Proteins as Microarrays for High-Throughput Function Determination. *Science* **2000**, *289*, 1760–1763. [CrossRef]
199. Mendoza, L.G.; McQuary, P.; Mongan, A.; Gangadharan, R.; Brignac, S.; Eggers, M. High-Throughput Microarray-Based Enzyme-Linked Immunosorbent Assay (ELISA). *BioTechniques* **1999**, *27*, 778–788. [CrossRef] [PubMed]
200. Pavlickova, P.; Schneider, E.M.; Hug, H. Advances in Recombinant Antibody Microarrays. *Clin. Chim. Acta* **2004**, *343*, 17–35. [CrossRef] [PubMed]

201. Talapatra, A.; Rouse, R.; Hardiman, G. Protein Microarrays: Challenges and Promises. *Pharmacogenomics* **2002**, *3*, 527–536. [CrossRef] [PubMed]
202. Templin, M.F.; Stoll, D.; Schrenk, M.; Traub, P.C.; Vohringer, C.F.; Joos, T.O. Protein Microarray Technology. *Trends Biotechnol.* **2002**, *20*, 160–166. [CrossRef]
203. Xu, Q.; Lam, K.S. Protein and Chemical Microarrays-Powerful Tools for Proteomics. *J. Biomed. Biotechnol.* **2003**, *2003*, 257–266. [CrossRef]
204. Seo, J.; Shin, J.-Y.; Leijten, J.; Jeon, O.; Camci-Unal, G.; Dikina, A.D.; Brinegar, K.; Ghaemmaghami, A.M.; Alsberg, E.; Khademhosseini, A. High-Throughput Approaches for Screening and Analysis of Cell Behaviors. *Biomaterials* **2018**, *153*, 85–101. [CrossRef] [PubMed]
205. Zhu, H.; Fohlerová, Z.; Pekárek, J.; Basova, E.; Neužil, P. Recent Advances in Lab-on-a-Chip Technologies for Viral Diagnosis. *Biosens. Bioelectron.* **2020**, *153*, 112041. [CrossRef]
206. Brambilla, D.; Chiari, M.; Gori, A.; Cretich, M. Towards Precision Medicine: The Role and Potential of Protein and Peptide Microarrays. *Analyst* **2019**, *144*, 5353–5367. [CrossRef]
207. Bowser, B.L.; Robinson, R.A.S. Enhanced Multiplexing Technology for Proteomics. *Annu. Rev. Anal. Chem.* **2023**, *16*, 379–400. [CrossRef]
208. Zhang, H.; Miller, B.L. Immunosensor-Based Label-free and Multiplex Detection of Influenza Viruses: State of the Art. *Biosens. Bioelectron.* **2019**, *141*, 111476. [CrossRef]
209. Panda, S.; Hajra, S.; Mistewicz, K.; Nowacki, B.; In-na, P.; Krushynska, A.; Mishra, Y.K.; Kim, H.J. A Focused Review on Three-Dimensional Bioprinting Technology for Artificial Organ Fabrication. *Biomater. Sci.* **2022**, *10*, 5054–5080. [CrossRef] [PubMed]
210. Brittain, W.J.; Brandsetter, T.; Prucker, O.; Rühe, J. The Surface Science of Microarray Generation–A Critical Inventory. *ACS Appl. Mater. Interfaces* **2019**, *11*, 39397–39409. [CrossRef] [PubMed]
211. Gao, S.; Guisán, J.M.; Rocha-Martin, J. Oriented Immobilization of Antibodies onto Sensing Platforms-A Critical Review. *Anal. Chim. Acta* **2022**, *1189*, 338907. [CrossRef]
212. Clancy, K.F.A.; Dery, S.; Laforte, V.; Shetty, P.; Juncker, D.; Nicolau, D.V. Protein Microarray Spots are Modulated by Patterning Method, Surface Chemistry and Processing Conditions. *Biosens. Bioelectron.* **2019**, *130*, 397–407. [CrossRef] [PubMed]
213. Phizicky, E.; Bastiaens, P.I.H.; Zhu, H.; Snyder, M.; Fields, S. Protein Analysis on a Proteomic Scale. *Nature* **2003**, *422*, 208–215. [CrossRef]
214. Anderson, K.S.; Ramachandran, N.; Wong, J.; Raphael, J.V.; Hainsworth, E.; Demirkan, G.; Cramer, D.; Aronzon, D.; Hodi, F.S.; Harris, L.; et al. Application of Protein Microarrays for Multiplexed Detection of Antibodies to Tumor Antigens in Breast Cancer. *J. Proteome Res.* **2008**, *7*, 1490–1499. [CrossRef] [PubMed]
215. Spatz, J.P.; Mössmer, S.; Hartmann, C.; Möller, M.; Herzog, T.; Krieger, M.; Boyen, H.-G.; Ziemann, P.; Kabius, B. Ordered Deposition of Inorganic Clusters from Micellar Block Copolymer Films. *Langmuir* **2000**, *16*, 407–415. [CrossRef]
216. Leung, B.O.; Hitchcock, A.P.; Cornelius, R.; Brash, J.L.; Scholl, A.; Doran, A. X-ray Spectromicroscopy Study of Protein Adsorption to a Polystyrene–Polylactide Blend. *Biomacromolecules* **2009**, *10*, 1838–1845. [CrossRef]
217. Li, L.; Hitchcock, A.P.; Cornelius, R.; Brash, J.L.; Scholl, A.; Doran, A. X-ray Microscopy Studies of Protein Adsorption on a Phase Segregated Polystyrene/Polymethylmethacrylate Surface. 2. Effect of pH on Site Preference. *J. Phys. Chem. B* **2008**, *112*, 2150–2158. [CrossRef]
218. Ngo, B.K.D.; Grunlan, M.A. Protein Resistant Polymeric Biomaterials. *ACS Macro Lett.* **2017**, *6*, 992–1000. [CrossRef]
219. Katyal, P.; Mahmoudinobar, F.; Montclare, J.K. Recent Trends in Peptide and Protein-Based Hydrogels. *Curr. Opin. Struct. Biol.* **2020**, *63*, 97–105. [CrossRef] [PubMed]
220. Binaymotlagh, R.; Chronopoulou, L.; Haghighi, F.H.; Fratoddi, I.; Palocci, C. Peptide-Based Hydrogels: New Materials for Biosensing and Biomedical Applications. *Materials* **2022**, *15*, 5871. [CrossRef] [PubMed]
221. Weinman, C.J.; Gunari, N.; Krishnan, S.; Dong, R.; Paik, M.Y.; Sohn, K.E.; Walker, G.C.; Kramer, E.J.; Fischer, D.A.; Ober, C.K. Protein Adsorption Resistance of Anti-Biofouling Block Copolymers Containing Amphiphilic Side Chains. *Soft Matter* **2010**, *6*, 3237–3243. [CrossRef]
222. Wu, H.-X.; Zhang, X.-H.; Huang, L.; Ma, L.-F.; Liu, C.-J. Diblock Polymer Brush (PHEAA-b-PFMA): Microphase Separation Behavior and Anti-Protein Adsorption Performance. *Langmuir* **2018**, *34*, 11101–11109. [CrossRef] [PubMed]

Disclaimer/Publisher's Note: The statements, opinions and data contained in all publications are solely those of the individual author(s) and contributor(s) and not of MDPI and/or the editor(s). MDPI and/or the editor(s) disclaim responsibility for any injury to people or property resulting from any ideas, methods, instructions or products referred to in the content.

Article

Biosynthesis of Polyhydroxyalkanoates in *Cupriavidus necator* B-10646 on Saturated Fatty Acids

Natalia O. Zhila [1,2,*], Kristina Yu. Sapozhnikova [1,2], Evgeniy G. Kiselev [1,2], Ekaterina I. Shishatskaya [1,2] and Tatiana G. Volova [1,2]

[1] Institute of Biophysics SB RAS, Federal Research Center "Krasnoyarsk Science Center SB RAS", 50/50 Akademgorodok, Krasnoyarsk 660036, Russia; kristina.sap@list.ru (K.Y.S.); evgeniygek@gmail.com (E.G.K.); shishatskaya@inbox.ru (E.I.S.); volova45@mail.ru (T.G.V.)

[2] Basic Department of Biotechnology, School of Fundamental Biology and Biotechnology, Siberian Federal University, 79 Svobodnyi Av., Krasnoyarsk 660041, Russia

* Correspondence: nzhila@mail.ru; Tel.: +7-931-290-5491; Fax: +7-391-243-3400

Abstract: It has been established that the wild-type *Cupriavidus necator* B-10646 strain uses saturated fatty acids (SFAs) for growth and polyhydroxyalkanoate (PHA) synthesis. It uses lauric (12:0), myristic (14:0), palmitic (16:0) and stearic (18:0) acids as carbon sources; moreover, the elongation of the C-chain negatively affects the biomass and PHA yields. When bacteria grow on C12 and C14 fatty acids, the total biomass and PHA yields are comparable up to 7.5 g/L and 75%, respectively, which twice exceed the values that occur on longer C16 and C18 acids. Regardless of the type of SFAs, bacteria synthesize poly(3-hydroxybutyrate), which have a reduced crystallinity (C_x from 40 to 57%) and a molecular weight typical for poly(3-hydroxybutyrate) (P(3HB)) (M_w from 289 to 465 kDa), and obtained polymer samples demonstrate melting and degradation temperatures with a gap of about 100 °C. The ability of bacteria to assimilate SFAs opens up the possibility of attracting the synthesis of PHAs on complex fat-containing substrates, including waste.

Keywords: degradable polyhydroxyalkanoates; PHAs; fatty acids; fatty acid mixture; biosynthesis; properties

Citation: Zhila, N.O.; Sapozhnikova, K.Y.; Kiselev, E.G.; Shishatskaya, E.I.; Volova, T.G. Biosynthesis of Polyhydroxyalkanoates in *Cupriavidus necator* B-10646 on Saturated Fatty Acids. *Polymers* **2024**, *16*, 1294. https://doi.org/10.3390/polym16091294

Academic Editors: Arash Moeini, Gabriella Santagata and Pierfrancesco Cerruti

Received: 25 March 2024
Revised: 20 April 2024
Accepted: 3 May 2024
Published: 5 May 2024

Copyright: © 2024 by the authors. Licensee MDPI, Basel, Switzerland. This article is an open access article distributed under the terms and conditions of the Creative Commons Attribution (CC BY) license (https://creativecommons.org/licenses/by/4.0/).

1. Introduction

The global production of synthetic, non-degradable plastics has reached 400 million tons per year [1,2]. The widespread accumulation of plastic waste is polluting the world's oceans, thereby negatively affecting water quality and threatening biota [3,4]. Even more worrying is the increasing accumulation of microplastics in the biosphere [5]. Solving the problem of plastic waste is associated with a gradual transition to a new generation of degradable polymer materials [6,7], among which a special place belongs to polymers of microbiological origin, such as polyhydroxyalkanoates (PHAs).

PHAs are a family of biodegradable thermoplastic polymers of different chemical structures with various physicochemical properties [8–11]. At present, these biopolymers are rightfully considered as the most promising material of the 21st century for a wide variety of applications—ranging from the municipal and agricultural to pharmacology and biomedicine [12–15].

The key problem for increasing production volumes and expanding the scope of PHA application is reducing their cost, which is often achieved primarily through the use of available carbon raw materials including waste. Potential raw materials for PHA synthesis can be various individual compounds, as well as, most importantly, in various types of waste (e.g., the products of processing and hydrolysis of plant materials, waste from food, pharmaceutical waste, alcohol waste, pulp and paper waste, etc.) [16–21].

The possibility of PHA synthesis from the waste of various origins is of great importance as it contributes to "The Circular Economy" [11,22–24]. The use of waste as substrates in biotechnological processes for obtaining target products is not only a way to reduce the volume of their accumulation in the biosphere, but also a way through which to increase the efficiency of industrial production and the complete use of raw materials in general [25].

Great potential is represented by fat-containing wastes, which are accumulated in huge quantities in the food industry due to the lack of rational and effective technologies for their processing. The amount of fat waste generated annually is about 29 million tons [26]. The waste of fat-containing raw materials, which are low-grade oils, waste cooking oils, as well as fish processing waste containing smoking components (phenolic compounds, polycyclic aromatic hydrocarbons, etc.) are unsuitable not only for humans, but also for use in animal and aquaculture feed. These wastes must be treated to prevent environmental pollution, or they should be disposed safely [27]. Such wastes often are transported to municipal solid waste landfills and burned, so the possibility of their use is a relevant solution. An environmentally significant area for the use of large-capacity fat waste can be the biotechnological synthesis of target products including degradable PHAs.

Despite the fact that interest in lipid substrates for PHA synthesis has emerged relatively recently, the results obtained are encouraging. The possibility of conducting PHA synthesis using vegetable oils of various origins [28], as well as the low-quality and waste fats of animal origin [29–31] and fat-containing fish processing waste [32–35], have been shown. It is important to note that the high-energy efficiency of the transformation of lipid substrates in microbial metabolism processes with a possible theoretical PHA yield is up to 0.7–0.8 g/g. This is practically twice as high when compared to the results for sugars [36–38].

However, a complicating factor in the synthesis of PHAs on complex fatty substrates may be the found in the selective consumption of fatty acids (FAs) by bacteria. Thus, during the PHA accumulation on complex fatty substrates, the uneven consumption of saturated, polyenoic and unsaturated FAs as a component of the C-substrate formed under the action of lipolytic enzymes was revealed [39].

The effect of uneven FA consumption was discovered in the culture of *Cupriavidus necator* B-10646 when the PHA synthesis on vegetable oils (sunflower and Siberian seed oils) in contrast to palm oil (the FA from which was utilized by bacteria evenly) was studied [40]. The aforementioned paper [41] showed that the *Ralstonia eutropha* H16 strain more actively metabolized palmitic, oleic and linoleic fatty acids when it was grown on soybean oil in contrast to linolenic acid, which was practically not used. When *C. necator* B-10646 was grown on the fish processing fatty waste obtained during the production of canned sprats and on the fat obtained from the heads and ridges of Atlantic mackerel, the bacteria primarily utilized polyenoic fatty acids (linoleic and linolenic acids) in contrast to saturated fatty acids and unsaturated monoenoic oleic acid, the content of which increased in the culture (which led to changes in the residual lipid saturation [42]). These results are comparable with the data of paper [43], which also noted the uneven consumption of FAs by bacteria during the PHA synthesis on pollock waste by six strains of *Pseudomonas*. The authors showed that these changes manifest themselves differently depending on the strain specificity of the bacteria, the synthesis route and PHA composition. Uneven FA utilization by bacteria and the accumulation of non-recyclable acids in the culture negatively affects the complete use of the substrate, thereby reducing the final yield of the product and the economics of the process as a whole.

At the same time, it is known that PHA producers metabolize many FAs as the sole carbon source, and their metabolism is associated with intracellular PHA synthesis.

The potential of fatty acids as a substrate for PHA synthesis by representatives of various taxa is illustrated in Table 1. The results were obtained under the same conditions of laboratory cultivation of PHA producers in glass flasks as a fermentation vessel, which allows for a comparison of the achieved indicators.

Table 1. Published data on the P(3HB) synthesis by various wild-type producers during growth on different fatty acids as a carbon source.

Strain	X, g/L	P(3HB), %	M_w, kDa	Đ	T_{melt}, °C	T_{degr}, °C	C_x, %	Reference
Lauric acid								
Alcaligenes sp. AK 201	~2.4	40%	304	1.9	102	-	-	[44]
Aeromonas hydrophila	2.7–4.3	19.4–41.1	-	-	-	-	-	[45,46]
Aeromonas salmonicida 741	2.8–3.8	13.3–30.0 *	-	-	-	-	-	[45]
Klebsiella pneumoniae ZMd31	0.6–0.7	11.0–18.3 *	-	-	-	-	-	[45]
Burkholderia sp.	1.4–2.1	8–69	-	-	-	-	-	[45]
Bacillus cereus SPV	0.51	61.81	-	-	-	-	-	[47]
Myristic acid								
Alcaligenes sp. AK 201	~2.9	55%	1416	3.1	106	-	-	[44]
C. necator DSM 545	1.43	-	-	-	-	-	-	[48]
D. tsuruhatensis Bet002	-	76.7	131	1.1	173.2	289.8	-	[49]
Burkholderia sp.	1.1–1.9	1–49	-	-	-	-	-	[50]
Palmitic acid								
Alcaligenes sp. AK 201	~3.2	55%	1442	2.5	108	-	-	[44]
C. necator DSM 545	1.97	-	-	-	-	-	-	[48]
D. tsuruhatensis Bet002	-	53.8	166	1.5	175.7	302.9	-	[49]
Burkholderia sp.	0.6–1.5	tr-9	-	-	-	-	-	[50]
Rhodococcus pyridinivorans KY703220	1.435 (OD)	40	-	-	-	-	-	[51]
Stearic acid								
Alcaligenes sp. AK 201	~2.3	30%	986	1.9	116	-	-	[44]
C. necator DSM 545	2.11	-	-	-	-	-	-	[48]
D. tsuruhatensis Bet002	-	45.0	188	1.8	177.4	391.8	-	[49]
Burkholderia sp.	0.5–1.0	tr-1	-	-	-	-	-	[50]

* The data on the PHA composition are not reported.

The most representative array of data was obtained from the study of the *Pseudomonas* strains, as well as the *Aeromonas hydrophila* [45,46,52], *Delftia* [49,53], *Burkholderia* sp. [50], *Klebsiella pneumonia* [45], *Bacillus cereus* [47], and *Rhodococcus pyridinivorans* [51] strains. There is very limited information regarding the ability of *Cupriavidus* strains to synthesize the PHAs on fatty acids (Table 1). The bacteria belonging to the genus *Cupriavidus* (formerly *Hydrogenomonas*, *Alcaligenes*, *Ralstonia*, and *Wautersia*) are promising producers of PHAs [54], both those under autotrophic conditions and on various organic substrates, including individual fatty acids [44,48,55,56]. The highly productive *Cupriavidus necator* B-10646 strain, which has broad organotrophic potential, is capable of synthesizing PHAs in high yields using mixtures of CO_2 and H_2 [57], sugars [58,59], and glycerol [60]. Regarding the ability of the *C. necator* B-10646 strain to synthesize PHAs, it has been mostly oleic acid that has been studied; moreover, the possibility of poly(3-hydroxybutyrate) and copolymers with a 3-hydroxyvalerate synthesis and the addition of valerate precursors to a medium has been shown [61,62]. It has also been shown that saturated fatty acids, as well as oleic acid, are poorly utilized by this strain when it is grown on complex fatty substrates [41,42].

The purpose of this work is to study the ability of the wild-type *Cupriavidus necator* B-10646 strain to metabolize saturated fatty acids as the sole carbon source for the synthesis of PHAs and the influence of the type of FAs on the composition and properties of polymers.

2. Materials and Methods

2.1. The PHA Producer Strain, Media and Cultivation Technique

In all experiments, the wild-type *Cupriavidus necator* B-10646 strain was used. It is registered in the Russian National Collection of Industrial Microorganisms (VKPM) [63]. It is able to synthesize PHAs using a wide range of carbon substrates (CO_2, sugars, glycerol, plant oils, etc.).

The nutrient medium used was Schlegel's mineral salt medium [64], which is based on a phosphate buffer (Na_2HPO_4, 9.1 g/L; KH_2PO_4, 1.5 g/L) containing a source of magnesium

(MgSO$_4$, 0.2 g/L), iron (Fe$_3$C$_6$H$_5$O$_7$, 0.025 g/L), nitrogen source (NH$_4$Cl, 0.7 g/L) and a set of microelements (3 mL solution/L of medium; solution composition (g/L): H$_3$BO$_3$ (0.228), CoCl$_2$ × 6H$_2$O (0.03), CuSO$_4$ × 5H$_2$O (0.008), MnCl$_2$ × 4H$_2$O (0.008), ZnSO$_4$ × 7H$_2$O (0.176), NaMoO$_4$ × 2H$_2$O (0.05), NiCl (0.008)). The carbon sources were constituted of the following: saturated fatty acids (lauric (C12:0), myristic (C14:0), palmitic (C16:0) and stearic (C18:0) (purity 97%, Acros Organics, Brussels, Belgium), monounsaturated oleic acid (18:1) (purity 98%, EKOS-1, Staraya Kupavna, Russia) and mixtures thereof were used, as described in the text.

Bacteria were grown under aerobic conditions at a temperature of 30 °C in a thermostatic incubator shaker "Incubator Shaker Innova" (New Brunswick Scientific, Edison, NJ, USA) in 0.5 L flasks. The bacteria were cultured in batch mode and in the conditions developed earlier for PHA biosynthesis. To obtain the inoculum, a museum culture of a bacterial strain stored on Schlegel agar medium at 5 °C was resuspended until a starting cell concentration of 0.1–0.2 g/L. The bacteria were cultured in Schlegel's salt medium in periodic mode: during the first 25–35 h of growth, 0.7 g/L of a nitrogen source was added to the medium, which served as a limiting factor and stimulated the production of PHAs. The source of nitrogen was exhausted in the subsequent hours of growth.

During the growth of the bacteria, samples were taken periodically (every 24 h) to analyze the accumulation of biomass and polymers in the cells. The dynamics of the cell growth in the culture were recorded by the optical density of the bacterial suspension at a wavelength of λ = 440 nm (UNICO 2100, Dayton, NJ, USA). The cell concentration (i.e., the yield of bacterial biomass X, g/L) was assessed by the gravimetric method by drying a twice-washed biomass at 105 °C for 24 h, which was centrifuged at 6000 rpm.

2.2. PHA Analysis

To determine the intracellular content of the PHAs in bacterial cells and its monomer composition, dry cell biomass and native polymers extracted from the cells and purified to a homogeneous state were used. The extraction of PHAs from a biomass was carried out in two stages: (1) the actual extraction of the polymer with dichloromethane and the concentration of the resulting extract on a rotary evaporator; and (2) ethanol precipitation. To purify the obtained samples, the polymer was dissolved and reprecipitated several times.

The purity and composition of the polymer was determined by a chromatography of the methyl esters of fatty acids after methanolysis of the purified polymer samples using a 7890A chromatograph-mass spectrometer (Agilent Technologies, Santa Clara, CA, USA) equipped with a 5975C mass detector (Agilent Technologies, Santa Clara, CA, USA) [65]. Benzoic acid was used as an internal standard to determine the total intracellular PHAs [66].

2.3. PHA Properties

The molecular weight and molecular weight distribution of the PHAs were examined using a size-exclusion chromatograph (Agilent Technologies 1260 Infinity, Waldbronn, Germany) with a refractive index detector, and this was achieved using an Agilent PLgel Mixed-C column, where the weight average, M_w, number average, M_n and polydispersity (Đ = M_w/M_n) were determined. The thermal properties of the polymer were analyzed using a DSC-1 differential scanning calorimeter (Mettler Toledo, Schwerzenbac, Switzerland). The melting points were determined from the exothermal peaks in the thermograms using the STARe software. Thermal degradation of the samples was investigated using a TGA2 thermal analysis system (Mettler Toledo, Schwerzenbac, Switzerland). The theoretical degree of crystallinity was calculated using the following formula:

$$C_x = (\Delta H_i)/(\Delta H_0), \qquad (1)$$

where ΔH_i is specific enthalpy of melting of the sample (J/g), and ΔH_0 is the specific enthalpy of a melted 100% crystallized P(3HB) (146 J/g) [67]. The methods for analyzing the physicochemical properties of PHAs have been previously described in detail [59].

2.4. Statistical Analysis

The statistical analysis of the results was performed by conventional methods using the standard software package of Microsoft Excel 2013 (Ver. 15.0.4420.1017). Each experiment was performed five times. The arithmetic means and standard deviations were found. To compare the groups, the Mann–Whitney test was used at the significance level of $p \leq 0.05$.

3. Results and Discussion

3.1. Growth and PHA Synthesis by Bacteria C. necator B-10646 on Individual Saturated Fatty Acids

The growth of *C. necator* B-10646 bacteria was conducted using one of following four saturated fatty acids as the sole carbon source: myristic (12:0), lauric (14:0), palmitic (16:0) and stearic (18:0) acids. These acids were studied for the first time. These fatty acids were chosen due to their wide distribution in vegetable oils and low-grade animal fats, which are currently being actively studied as a carbon substrate for the production of PHAs [29,68–70]. The results of assessing the ability of *C. necator* B-10646 bacteria to grow and synthesize polymers using individual saturated fatty acids as the sole carbon substrate are presented in Figure 1.

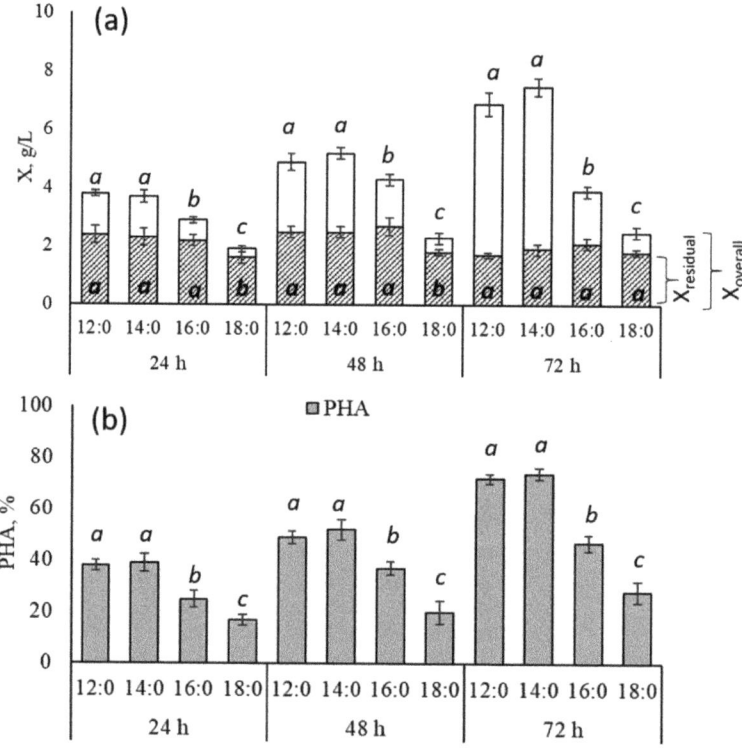

Figure 1. The yield of the overall (X, g/L) and residual bacterial biomass (X_{res}, g/L) (**a**) and the intracellular PHA content (**b**) in the culture of the wild-type *C. necator* B-10646 strain during growth on the saturated FAs of various structures. The letters indicate the significance of the differences when comparing groups according to the Mann–Whitney test at a level of $p \leq 0.05$, where identical letters indicate no significant differences.

All the studied saturated fatty acids (FAs) were supported the growth of bacteria *C. necator* B-10646 and were accompanied by the synthesis and accumulation of the polymer in cells (Figure 1a). The highest yield of the bacterial biomass (X = 7.5 ± 0.4 g/L) was obtained when the bacteria were grown on myristic acid as the only carbon source; using lauric acid resulted in a slightly lower (6.7 ± 0.5 g/L) growth, but these differences were not statistically significant. In the aforementioned cases, the values of the residual bacterial biomass (X_{res}) were 1.9 ± 0.1 and 1.7 ± 0.3 g/L, respectively, which also did not differ significantly. The lowest bacterial biomass yield (X = 2.5 ± 0.3 g/L, X_{res} = 1.8 ± 0.2 g/L) was recorded for stearic acid. The growth of the bacteria on palmitic acid was comparable to the results obtained on the myristic and lauric acid-grown FAs, but this was only found during the first 48 h of bacterial cultivation. Further, there was no increase in the X value; moreover, a decrease value was noted by 72 h, which indicates the death of some cells in the culture.

The intracellular content of the polymer in the culture of *C. necator* B-10646 during growth on the saturated FAs also varied (Figure 1b). The PHA content was comparable and the highest at 74 ± 2 and 72 ± 2% when bacteria were grown on myristic and lauric acids, respectively. The intracellular content of PHAs on palmitic acid was significantly lower (47 ± 3%); in the case of stearic acid, it was very low (28 ± 4%). The results of a comparative analysis of the production indicators of the *C. necator* B-10646 culture when bacteria were grown on the saturated FAs of various structures are presented in Table 2.

Table 2. The production indicators in the culture of the wild-type *C. necator* B-10646 strain on the saturated FAs of various structures.

Fatty Acid (FA)	X, g/L	PHAs, g/L	X_{res}, g/L	PHAs, %	Y_X, g/g	Y_{PHA}, g/g	Biomass Productivity		PHA Productivity, P_{PHA}, g/L·h	Degree of Use of FAs, %
							P_X, g/L·h	P_{Xres}, g/L·h		
Lauric C12:0	6.7 [a]	5.0 [a]	1.7 [a]	75 [a]	0.69 [a]	0.52 [a]	0.093 [a]	0.024 [a]	0.069 [a]	64.7 [a]
Myristic C14:0	7.5 [a]	5.6 [a]	1.9 [a]	74 [a]	0.83 [b]	0.62 [a]	0.104 [a]	0.026 [a]	0.078 [a]	60.0 [a]
Palmitic C16:0	3.9 [b]	1.8 [b]	2.1 [a]	47 [b]	0.67 [a]	0.31 [b]	0.054 [b]	0.029 [a]	0.025 [b]	38.7 [b]
Stearic C18:0	2.5 [c]	0.7 [c]	1.8 [a]	28 [c]	0.66 [a]	0.18 [c]	0.035 [c]	0.025 [a]	0.010 [c]	25.3 [c]

The letters indicate the significance of the differences when comparing groups according to the Mann–Whitney test at the level of $p \leq 0.05$, where identical letters indicate no significant differences.

The highest values of all the studied indicators were obtained when myristic acid was used as the C-substrate. Thus, the productivity of the *C. necator* B-10646 culture calculated for the total growth period (72 h) in terms of the overall bacterial biomass (X, g/L) and polymer (PHAs) were 0.104 and 0.078 g/L·h, respectively; moreover, the economic coefficients (Y_X and Y_{PHA}), were 0.83 and 0.62 g/g, respectively. The productivity indicators obtained for the bacterial growth on lauric acid were close to the results obtained from the growth on myristic acid, but the economic coefficients were significantly inferior. The biomass and polymer productivity decreased relative to myristic acid by two and three times, respectively, and the economic ratios also dropped significantly when the bacteria were grown on palmitic acid. The least productive was the process on stearic acid. When stearic acid was used as the only C-substrate, all of the calculated indicators decreased even more significantly than those obtained on palmitic acid (Table 2).

An analysis of the literature data showed that the yield of bacterial biomass and polymers when the bacteria of various species are grown on saturated fatty acids with a C-chain length from C12 to C18 also vary widely (Table 1). The wild-type and genetically modified *Pseudomonas* strains have been the most studied in terms of being grown on saturated FAs. This is in addition to the P(3HB) homopolymer PHA copolymers also have been synthesized containing medium-chain monomers of various structures. However, relatively low yields of bacterial biomass (from 1.0 to 5.5 g/L) and PHA yields of no more than 50% [71–82] have been obtained in most studies. This is also typical for other taxa, such as *Delftia* [49,53], *Burkholderia* sp. [50], *Klebsiella pneumonia* [45], *Bacillus cereus* [47] and *Rhodococcus pyridinivorans* [51]. A number of studies have described *Alcaligenes* (later *Cupriavidus*) strains as capable of accumulating up to 3.2 g/L of biomass that contain up to 55% of PHAs when they were grown on saturated the fatty acids of C12-C18 [44].

The yield of the biomass and PHAs obtained in the culture of *C. necator* B-10646 when using myristic acid exceeded the data available in the literature for most of the bacteria of various taxa [53,77,80], as well as for the closely related strains of *Alcaligenes* sp. AK 201 [44] and *C. necator* DSM 454 [48]. The quantitative indicators obtained in this paper when *C. necator* B-10646 was cultivated on lauric acid also exceed the data on the accumulation of bacterial biomass and PHAs in cultures of *A. hydrophila* and *A. salmocida* [45,46]. The published data on bacterial growth and PHA synthesis on palmitic and stearic fatty acids are limited. Furthermore, in general and in terms of production, the indicators regarding the most of all strains [48,50,78] have also been found to be inferior to the results obtained in this paper.

An important indicator of the efficiency of biotechnological processes that affects the economy of process is the completeness of the substrate being used, as well as, above all, carbon, the costs of which can reach 45–50% during the synthesis of PHAs [83]. It has been shown that the completeness of the studied saturated FAs when used by the *C. necator* B-10646, similar to the achieved production indicators (Table 2), varies significantly. At the same starting concentration of FAs in the nutrient medium (15.0 g/L), the amount of these substrates used by the bacterial culture during growth was 9.7, 9.0, 5.8 and 3.8 g/L, for lauric, myristic, palmitic and stearic FAs, respectively. The lauric and myristic fatty acids were completely utilized the most (by 64.7 and 60.0%, respectively). This indicator for palmitic and stearic fatty acids was almost two times lower and amounted to 38.7 and 25.3%, respectively (Table 2).

Thus, all of the saturated fatty acids studied were actively consumed and metabolized by *C. necator*. This is in contrast to the process on complex fatty substrates, where the fatty acids used were poor in terms of bacteria growth and accumulation in the culture. It is known that microbial metabolism depends on the activity of lipolytic enzymes, under the influence of which triacylglycerols (TAGs) are hydrolyzed to diacylglycerols (DAGs) and monoacylglycerols (MAGs) with the formation of the mixtures of glycerol and free fatty acids at the interface between lipids and water [84,85]. It is known that some FAs are freely transported into cells in a non-dissociated form as a result of nonionic diffusion [86]. In the cytoplasm of cells, fatty acids are metabolized via the β-oxidation pathway to form (R)-3-hydroxyacyl-CoA, which is a building block for PHA monomers [87]. The process of PHA synthesis in bacterial cells through the β-oxidation cycle of FAs, which are formed from TAG in complex fatty substrates, is shown in Figure 2.

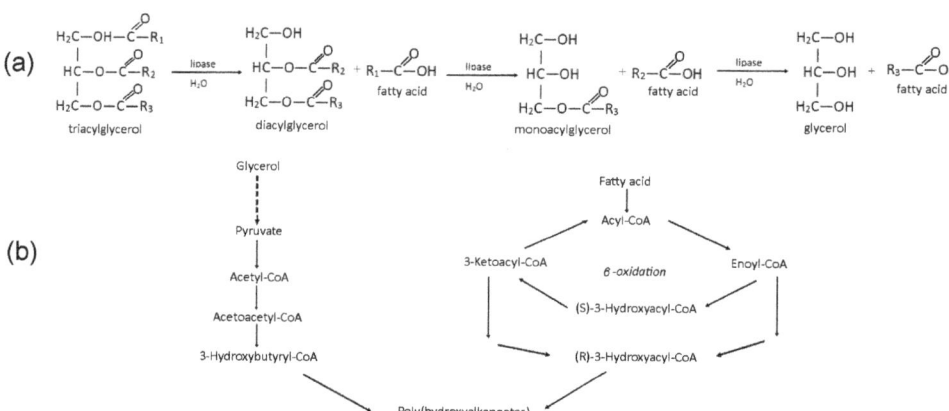

Figure 2. The scheme of the PHA synthesis from fatty acids: hydrolysis of the triacylglycerols to glycerol and free fatty acids (**a**) and pathways for the synthesis of PHAs from glycerol and fatty acids (**b**).

The uneven consumption of FAs by bacteria is described in papers where the growth and synthesis of PHAs on vegetable and animal fats were studied [29,30,87]. The authors of these papers, via analyzing the process of lipid hydrolysis under the influence of microbial extracellular lipases, took into account the dynamics of the conversion of triacylglycerols into DAGs, MAGs and FFAs. It was then concluded that FAs with different carbon chain lengths, which are released from complex lipids, penetrate into cells and are metabolized differently. It is possible that the lipase activity of the studied *C. necator* B-10646 strain (determined at a level of 6.6–11.5 U/mL) is not active enough and does not ensure a complete hydrolysis of the lipid component of complex substrates, nor does it ensure the release of all acids into the medium, thus making them accessible to cells. This prompted us to investigate how the studied saturated FAs, when they are added simultaneously to the nutrient medium, are utilized in the mixtures.

3.2. Growth and PHA Synthesis by Bacteria C. necator B-10646 on Mixtures of Fatty Acids

The PHA synthesis on the mixtures of saturated FAs was studied for the first time. For the purposes of research, several variants of the mixtures of saturated fatty acids, differing in the set of fatty acids and their ratio (Table 3), were prepared.

Table 3. Composition of the FA mixtures as a carbon substrate for the growth of the wild-type *C. necator* B-10646 strain.

Mixture Number	FA Composition in the Mixture	FA Concentration, g/L	Total of FA, g/L	FA Ratio
1	Lauric C12:0 Myristic C14:0 Palmitic C16:0 Stearic C18:0	3.75 3.75 3.75 3.75	15.0	C12:0/C14:0/C16:0/ C18:0 = 1.0/1.0/1.0/1.0
2	Myristic C14:0 Palmitic C16:0 Stearic C18:0	5.0 5.0 5.0	15.0	C14:0/C16:0/C18:0 = 1.0/1.0/1.0
3	Myristic C14:0 Palmitic C16:0 Stearic C18:0	1.5 11.7 1.8	15.0	C14:0/C16:0/C18:0 = 1.0/7.8/1.2
4	Lauric C12:0 Myristic C14:0 Palmitic C16:0 Stearic C18:0 Oleic C18:1ω9	4.0 4.0 4.0 4.0 4.0	20.0	C12:0/C14:0/C16:0/C18:0/ C18:1ω9 = 1.0/1.0/1.0/1.0/1.0
5	Myristic C14:0 Palmitic C16:0 Stearic C18:0 Oleic C18:1ω9	1.5 9.0 1.5 8.0	20.0	C14:0/C16:0/C18:0/ C18:1ω9 = 1.0/6.0/1.0/5.3

The first mixture contained four saturated FAs that are characteristic for vegetable oils [88]: lauric, myristic, palmitic and stearic. The concentration of all acids was the same, i.e., 3.75 g/L, and the ratio used was 1.0/1.0/1.0/1.0 (at a total of 15.0 g/L). The second mixture contained three saturated acids, myristic, palmitic and stearic, which were found in the fat obtained from fish processing waste (in which lauric acid was not detected [35]). The total concentration of the three acids and their ratio in the medium were similar to the first mixture. The third mixture contained three FAs similar to Mixture 2, but the concentration of acids in the mixture was close to the content in the so-called sprat oil [42] and was 1.5, 11.7 and 1.8 g/L (total 15.0 g/L) for the myristic, palmitic and stearic acids, respectively, with an FA ratio of C14:0/C16:0/C18:0 = 1.0/7.8/1.2.

The results of the growth and synthesis of the PHAs in the bacteria *C. necator* B-10646 strain using mixtures of saturated FAs as a carbon substrate are presented in Figure 3 and Table 4.

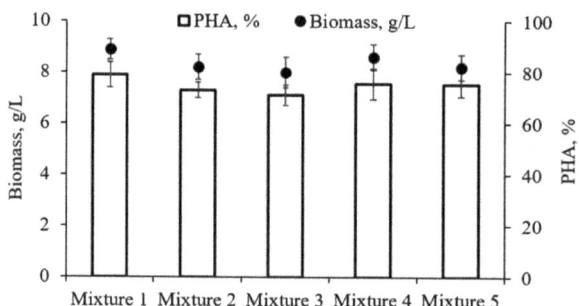

Figure 3. The bacterial biomass yield and intracellular PHA content in the culture of the wild-type *C. necator* B-10646 strain on the mixtures of fatty acids. The composition of the FA mixtures is detailed in Table 3.

Table 4. The production indicators of the wild-type *C. necator* B-10646 strain on mixtures of FAs of various content.

FA Mixture Number	X, g/L	PHAs, g/L	PHAs, %	X_{res}, g/L	Y_X, g/g	Y_{PHA}, g/g	Biomass Productivity P_X, g/L·h	P_{Xres}, g/L·h	PHA Productivity, P_{PHA}, g/L·h	Degree of the Use of FAs, %
						Saturated FAs				
1	8.9 [a]	7.0 [a]	79 [a]	1.9 [a]	0.63 [ab]	0.49 [ac]	0.124 [a]	0.026 [a]	0.097 [a]	94.9 [ab]
2	8.2 [ab]	6.0 [b]	73 [a]	2.2 [a]	0.59 [a]	0.43 [b]	0.114 [ab]	0.031 [a]	0.083 [b]	92.7 [a]
3	8.0 [b]	5.7 [b]	71 [a]	2.4 [a]	0.62 [ab]	0.44 [abc]	0.111 [b]	0.033 [a]	0.079 [b]	86.1 [bcd]
						Saturated FAs + Oleic acid				
4	8.6 [ab]	6.5 [ab]	76 [a]	2.1 [a]	0.69 [b]	0.52 [ac]	0.119 [ab]	0.029 [a]	0.090 [ab]	83.1 [c]
5	8.2 [ab]	6.1 [ab]	75 [a]	2.1 [a]	0.61 [ab]	0.45 [abc]	0.114 [ab]	0.029 [a]	0.085 [ab]	90.4 [ad]

The composition of the FA mixtures corresponds to Table 3. The letters indicate the significance of differences when comparing groups according to the Mann–Whitney test at the level of $p \leq 0.05$, where identical letters indicate no significant differences.

The bacterial biomass yields obtained by cultivating *C. necator* B-10646 on mixtures of fatty acids in most of the cases did not statistically differ significantly (X = 7.9–8.9 g/L when X_{res} was 1.9–2.4 g/L). Moreover, they slightly exceeded the results that were obtained when using individual fatty acids. Mixture 1, which contained the four studied acids, was the most effective. It provided a biomass and polymer productivity of 0.124 and 0.097 g/L*h, respectively. The use of Mixtures 2 and 3 ensured productivity in terms of the overall biomass and polymer at 0.110–0.114 and 0.076–0.083 g/L·h, respectively. The completeness of the utilization of the fatty acids in Mixtures 1 and 2 was close (92.7–94.9%); however, for Mixture 3, it was significantly lower (86.1%) than for Mixture 2. Meanwhile, it was noted that, in all cases, the completeness of the utilization of the fatty acids exceeded the results obtained with individual fatty acids (25.0–64.7%) (Table 4).

It is not possible to compare the results with published ones since no studies on the synthesis of PHAs on mixtures of saturated FAs can be found in the available literature.

The utilization of individual fatty acids from the studied mixtures is illustrated in Figure 4. A comparison of the initial proportions of the FAs in the nutrient medium with the residual ones in the culture showed no changes. This is an indicator of the uniform utilization of fatty acids by the bacteria during growth on the studied mixtures of FAs, which is in contrast to the uneven consumption of these FAs from complex fat-containing substrates.

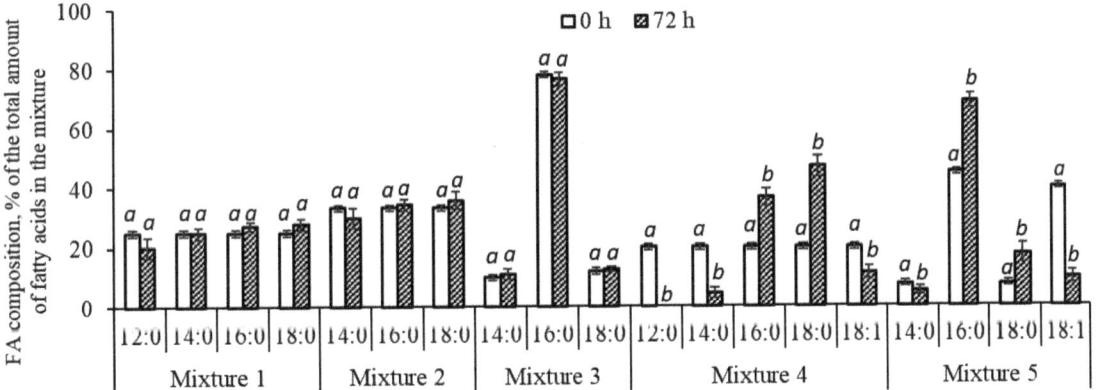

Figure 4. The ratio of the fatty acids in the initial nutrient medium and in the culture of the wild-type bacteria *C. necator* B-10646 strain at the end of the growth (72 h). The letters indicate the significance of differences when comparing the groups according to a Mann–Whitney test at the level of $p \leq 0.05$, where identical letters indicate no significant differences.

The revealed effects of the uneven consumption of FAs by bacteria when grown on substrates of complex composition are, apparently, because natural fat-containing sources contain a wide range of fatty acids of various structures, including saturated and unsaturated acids, mono- and polyenoic, branched, etc. Seemingly, the process of FA transport and the affinity for bacteria as a growth substrate are different. It has been shown that bacteria primarily utilize polyenoic fatty acids, while monoenoic and saturated fatty acids are poorly or not metabolized at all, so their content in the residual substrate increases [41,42,89]. It is also possible that there is competition between FAs for the transporter in the processes of the active transport of fatty acids into cells.

Therefore, it was considered appropriate to study the growth of bacteria and the dynamics of fatty acid consumption if the composition of the mixtures became more complex. FA mixtures were prepared and studied (Table 3), to which oleic acid (unsaturated fatty acid) was added. Oleic acid, as a rule, dominates in the composition of natural fats of various origins and can account for 20–25% or more in relation to the number of FAs. Mixture 4 was based on Mixture 1, i.e., oleic acid was added to four unsaturated FAs, the ratio of all FAs was the same and their total concentration was 20 g/L. Mixture 5 contained a double concentration of oleic acid, but the shortest acid, lauric acid, was excluded from it, that is, palmitic and oleic fatty acids were found to dominate in this mixture. This corresponds to the FA content in sprat oil [42].

The *C. necator* B-10646 cultivation in Mixture 4 (which contained, in addition to saturated FAs, unsaturated oleic acid in equal parts) did not significantly affect the production indicators of the culture (Table 4), which, in terms of biomass and PHA yields, were comparable to the results obtained on FA mixtures without oleic acid. At the same time, it was revealed that the consumption of fatty acids by bacteria from the four- and five-component mixtures with oleic acid changes (Figure 4). The presence of oleic acid in both variants negatively affected the consumption of palmitic and stearic fatty acids by the bacteria, the proportion of which among the residual FAs in the culture increased but did not affect the utilization of lauric and myristic acids. Oleic acid, in contrast to the growth of this strain on complex fat-containing substrates (as we previously showed [42]), was consumed by bacteria. Its concentration in the culture compared to the initial concentration in the nutrient medium significantly decreased from 4.0 to 2.3 g/L in Mixture 4 and, more significantly, from 8.0 to 1.9 g/L in Mixture 5.

The obtained results showing a decrease in the consumption of individual saturated fatty acids by bacteria in the presence of oleic acid, in our opinion, may indicate support for the assumption of competitive relationships between FAs in the processes of their transport into the cell. According to the study of [90], the transport of free fatty acids into the cell, depending on the length of the carbon chain, can be passive or active. Therefore, short-chain (C4–C6) and medium-chain fatty acids (C7–C11) can enter the cell by free diffusion, while longer fatty acids (>C12) require specialized transporters. It has been established that, in the cells of Gram-negative bacteria, the FA transport protein is FadL, which is located on the outer side of the cell membrane [91]. Long-chain fatty acids penetrate the outer membrane through FadL, which then pass through the periplasmic space and enter the inner membrane. It is believed that, after being transported across the membrane, fatty acids enter the periplasmic space. This is where, changing their orientation, they penetrate the inner membrane layer and are activated by the cytosolic acyl-CoA synthase FadD after that enter the β-oxidation cycle [90]. It can be assumed that a similar mechanism of FA transport is also implemented in most Gram-negative bacteria, including representatives of *C. necator*. Due to the fact that the issue of the transport of fatty acids from the periplasmic space through the inner cell membrane has not yet been fully studied, the reason for the uneven utilization of fatty acids may lie precisely in this aspect of FA transport into the cell. However, this issue requires additional and special research.

3.3. The Composition and Physicochemical Properties of PHAs Synthesized by C. necator B-10646 on Individual Fatty Acids and Their Mixtures

PHAs are a family of polymers with a different set and ratio of monomers, so their properties vary significantly. The composition of PHAs that are synthesized depends on many factors and, above all, on the physiological and biochemical specificity of the producer strains and carbon nutrition conditions. The published data on the properties of PHAs synthesized by the bacteria of various taxa on various fatty acids as the sole carbon source are very limited (Table 1). The information in Table 1 shows that the available data on the properties of PHAs synthesized on FAs are extremely limited. At the same time, the need and importance for studying the properties of PHAs that are synthesized on new substrates or new strains is due to the fact that the value of the samples is determined not only by the yield of the polymer, but also by the manufacturability of these polymers, that is, the possibility of processing and obtaining products from them. This is determined by the physicochemical properties of polymers, which primarily involve molecular weight and temperature characteristics. These situations are possible when a particular producer is capable of PHA synthesis; however, properties such as, for example, the degree of polymerizability or thermal behavior are such that the processing and obtaining of products from these polymers using accessible methods is difficult or impossible.

In the presented paper, the properties of PHAs synthesized on saturated fatty acids was studied for the first time. At the same time, the composition and properties of polymers synthesized by *C. necator* B-10646 were studied both on individual saturated FAs and on their mixtures (Table 5).

According to gas chromatography data, all of the synthesized PHA samples represent a poly(3-hydroxybutyrate) homopolymer regardless of whether fatty acids were supplied as a monosubstrate or as a part of mixtures. As an example, Figure 5 shows an ion chromatogram and mass spectrum, wherein the monomeric composition of P(3HB) synthesized by *C. necator* B-10646 using lauric acid is illustrated.

The molecular weight characteristics of the P(3HB) samples synthesized by *C. necator* B-10646 on the individual saturated FAs and mixtures of FAs differed (Figure 6). One can note a tendency for the average molecular weight of the polymer to increase and for the polydispersity to decrease with the lengthening of the carbon chain of the studied saturated FAs as the sole C-substrate.

Table 5. The properties of the P(3HB) samples synthesized by the wild-type *C. necator* B-10646 strain on the individual saturated FAs and mixtures of FAs.

P(3HB) Sample Number	Substrate	M_w, kDa	Đ	C_x, %	T_{melt}, °C	H_{melt}, J/g	T_{degr}, °C	T_g, °C	T_{cryst}, °C
				Saturated FAs					
1	Lauric (12:0)	305.5	3.74	49.7	169.3	72.6	127.1 (19.5%) 284.7	-	65.2 47.5
2	Myristic (14:0)	364.9	3.71	52.8	169.2	77.1	113.0 (19.4%) 284.9	1.8	67.4 45.5
3	Palmitic (16:0)	423.7	3.46	50.7	170.1	74.1	113.0 (21.4%) 284.7	2.0	60.7 48.3
4	Stearic (18:0)	447.1	2.88	47.5	166.0	69.3	282.9	-	53.2
				FA Mixture *					
5	Mixture 1	289.3	2.72	39.2	157.4 168.6	7.9 49.3	92.4 (14.9%) 264.5	-	74.1
6	Mixture 2	333.4	2.74	53.2	171.5	77.7	127.5 (9.2%) 295.7	3.0	62.1
7	Mixture 3	464.5	3.48	56.0	171.0	81.8	115.3 (17%) 267.6	-	66.0
8	Mixture 4	325.0	3.22	57.1	169.8	83.4	111.3 (22.8%) 281.1	-	65.3
9	Mixture 5	315.5	3.32	49.4	150.3 163.1	10.4 61.7	77.1 (3.6%) 138.5 (8.1%) 252.4	-	64.4

* The composition of FA mixtures corresponds to Table 3.

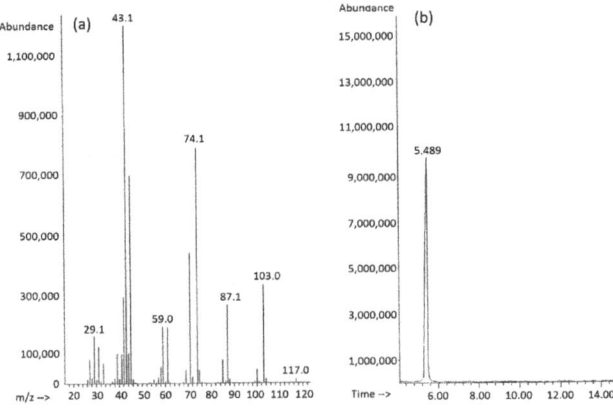

Figure 5. The mass spectrum (a) and ion chromatogram (b) of the P(3HB) sample synthesized by the wild-type *C. necator* B-10646 strain on saturated FAs. The retention time of the 3HB unit is 5.489 min.

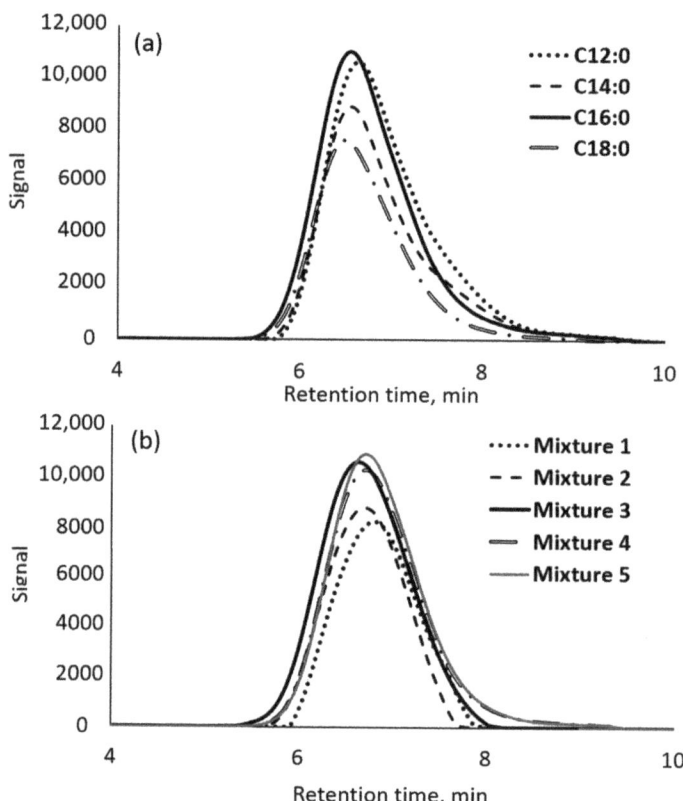

Figure 6. GPC chromatogram of the P(3HB) samples synthesized by the wild-type *C. necator* B-10646 strain on individual saturated FAs (**a**) and on the mixtures of FAs (**b**). The composition of FA mixtures corresponds to Table 3.

The average molecular weight (M_w) of the P(3HB) samples increased almost 1.5 times—from 305.5 to 447.1 kDa—with an increase in the length of the carbon chain in the fatty acids (from 12 to 18 carbon atoms in the lauric and stearic fatty acids, respectively) against the background of a decrease, from 3.74 to 2.88, in the polydispersity (Đ). The molecular weight (M_w) of the sample obtained by *C. necator* B-10646 cultivation on the FA Mixture 1 was somewhat lower than this indicator when individual fatty acids were used, whereby it amounted to 289.3 kDa with a polydispersity value of 2.72. When bacteria were grown on the FA Mixture 3, the P(3HB) samples had a slightly increased molecular weight compared to individual fatty acids (M_w = 464.5 kDa at Đ = 3.48). The polymer samples synthesized on all other mixtures (FA mixtures 2 and 4–5) were comparable in their molecular weight characteristics to those obtained using individual fatty acids.

An analysis of publications showed that the effect of fatty acids on the molecular weight characteristics of PHAs was the most studied issue. The results of a study of the average molecular weight of P(3HB) synthesized by *C. necator* B-10646 using lauric fatty acid were found to be consistent with the results obtained in [44], where the M_w value of the polymer samples synthesized by the *Alcaligenes* sp. AK 201 strain on lauric acid was 304 kDa. When lauric acid was replaced with fatty acids that have a longer C-chain length, an increase in M_w values was found, and these changes were more pronounced [44]. The average molecular weight of the polymer increased to 1442 kDa when the bacteria was grown on myristic and palmitic acids, and it was somewhat less, to 986 kDa, on stearic acid [44].

The temperature properties of the P(3HB) samples synthesized on individual saturated FAs and mixtures of FAs are presented in Table 5 and Figure 7.

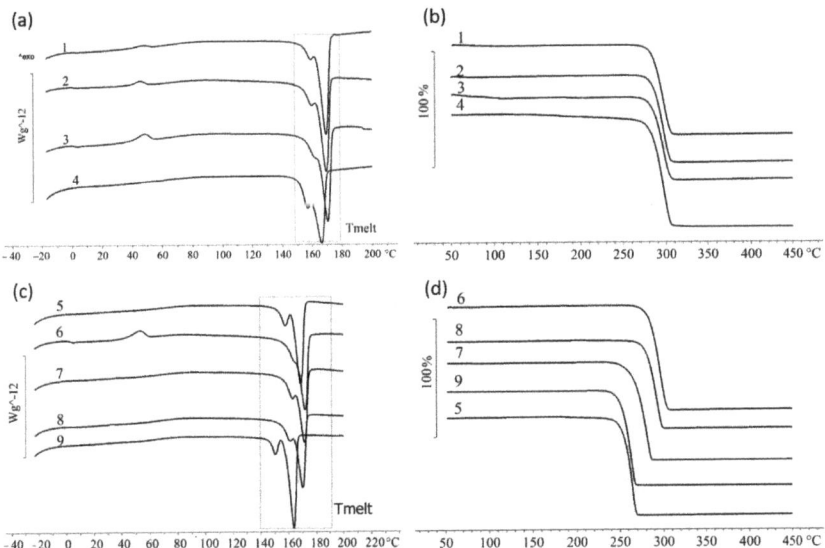

Figure 7. The temperature characteristics of P(3HB) samples synthesized by the wild-type *C. necator* B-10646 strain on individual saturated FA–DSC curves (**a**) and thermal stability (TGA) (**b**). The temperature characteristics of the P(3HB) samples synthesized on FA mixture–DSC curves (**c**) and thermal stability (TGA) (**d**). The numbering of polymer samples corresponds to Table 5.

It was shown that all of the samples of polymers synthesized on the studied saturated fatty acids had a similar melting point (T_{melt}), 166–170 °C, which is typical for P(3HB). The melting peak in the thermograms (Figure 7) was found to be narrow and had a small shoulder on the low-temperature side. The samples synthesized on lauric, myristic and palmitic acids were characterized by the presence of two crystallization peaks: the first peak was when the samples were cooled (60–67 °C), and the second peak was when they were reheated (45–48 °C). The samples of polymers synthesized using the fatty acids studied demonstrated good thermal stability. The gap between the melting point and the thermal degradation temperature (T_{degr}) was 114–118 °C, and the average thermal degradation temperature of the samples was 284 ± 1 °C.

The thermograms of the P(3HB) samples synthesized on the mixtures of fatty acids were characterized by narrow melting peaks. The samples synthesized on Mixtures 1 and 5 had two melting peaks; for the other samples, this peak degenerated into a small shoulder at the main melting peak on the low-temperature side. For these polymer samples (which were obtained when Mixtures 1 and 5 were used), the lowest melting point values were recorded at 157.4 and 168.6 °C and 150.3 and 163.1 °C accordingly. For the samples obtained on Mixtures 2–4, the melting point was 170–171 °C, which is close to the melting point of the polymer samples synthesized on individual fatty acids.

The samples of the polymers synthesized on mixtures of fatty acids demonstrated varying thermal stability, and these were slightly different from the samples obtained on separate fatty acids. The highest temperature of thermal degradation was recorded for those polymers obtained using Mixture 2 (295.7 °C). The lowest temperature of thermal degradation and, consequently, the lowest thermostability was characteristic of the samples obtained using Mixtures 1, 3 and 5. In those cases, the T_{degr} values were 267.6, 264.5 and 252.4 °C, respectively.

In the available literature, the data on the temperature characteristics of P(3HB) that were obtained on individual fatty acids are limited. Most of the data published relate to multicomponent copolymers that are synthesized by representatives of *Pseudomonas*, where the T_{melt} lies in the range of 52–77 °C [46–51,71–77]. The results obtained in this work are close to the temperature characteristics of the P(3HB) synthesized by the *D. tsuruhatensis* Bet002 strain on individual fatty acids (T_{melt} 173.2–177.4 °C; T_{degr} 289.8–391.8 °C) [49].

No data on the degree of the crystallinity of PHAs synthesized on saturated fatty acids with a carbon chain length of 12–18 carbon atoms have been found in the available literature. The degree of the crystallinity of the P(3HB) samples synthesized by the *C. necator* B-10646 strain, except for the sample that was obtained using Mixture 1, was determined from thermograms, and it was found to be close and ranged from 47 to 57%. This value is below the known data, where the C_x of a P(3HB) is usually determined higher (from 60% and above). The lowest degree of crystallinity (39%) was recorded for the polymer sample that was obtained from Mixture 1 (Table 5), which is unusual for a P(3HB) homopolymer.

The novelty and significance of the results obtained is due, firstly, to the new data on the properties of PHAs, which were synthesized on a little-studied carbon substrate, i.e., saturated FAs. The study showed that, in general, the basic properties (i.e., the molecular mass and temperature characteristics) generally correspond to the data, and they are characteristic of the polymer samples synthesized on many other generally used C-substrates. The revealed reduced values of the degree of crystallinity can be regarded as a positive point since a decrease in crystallinity makes a PHA sample more technological. Secondly, the results obtained allow us to consider saturated fatty acids as a C-substrate for the synthesis of PHAs by the highly productive wild-type *C. necator* B-10646 strain, and they also indicates an increase in the completeness of the use of complex fat-containing waste, of which saturated acids, as a rule, are poorly utilized by bacteria.

4. Conclusions

It was shown that saturated FAs with different C-chain lengths (C12, C14, C16 and C18), individually and in mixtures, ensure the growth and synthesis of PHAs by the wild-type *Cupriavidus necator* B-10646 strain. The length of the C-chain of SFAs influences the overall yield of the biomass and PHAs, which are two times higher when bacteria are grown on C12 and C14 FAs compared to longer C16 and C18 acids. On all of the types of SFAs, the bacteria synthesized poly(3-hydroxybutyrate) with typical molecular weight and temperature characteristics but reduced crystallinity. The ability of bacteria to assimilate SFAs opens up the possibility of attracting the synthesis of PHAs to complex fat-containing substrates, including waste.

Author Contributions: PHA synthesis on FA, N.O.Z.; PHA synthesis on FA mixtures, K.Y.S.; DTA-DSC investigation, E.G.K.; molecular mass research, E.I.S.; conceptualization, results analysis and writing, T.G.V. All authors have read and agreed to the published version of the manuscript.

Funding: This research was funded by the Russian Science Foundation, grant number 23-64-10007.

Institutional Review Board Statement: Not applicable.

Data Availability Statement: All the data are available in the paper.

Acknowledgments: The authors would like to express their special thanks to the Krasnoyarsk Regional Center of Research Equipment of Federal Research Center "Krasnoyarsk Science Center SB RAS" for providing equipment to ensure the accomplishment of this project.

Conflicts of Interest: The authors declare no conflicts of interest.

References

1. Geyer, R. Production, use, and fate of synthetic polymers. In *Plastic Waste and Recycling: Environmental Impact, Societal Issues, Prevention, and Solutions*; Letcher, T.M., Ed.; Academic Press: London, UK, 2020; pp. 13–32. [CrossRef]
2. Plastics Europe. Available online: https://plasticseurope.org/knowledge-hub/plastics-the-fast-facts-2023/ (accessed on 18 March 2024).
3. Napper, I.E.; Davies, B.F.R.; Clifford, H.; Elvin, S.; Koldewey, H.J.; Mayewski, P.A.; Miner, K.R.; Potocki, M.; Elmore, A.C.; Gajurel, A.P.; et al. Reaching new heights in plastic pollution—Preliminary findings of microplastics on mount everest. *One Earth* **2020**, *3*, 621–630. [CrossRef]
4. Jamieson, A.J.; Onda, D.F.L. Lebensspuren and müllspuren: Drifting plastic bags alter microtopography of seafloor at full ocean depth (10,000 m, Philippine Trench). *Cont. Shelf Res.* **2022**, *250*, 104867. [CrossRef]
5. González-Pleiter, M.; Edo, C.; Aguilera, Á.; Viúdez-Moreiras, D.; Pulido-Reyes, G.; González-Toril, E.; Osuna, S.; de Diego-Castilla, G.; Leganés, F.; Fernández-Piñas, F.; et al. Occurrence and transport of microplastics sampled within and above the planetary boundary layer. *Sci. Total Environ.* **2021**, *761*, 143213. [CrossRef]
6. Awasthi, S.K.; Kumar, M.; Kumar, V.; Sarsaiya, S.; Anerao, P.; Ghosh, P.; Singh, L.; Liu, H.; Zhang, Z.; Awasthi, M.K. A comprehensive review on recent advancements in biodegradation and sustainable management of biopolymers. *Environ. Pollut.* **2022**, *307*, 119600. [CrossRef] [PubMed]
7. Tan, J.; Jia, S.; Ramakrishna, S. Accelerating plastic circularity: A critical assessment of the pathways and processes to circular plastics. *Processes* **2023**, *11*, 1457. [CrossRef]
8. Chen, G.Q.; Chen, X.Y.; Wu, F.Q.; Chen, J.C. Polyhydroxyalkanoates (PHA) toward cost competitiveness and functionality. *Adv. Ind. Eng. Polym. Res.* **2020**, *3*, 1–7. [CrossRef]
9. Mitra, R.; Xu, T.; Chen, G.Q.; Xiang, H.; Han, J. An updated overview on the regulatory circuits of polyhydroxyalkanoates synthesis. *Microb. Biotechnol.* **2022**, *15*, 1446–1470. [CrossRef] [PubMed]
10. Tan, D.; Wang, Y.; Tong, Y.; Chen, G.Q. Grand challenges for industrializing polyhydroxyalkanoates (PHAs). *Trends Biotechnol.* **2021**, *39*, 953–963. [CrossRef]
11. Koller, M.; Mukherjee, A. A new wave of industrialization of PHA biopolyesters. *Bioengineering* **2022**, *9*, 74. [CrossRef]
12. Koller, M.; Mukherjee, A. Polyhydroxyalkanoates—Linking properties, applications, and end-of-life options. *Chem. Biochem. Eng. Q.* **2020**, *34*, 115–129. [CrossRef]
13. Markets and Markets. Available online: https://www.marketsandmarkets.com/Market-Reports/pha-market-395.html (accessed on 18 March 2024).
14. Dalton, B.; Bhagabati, P.; De Micco, J.; Padamati, R.B.; O'Connor, K. A review on biological synthesis of the biodegradable polymers polyhydroxyalkanoates and the development of multiple applications. *Catalysts* **2022**, *12*, 319. [CrossRef]
15. Palmeiro-Sánchez, T.; O'Flaherty, V.; Lens, P.N.L. Polyhydroxyalkanoate bio-production and its rise as biomaterial of the future. *J. Biotechnol.* **2022**, *348*, 10–25. [CrossRef]
16. Wang, Y.; Huang, J.; Liang, X.; Wei, M.; Liang, F.; Feng, D.; Xu, C.; Xian, M.; Zou, H. Production and waste treatment of polyesters: Application of bioresources and biotechniques. *Crit. Rev. Biotechnol.* **2023**, *43*, 503–520. [CrossRef]
17. Naitam, M.G.; Tomar, G.S.; Kaushik, R.; Singh, S.; Nain, L. Agro-industrial waste as potential renewable feedstock for biopolymer polyhydroxyalkanoates (PHA) production. *Enzym. Eng.* **2022**, *11*, 190–206. Available online: https://www.longdom.org/open-access/agroindustrial-waste-as-potential-renewable-feedstock-for-biopolymer-polyhydroxyalkanoates-pha-production-93784.html#ai (accessed on 2 May 2024).
18. Kannah, R.Y.; Kumar, M.D.; Kavitha, S.; Banu, J.R.; Tyagi, V.K.; Rajaguru, P.; Kumar, G. Production and recovery of polyhydroxyalkanoates (PHA) from waste streams—A review. *Bioresour. Technol.* **2022**, *366*, 128203. [CrossRef] [PubMed]
19. Mahato, R.P.; Kumar, S.; Singh, P. Production of polyhydroxyalkanoates from renewable resources: A review on prospects, challenges and applications. *Arch. Microbiol.* **2023**, *205*, 172. [CrossRef]
20. Che, L.; Jin, W.; Zhou, X.; Han, W.; Chen, Y.; Chen, C.; Jiang, G. Current status and future perspectives on the biological production of polyhydroxyalkanoates. *Asia-Pac. J. Chem. Eng.* **2023**, *18*, e2899. [CrossRef]
21. de Mello, A.F.M.; de Souza Vandenberghe, L.P.; Machado, C.M.B.; Brehmer, M.S.; de Oliveira, P.Z.; Binod, P.; Sindhu, R.; Soccol, C.R. Polyhydroxyalkanoates production in biorefineries: A review on current status, challenges and opportunities. *Bioresour. Technol.* **2023**, *393*, 130078. [CrossRef]
22. Adeleye, A.T.; Odoh, C.K.; Enudi, O.C.; Banjoko, O.O.; Osiboye, O.O.; Odediran, E.T.; Louis, H. Sustainable synthesis and applications of polyhydroxyalkanoates (PHAs) from biomass. *Process Biochem.* **2020**, *96*, 174–193. [CrossRef]
23. Parlato, M.C.; Valenti, F.; Porto, S.M. Covering plastic films in greenhouses system: A GIS-based model to improve post use suistainable management. *J. Environ. Manag.* **2020**, *263*, 110389. [CrossRef]
24. Mukherjee, A.; Koller, M. Polyhydroxyalkanoate (PHA) bio-polyesters–circular materials for sustainable development and growth. *Chem. Biochem. Eng. Q.* **2022**, *36*, 273–293. [CrossRef]
25. Choi, S.Y.; Lee, Y.; Yu, H.E.; Cho, I.J.; Kang, M.; Lee, S.Y. Sustainable production and degradation of plastics using microbes. *Nat. Microbiol.* **2023**, *8*, 2253–2276. [CrossRef] [PubMed]
26. Maddikeri, G.L.; Pandit, A.B.; Gogate, P.R. Adsorptive removal of saturated and unsaturated fatty acids using ion-exchange resins. *Ind. Eng. Chem. Res.* **2012**, *51*, 6869–6876. [CrossRef]

27. AMEC. *Management of Wastes from Atlantic Seafood Processing Operations*; AMEC Earth and Environment Limited: Dartmouth, NS, Canada, 2003.
28. Bong, C.P.C.; Alam, M.N.H.Z.; Samsudin, S.A.; Jamaluddin, J.; Adrus, N.; Yusof, A.H.M.; Muis, Z.A.; Hashim, H.; Salleh, M.M.; Abdullah, A.R.; et al. A review on the potential of polyhydroxyalkanoates production from oil-based substrates. *J. Environ. Manag.* **2021**, *298*, 113461. [CrossRef] [PubMed]
29. Riedel, S.L.; Jahns, S.; Koenig, S.; Bock, M.C.; Brigham, C.J.; Bader, J.; Stahl, U. Polyhydroxyalkanoates production with *Ralstonia eutropha* from low quality waste animal fats. *J. Biotechnol.* **2015**, *214*, 119–127. [CrossRef] [PubMed]
30. Saad, V.; Gutschmann, B.; Grimm, T.; Widmer, T.; Neubauer, P.; Riedel, S.L. Low-quality animal by-product streams for the production of PHA-biopolymers: Fats, fat/protein-emulsions and materials with high ash content as low-cost feedstocks. *Biotechnol. Lett.* **2021**, *43*, 579–587. [CrossRef] [PubMed]
31. Gutschmann, B.; Maldonado Simões, M.; Schiewe, T.; Schröter, E.S.; Münzberg, M.; Neubauer, P.; Bockisch, A.; Riedel, S.L. Continuous feeding strategy for polyhydroxyalkanoate production from solid waste animal fat at laboratory-and pilot-scale. *Microb. Biotechnol.* **2023**, *16*, 295–306. [CrossRef]
32. Thuoc, D.V.; My, D.N.; Loan, T.T.; Sudesh, K. Utilization of waste fish oil and glycerol as carbon sources for polyhydroxyalkanoate production by *Salinivibrio* sp. M318. *Int. J. Biol. Macromol.* **2019**, *141*, 885–892. [CrossRef] [PubMed]
33. Thuoc, D.V.; Anh, V.T. Bioconversion of crude fish Oil into poly-3-hydroxybutyrate by *Ralstonia* sp. M91. *Appl. Biochem. Microbiol.* **2021**, *57*, 219–225. [CrossRef]
34. Loan, T.T.; Trang, D.T.Q.; Huy, P.Q.; Ninh, P.X.; Thuoc, D.V. A fermentation process for the production of poly (3-hydroxybutyrate) using waste cooking oil or waste fish oil as inexpensive carbon substrate. *Biotechnol. Rep.* **2022**, *33*, e00700. [CrossRef]
35. Zhila, N.O.; Kiselev, E.G.; Volkov, V.V.; Mezenova, O.Y.; Sapozhnikova, K.Y.; Shishatskaya, E.I.; Volova, T.G. Properties of Degradable Polyhydroxyalkanoates Synthesized from New Waste Fish Oils (WFOs). *Int. J. Mol. Sci.* **2023**, *24*, 14919. [CrossRef] [PubMed]
36. Akiyama, M.; Tsuge, T.; Doi, Y. Environmental life cycle comparison of polyhydroxyalkanoates produced from renewable carbon resources by bacterial fermentation. *Polym. Degrad. Stab.* **2003**, *80*, 183–194. [CrossRef]
37. Park, D.H.; Kim, B.S. Production of poly (3-hydroxybutyrate) and poly (3-hydroxybutyrate-co-4-hydroxybutyrate) by *Ralstonia eutropha* from soybean oil. *New Biotechnol.* **2011**, *28*, 719–724. [CrossRef] [PubMed]
38. Jiang, G.; Hill, D.J.; Kowalczuk, M.; Johnston, B.; Adamus, G.; Irorere, V.; Radecka, I. Carbon sources for polyhydroxyalkanoates and an integrated biorefinery. *Int. J. Mol. Sci.* **2016**, *17*, 1157. [CrossRef] [PubMed]
39. Kumar, V.; Kumar, S.; Singh, D. Microbial polyhydroxyalkanoates from extreme niches: Bioprospection status, opportunities and challenges. *Int. J. Biol. Macromol.* **2020**, *147*, 1255–1267. [CrossRef] [PubMed]
40. Volova, T.; Sapozhnikova, K.; Zhila, N. *Cupriavidus necator* B-10646 growth and polyhydroxyalkanoates production on different plant oils. *Int. J. Biol. Macromol.* **2020**, *164*, 121–130. [CrossRef] [PubMed]
41. Kahar, P.; Tsuge, T.; Taguchi, K.; Doi, Y. High yield production of polyhydroxyalkanoates from soybean oil by *Ralstonia eutropha* and its recombinant strain. *Polym. Degrad. Stab.* **2004**, *83*, 79–86. [CrossRef]
42. Zhila, N.O.; Sapozhnikova, K.Y.; Kiselev, E.G.; Shishatskaya, E.I.; Volova, T.G. Synthesis and properties of polyhydroxyalkanoates on waste fish oil from the production of canned sprats. *Processes* **2023**, *11*, 2113. [CrossRef]
43. Ashby, R.D.; Solaiman, D.K. Poly(hydroxyalkanoate) biosynthesis from crude Alaskan pollock (*Theragra chalcogramma*) oil. *J. Polym. Environ.* **2008**, *16*, 221–229. [CrossRef]
44. Akiyama, M.; Taima, Y.; Doi, Y. Production of poly (3-hydroxyalkanoates) by a bacterium of the genus *Alcaligenes* utilizing long-chain fatty acids. *Appl. Microbiol. Biotechnol.* **1992**, *37*, 698–701. [CrossRef]
45. Chen, B.Y.; Shiau, T.J.; Wei, Y.H.; Chen, W.M. Feasibility study on polyhydroxybutyrate production of dye-decolorizing bacteria using dye and amine-bearing cultures. *J. Taiwan Inst. Chem. Eng.* **2012**, *43*, 241–245. [CrossRef]
46. Chen, B.Y.; Hung, J.Y.; Shiau, T.J.; Wei, Y.H. Exploring two-stage fermentation strategy of polyhydroxyalkanoate production using *Aeromonas hydrophila*. *Biochem. Eng. J.* **2013**, *78*, 80–84. [CrossRef]
47. Valappil, S.P.; Peiris, D.; Langley, G.J.; Herniman, J.M.; Boccaccini, A.R.; Bucke, C.; Roy, I. Polyhydroxyalkanoate (PHA) biosynthesis from structurally unrelated carbon sources by a newly characterized *Bacillus* spp. *J. Biotechnol.* **2007**, *127*, 475–487. [CrossRef] [PubMed]
48. Povolo, S.; Basaglia, M.; Fontana, F.; Morelli, A.; Casella, S. Poly(hydroxyalkanoate) production by *Cupriavidus necator* from fatty waste can be enhanced by phaZ1 inactivation. *Chem. Biochem. Eng. Q.* **2015**, *29*, 67–74. [CrossRef]
49. Gumel, A.M.; Annuar, M.S.M.; Heidelberg, T. Effects of carbon substrates on biodegradable polymer composition and stability produced by *Delftia tsuruhatensis* Bet002 isolated from palm oil mill effluent. *Polym. Degrad. Stab.* **2012**, *97*, 1224–1231. [CrossRef]
50. Chee, J.Y.; Tan, Y.; Samian, M.R.; Sudesh, K. Isolation and characterization of a *Burkholderia* sp. USM (JCM15050) capable of producing polyhydroxyalkanoate (PHA) from triglycerides, fatty acids and glycerols. *J. Polym. Environ.* **2010**, *18*, 584–592. [CrossRef]
51. Khan, N.; Jamil, N. Biosynthesis of poly-3-hydroxybutyrate by *Rhodococcus pyridinivorans* using unrelated carbon sources. *Adv. Life Sci.* **2021**, *8*, 128–132.
52. Chen, G.; Zhang, G.; Park, S.; Lee, S. Industrial scale production of poly (3-hydroxybutyrate-co-3-hydroxyhexanoate). *Appl. Microbiol. Biotechnol.* **2001**, *57*, 50–55. [CrossRef] [PubMed]

53. Romanelli, M.G.; Povolo, S.; Favaro, L.; Fontana, F.; Basaglia, M.; Casella, S. Engineering *Delftia acidovorans* DSM39 to produce polyhydroxyalkanoates from slaughterhouse waste. *Int. J. Biol. Macromol.* **2014**, *71*, 21–27. [CrossRef]
54. Morlino, M.S.; García, R.S.; Savio, F.; Zampieri, G.; Morosinotto, T.; Treu, L.; Campanaro, S. *Cupriavidus necator* as a platform for PHA production: An overview of strains, metabolism, and modeling approaches. *Biotechnol. Adv.* **2023**, *69*, 108264. [CrossRef]
55. Grigull, V.H.; Domingos da Silva, D.; Formolo Garcia, M.C.; Furlan, S.A.; Testa Pezzin, A.P.; Lima dos Santos Schneider, A.; Falcão Aragão, G. Production and characterization of poly (3-hydroxybutyrate) from oleic acid by *Ralstonia eutropha*. *Food Technol. Biotechnol.* **2008**, *46*, 223–228.
56. Schneider, A.L.S.; Silva, D.D.; Garcia, M.C.F.; Grigull, V.H.; Mazur, L.P.; Furlan, S.A.; Pezzin, A.P.T. Biodegradation of poly (3-hydroxybutyrate) produced from *Cupriavidus necator* with different concentrations of oleic acid as nutritional supplement. *J. Polym. Environ.* **2010**, *18*, 401–406. [CrossRef]
57. Volova, T.G.; Kiselev, E.G.; Shishatskaya, E.I.; Zhila, N.O.; Boyandin, A.N.; Syrvacheva, D.A.; Vinogradova, O.N.; Kalacheva, G.S.; Vasiliev, A.D.; Peterson, I.V. Cell growth and accumulation of polyhydroxyalkanoates from CO_2 and H_2 of a hydrogen-oxidizing bacterium, *Cupriavidus eutrophus* B-10646. *Bioresour. Technol.* **2013**, *146*, 215–222. [CrossRef] [PubMed]
58. Volova, T.; Kiselev, E.; Vinogradova, O.; Nikolaeva, E.; Chistyakov, A.; Sukovatiy, A.; Shishatskaya, E. A glucose-utilizing strain, *Cupriavidus eutrophus* B-10646: Growth kinetics, characterization and synthesis of multicomponent PHAs. *PLoS ONE* **2014**, *9*, e87551. [CrossRef] [PubMed]
59. Volova, T.; Kiselev, E.; Nemtsev, I.; Lukyanenko, A.; Sukovatyi, A.; Kuzmin, A.; Shishatskaya, E. Properties of degradable polyhydroxyalkanoates with different monomer compositions. *Int. J. Biol. Macromol.* **2021**, *182*, 98–114. [CrossRef] [PubMed]
60. Volova, T.; Kiselev, E.; Zhila, N.; Shishatskaya, E. Synthesis of PHAs by Hydrogen Bacteria in a Pilot Production Process. *Biomacromolecules* **2019**, *20*, 3261–3270. [CrossRef] [PubMed]
61. Zhila, N.O.; Kalacheva, G.S.; Kiselev, E.G.; Volova, T.G. Synthesis of polyhydroxyalkanoates from oleic acid by *Cupriavidus necator* B-10646. *J. Sib. Fed. Univ. Biol.* **2020**, *13*, 208–217. [CrossRef]
62. Zhila, N.O.; Kalacheva, G.S.; Fokht, V.V.; Bubnova, S.S.; Volova, T.G. Biosynthesis of Poly (3-Hydroxybutyrate-co-3-Hydroxyvalerate) by *Cupriavidus necator* B-10646 from Mixtures of Oleic Acid and 3-Hydroxyvalerate Precursors. *J. Sib. Fed. Univ. Biol.* **2020**, *13*, 331–341. [CrossRef]
63. Volova, T.G.; Shishatskaya, E.I. *Cupriavidus eutrophus* Bacterial Strain VKPM B-10646-A Producer of Polyhydroxyalkanoates and a Method of Their Production (*Cupriavidus eutrophus* Shtamm Bakterii VKPM B-10646-Produtsent Poligidroksialkanoatov i Sposob Ikh Polucheniya). RU2439143C1, 10 January 2012. (In Russian)
64. Schlegel, H.G.; Kaltwasser, H.; Gottschalk, G. A submersion method for culture of hydrogen-oxidizing bacteria: Growth physiological studies. *Arch. Mikrobiol.* **1961**, *38*, 209–222. [CrossRef]
65. Braunegg, G.; Sonnleitner, B.Y.; Lafferty, R.M. A rapid gas chromatographic method for the determination of poly-β-hydroxybutyric acid in microbial biomass. *Eur. J. Appl. Microbiol. Biotechnol.* **1978**, *6*, 29–37. [CrossRef]
66. Brandl, H.; Gross, R.A.; Lenz, R.W.; Fuller, R.C. *Pseudomonas oleovorans* as a source of poly(β-hydroxyalkanoates) for potential applications as biodegradable polyesters. *Appl. Environ. Microbiol.* **1988**, *54*, 1977–1982. [CrossRef] [PubMed]
67. Barham, P.J.; Keller, A.; Otun, E.L.; Holmes, P.A. Crystallization and morphology of a bacterial thermoplastic: Poly-3-hydroxybutyrate. *J. Mater. Sci.* **1984**, *19*, 2781–2794. [CrossRef]
68. Budde, C.F.; Riedel, S.L.; Hübner, F.; Risch, S.; Popović, M.K.; Rha, C.; Sinskey, A.J. Growth and polyhydroxybutyrate production by *Ralstonia eutropha* in emulsified plant oil medium. *Appl. Microbiol. Biotechnol.* **2011**, *89*, 1611–1619. [CrossRef] [PubMed]
69. Kek, Y.K.; Lee, W.H.; Sudesh, K. Efficient bioconversion of palm acid oil and palm kernel acid oil to poly (3-hydroxybutyrate) by *Cupriavidus necator*. *Can. J. Chem.* **2008**, *86*, 533–539. [CrossRef]
70. Obruca, S.; Marova, I.; Snajdar, O.; Mravcova, L.; Svoboda, Z. Production of poly (3-hydroxybutyrate-3-hydroxyvalerate) from waste rapeseed oil using propanol as a precursor of 3-hydroxyvalerate. *Biotechnol. Lett.* **2010**, *32*, 1925–1932. [CrossRef] [PubMed]
71. Anis, S.N.S.; Mohd Annuar, M.S.; Simarani, K. Microbial biosynthesis and in vivo depolymerization of intracellular medium-chain-length poly-3-hydroxyalkanoates as potential route to platform chemicals. *Biotechnol. Appl. Biochem.* **2018**, *65*, 784–796. [CrossRef] [PubMed]
72. Chan, P.L.; Yu, V.; Wai, L.; Yu, H.F. Production of medium-chain-length polyhydroxyalkanoates by *Pseudomonas aeruginosa* with fatty acids and alternative carbon sources. *Appl. Biochem. Biotechnol.* **2006**, *132*, 933–941. [CrossRef]
73. Chung, A.L.; Jin, H.L.; Huang, L.J.; Ye, H.M.; Chen, J.C.; Wu, Q.; Chen, G.Q. Biosynthesis and characterization of poly (3-hydroxydodecanoate) by β-oxidation inhibited mutant of *Pseudomonas entomophila* L48. *Biomacromolecules* **2011**, *12*, 3559–3566. [CrossRef]
74. Ouyang, S.P.; Liu, Q.; Fang, L.; Chen, G.Q. Construction of PHA-operon-defined knockout mutants of *Pseudomonas putida* KT2442 and their applications in poly(hydroxyalkanoate) production. *Macromol. Biosci.* **2007**, *7*, 227–233. [CrossRef]
75. Ouyang, S.P.; Luo, R.C.; Chen, S.S.; Liu, Q.; Chung, A.; Wu, Q.; Chen, G.Q. Production of polyhydroxyalkanoates with high 3-hydroxydodecanoate monomer content by *fadB* and *fadA* knockout mutant of *Pseudomonas putida* KT2442. *Biomacromolecules* **2007**, *8*, 2504–2511. [CrossRef]
76. Liu, Q.; Luo, G.; Zhou, X.R.; Chen, G.Q. Biosynthesis of poly (3-hydroxydecanoate) and 3-hydroxydodecanoate dominating polyhydroxyalkanoates by β-oxidation pathway inhibited *Pseudomonas putida*. *Metabol. Eng.* **2011**, *13*, 11–17. [CrossRef]

77. Tan, I.K.P.; Kumar, K.S.; Theanmalar, M.; Gan, S.N.; Gordon Iii, B. Saponified palm kernel oil and its major free fatty acids as carbon substrates for the production of polyhydroxyalkanoates in *Pseudomonas putida* PGA1. *Appl. Microbiol. Biotechnol.* **1997**, *47*, 207–211. [CrossRef]
78. Impallomeni, G.; Guglielmino, S.P.; Carnazza, S.; Ferreri, A.; Ballistreri, A. Tween 20 and its major free fatty acids as carbon substrates for the production of polyhydroxyalkanoates in *Pseudomonas aeruginosa* ATCC 27853. *J. Polym. Environ.* **2000**, *8*, 97–102. [CrossRef]
79. Gumel, A.M.; Annuar, M.S.M.; Heidelberg, T. Growth kinetics, effect of carbon substrate in biosynthesis of mcl-PHA by *Pseudomonas putida* Bet001. *Braz. J. Microbiol.* **2014**, *45*, 427–438. [CrossRef]
80. Yao, J.; Zhang, G.; Wu, Q.; Chen, G.Q.; Zhang, R. Production of polyhydroxyalkanoates by *Pseudomonas nitroreducens*. *Antonie Leeuwenhoek* **1999**, *75*, 345–349. [CrossRef] [PubMed]
81. Tian, W.; Hong, K.; Chen, G.Q.; Wu, Q.; Zhang, R.Q.; Huang, W. Production of polyesters consisting of medium chain length 3-hydroxyalkanoic acids by *Pseudomonas mendocina* 0806 from various carbon sources. *Antonie Leeuwenhoek* **2000**, *77*, 31–36. [CrossRef] [PubMed]
82. Lee, S.Y.; Wong, H.H.; Choi, J.I.; Lee, S.H.; Lee, S.C.; Han, C.S. Production of medium-chain-length polyhydroxyalkanoates by high-cell-density cultivation of *Pseudomonas putida* under phosphorus limitation. *Biotechnol. Bioeng.* **2000**, *68*, 466–470. [CrossRef]
83. Lee, S.Y. Plastic bacteria? Progress and prospects for polyhydroxyalkanoate production in bacteria. *Trends Biotechnol.* **1996**, *14*, 431–438. [CrossRef]
84. Lu, J.; Brigham, C.J.; Rha, C.; Sinskey, A.J. Characterization of an extracellular lipase and its chaperone from *Ralstonia eutropha* H16. *Appl. Microbiol. Biotechnol.* **2013**, *97*, 2443–2454. [CrossRef]
85. Brigham, C.J.; Budde, C.F.; Holder, J.W.; Zeng, Q.; Mahan, A.E.; Rha, C.; Sinskey, A.J. Elucidation of β-oxidation pathways in *Ralstonia eutropha* H16 by examination of global gene expression. *J. Bacteriol.* **2010**, *192*, 5454–5464. [CrossRef]
86. Salmond, C.V.; Kroll, R.G.; Booth, I.R. The effect of food preservatives on pH homeostasis in *Escherichia coli*. *Microbiology* **1984**, *130*, 2845–2850. [CrossRef] [PubMed]
87. Riedel, S.L.; Lu, J.; Stahl, U.; Brigham, C.J. Lipid and fatty acid metabolism in *Ralstonia eutropha*: Relevance for the biotechnological production of value-added products. *Appl. Microbiol. Biotechnol.* **2014**, *98*, 1469–1483. [CrossRef] [PubMed]
88. Kostik, V.; Memeti, S.; Bauer, B. Fatty acid composition of edible oils and fats. *J. Hyg. Eng. Des.* **2013**, *4*, 112–116.
89. Riedel, S.L.; Bader, J.; Brigham, C.J.; Budde, C.F.; Yusof, Z.A.M.; Rha, C.; Sinskey, A.J. Production of poly (3-hydroxybutyrate-co-3-hydroxyhexanoate) by *Ralstonia eutropha* in high cell density palm oil fermentations. *Biotechnol. Bioeng.* **2012**, *109*, 74–83. [CrossRef] [PubMed]
90. Pavoncello, V.; Barras, F.; Bouveret, E. Degradation of exogenous fatty acids in *Escherichia coli*. *Biomolecules* **2022**, *12*, 1019. [CrossRef]
91. Black, P.N.; DiRusso, C.C. Transmembrane movement of exogenous long-chain fatty acids: Proteins, enzymes, and vectorial esterification. *Microbiol. Mol. Biol. Rev.* **2023**, *67*, 454–472. [CrossRef]

Disclaimer/Publisher's Note: The statements, opinions and data contained in all publications are solely those of the individual author(s) and contributor(s) and not of MDPI and/or the editor(s). MDPI and/or the editor(s) disclaim responsibility for any injury to people or property resulting from any ideas, methods, instructions or products referred to in the content.

Article

Upcycling of Poly(Lactic Acid) by Reactive Extrusion with Recycled Polycarbonate: Morphological and Mechanical Properties of Blends

Vito Gigante [1,2,†], Laura Aliotta [1,2,†], Maria-Beatrice Coltelli [1,2,*] and Andrea Lazzeri [1,2,*]

1 Department of Civil and Industrial Engineering, University of Pisa, 56122 Pisa, Italy
2 National Interuniversity Consortium of Materials Science and Technology (INSTM), 50121 Florence, Italy
* Correspondence: maria.beatrice.coltelli@unipi.it (M.-B.C.); andrea.lazzeri@unipi.it (A.L.)
† These authors contributed equally to this work.

Abstract: Poly(lactic acid) (PLA) is one of the most promising renewable polymers to be employed to foster ecological and renewable materials in many fields of application. To develop high-performance products, however, the thermal resistance and the impact properties should be improved. At the same time, it is also necessary to consider the end of life through the exploration of property assessment, following reprocessing. In this context the aim of the paper is to develop PLA/PC blends, obtained from recycled materials, in particular scraps from secondary processing, to close the recycling loop. Indeed, the blending of PLA with polycarbonate (PC) was demonstrated to be a successful strategy to improve thermomechanical properties that happens after several work cycles. The correlation between the compositions and properties was then investigated by considering the morphology of the blends; in addition, the reactive extrusions resulting in the formation of a PLA-PC co-polymer were investigated. The materials obtained are then examined by means of a dynamic-mechanical analysis (DMTA) to study the relaxations and transitions.

Keywords: reactive extrusion; mechanical behavior; recycled polymers; blending

Citation: Gigante, V.; Aliotta, L.; Coltelli, M.-B.; Lazzeri, A. Upcycling of Poly(Lactic Acid) by Reactive Extrusion with Recycled Polycarbonate: Morphological and Mechanical Properties of Blends. *Polymers* 2022, 14, 5058. https://doi.org/10.3390/polym14235058

Academic Editor: Arash Moeini

Received: 26 October 2022
Accepted: 17 November 2022
Published: 22 November 2022

Publisher's Note: MDPI stays neutral with regard to jurisdictional claims in published maps and institutional affiliations.

Copyright: © 2022 by the authors. Licensee MDPI, Basel, Switzerland. This article is an open access article distributed under the terms and conditions of the Creative Commons Attribution (CC BY) license (https://creativecommons.org/licenses/by/4.0/).

1. Introduction

Recycling plastics is one of the best opportunities available to reduce pollution, save raw materials, store carbon and protect ourselves from the negative effects of waste dispersion in nature and in the sea [1]. This assumption is also confirmed by the European Community directives that, starting from 2015, adopted an action plan aimed at fostering Europe's transition into a circular economy, in which waste is not simply disposed of but transformed into a valuable secondary raw material for further production [2]. There are a lot of reasons to incentivize and encourage plastic recycling, including limiting the use of landfills, optimising resources, limiting CO_2 emissions related to plastic production processes into the atmosphere and nurturing virtuous supply chains that can create sustainable employment [3].

In this framework, it is of great interest, both scientifically and in terms of industrial upscaling, to succeed in designing 'circular-by-design' materials, i.e., materials that can have such characteristics as to be recyclable after use, especially for high value-added applications. In addition, a topical challenge is to introduce in the recycling lines, renewable polymers, making them attractive for even high-performance applications through modifications and blending with other polymers. In this respect, the feasibility of a large window of production processes has guaranteed that poly(lactic acid) (PLA) is attractive for this purpose [4]. However, PLA is an extremely brittle material, it has a low toughness, it shows a low thermal resistance and it is relatively hydrophobic (with a static contact angle with water of 80°) [5]. To compensate for these defects (and above all to increase the impact properties) the incorporation of additives, such as inorganic particles or blending with tough polymers can be found in the literature [6–8]. Indeed, blending PLA with engineering plastics, such as polycarbonate (PC), because of its high Tg, high thermal stability, high tensile strength, and elongation at break, is a successful strategy to improve PLA properties, as confirmed by several studies [9–13]. The processing of PLA/PC blends can occur at temperatures lower than those typical of pure PC processing and this represents an opportunity to recycle post-consumer PC with a reduced energy consumption. Nevertheless, it has been demonstrated that PLA/PC blends are immiscible, so the adhesion between the two polymers is weak due to the high surface tension; over time, different methods of compatibility have been investigated: through the use of poly(butylene succinate-co-lactate) (PBSLA) [14] or poly(ethylene-co-vinyl) EVA [15] to form ternary systems. Ikehara et al. [16] used a biodegradable semi-crystalline polycarbonate called PEC (polyester carbonate) to verify that spherulites of the two polymers can interpenetrate, Wang et al. blended PC and PLA with an epoxidizing catalyst, which improved the toughness through a better compatibility [17].

The study of the effect of various catalysts (zinc borate, titanium, tetrabutyltitanate pigments) [18,19] on the interchange reactions of PLA and PC was investigated. The idea was to follow the path of PC/PBT [20] or PC/PET [21] blends where copolyesters were formed by a transesterification reaction and improved the compatibility of the two polymers.

The increase in the interfacial adhesion between PLA and PC by using catalysts that favor interchange reactions (capable of providing a new type of copolymer with high mechanical properties and an excellent thermal resistance), is the approach also used by Phuong et al. [22,23]. but with the step ahead to minimize the reaction times, making this solution exploitable, even in industrial processes where residence times in the extruder are limited.

As described above, the recycling of waste and discarded plastics is desirable for environmental and economic reasons. However, recycling of the PLA/PC blends is still in its early stages. Post-consumer recycling of the PLA/PC blends has been simulated in a few works [24,25], and the results clearly showed that aging corresponding to one year of use leads to the significant degradation of PLA, resulting in a reduced elongation at break. In addition, the content of PLA as a biopolymer in the PLA/PC blends should be as high as possible once performance requirements are met, which is helpful in reducing the environmental impact [26].

For all these reasons, the purpose of the present work is to implement knowledge on the recyclability of these blends, starting with a comparison of the process and mechanical properties of the PLA and PC scraps recycled from thermoforming and virgin polymers. In addition, it was sought to understand whether a catalytic system consisting of Triacetin (TA) and TetraButylAmmonium TetraPhenylBorate (TBATPB) patented by some of the authors of this paper [27] could also act as a compatibilizer for the recycled PLA (R-PLA)/recycled PC (R-PC) blends, by adding the catalyst during the extrusion to analyze the possibility of the improved adhesion between the components. The presence of the PLA-PC copolymers can be detected by means of a thermal-dynamic-mechanical analysis (DMTA) [28]: a new glass transition temperature intermediate between those of pure PLA and PC will demonstrate the formation of a new species.

With the intention of including as much PLA as possible, but at the same time achieving thermomechanical properties comparable with petro-based blends for durable applications, in this work, the recycled PLA/PC co-continuous blends were developed and produced [29,30], i.e., to compare the PLA matrix blends with the PC dispersed phase, and vice versa, while also investigating the differences in the presence and absence of the catalytic system. Through this study, therefore, the aim is to combine the need to increase knowledge of the thermomechanical properties of recycled polymeric materials, including renewable polyesters, and the opportunity to improve their interfacial adhesion.

2. Materials and Methods

2.1. Materials

PLA: Poly(lactic acid) 2003D (Natureworks LLC, Minnetonka, MN, USA) is a thermoplastic resin, derived from renewable resources (corn starch or sugar cane), transparent, with a high PM (around 200,000 g/mol) density of 1.24 g/cm^3, processable by extrusion, injection molding and thermoforming. It has an amount of around 4 percent of D units that introduce imperfections in the helical conformation of the polymer and defects in the crystal arrangement. It is routinely used as part of polymer blends.

PC: Bisphenol A Polycarbonate S3000 (Mitsubishi Chemical Co., Tokyo, Japan) with a density of 1.20 g/cm^3 and a molecular weight of 56,000 g/mol.

R-PLA: In this paper, reprocessed PLA2003D will be defined with the name of R-PLA. The regrinding of the PLA scraps has been carried out by Romei s.r.l (Florence, Italy).

R-PC: With the label R-PC, the post-industrial Bisphenol A Polycarbonate S3000, consisting of scraps of grey color, has been defined. The regrinding of the PC scraps has been accomplished by Romei s.r.l (Florence, Italy).

CATA: Triacetin (TA) and TetraButylAmmonium TetraPhenylBorate (TBATPB), both purchased from Sigma-Aldrich (Merk Life Science S.r.l., Milano, Italy) were used as catalysts for the interchange reactions, following the process described in [22].

Firstly, virgin PLA and virgin PC were compared with their recycled counterparts; thereafter the blends with a co-continuous morphology were produced and characterised, to evaluate whether the catalytic system acts as compatibilizer (Table 1).

Table 1. Blend compositions.

Blend Name	PLA (wt. %)	PC (wt. %)	R-PLA (wt. %)	R-PC (wt. %)	TA (wt. %)	TBATPB (wt. %)
PLA	100	-	-	-	-	-
PC	-	100	-	-	-	-
R-PLA	-	-	100	-	-	-
R-PC	-	-	-	100	-	-
R-PLA60/R-PC40	-	-	60	40	-	-
R-PLA60/R-PC40 + CATA	-	-	56.8	38	5	0.2
R-PLA40/R-PC60	-	-	40	60	-	-
R-PLA40/R-PC60 + CATA	-	-	38	56.8	5	0.2

2.2. Processing

A conic twin-screw micro compounder (ThermoScientific HAAKE MiniLab II, Karlsruhe, Germany) was used to process and extrude the polymeric blends with and without the catalytic system. The melt filament was collected by a heated cylinder piston and fed into a mini-injection molding machine (Thermo Scientific HAAKE Minijet II, Karlsruhe, Germany), to produce specimens for the tensile tests (25 × 5 × 1 mm) and for the impact/fracture properties (80 × 10 × 4 mm). The processing temperature selected for the blends with and without CATA was 235 °C, the mold was held at 60 °C for an injection cycle of 25 s. Regarding PLA and R-PLA, the extrusion temperature was set at 190 °C, whereas for PC and R-PC, it was set at 280 °C.

2.3. Testing Methodologies

Firstly, during the micro compounding process, the torque trend over time, was closely related to the viscosity of the fluid itself, and was evaluated to understand the variations in the melt strength among the various formulations tested.

The quasi-static tensile tests were carried out at room temperature on Haake type III dog-bone tensile bars (size: 25 × 5 × 1.5 mm), at a crosshead speed of 10 mm/min by an Instron 5500R universal testing machine (Canton, MA, USA), equipped with a 1 kN load cell and interfaced with a computer, running MERLIN software (INSTRON version 4.42 S/N–14733H).

The impact tests were performed on V-notched 80 × 10 × 4 mm specimens, using a 15 J Instron CEAST 9050 Charpy pendulum (INSTRON, Canton, MA, USA) following the standard procedure ISO 179.

The dynamic mechanical thermal analysis (DMTA) was performed on a on a Gabo Eplexor® 100N (Gabo Qualimeter GmbH, Ahlden, Germany). The test bars were of a size of 10 × 5 × 1.5 mm and placed on a tensile geometry configuration. The temperature used in the experiment ranged from −100 °C to 200 °C with a heating rate of 2 °C/min and a frequency of 1 Hz. The properties measured under this oscillating loading are the storage modulus (E′) and tan δ. The E′ value represents the stiffness of a viscoelastic material and is proportional to the energy stored during a loading cycle; tan delta is the ratio between the loss and storage modulus.

The morphology of the composites was studied by scanning electron microscopy (SEM) using JSM-5600LV (JEOL, Tokyo, Japan) and by analyzing the fractured surfaces of the samples obtained by breaking them in liquid nitrogen. Prior to the SEM analysis, all of the surfaces were sputtered with gold.

3. Results

Firstly, a preliminary study was performed, aimed at understanding whether PLA and PC recycled (R-PLA and R-PC) from industrial scraps kept rheological, processed and mechanical properties similar to virgin polymers. For this purpose, the measurements of the torque during the mixing time, the quasi-static tensile tests and the impact tests were carried out and discussed.

3.1. Comparison PLA/R-PLA and PC/R-PC

3.1.1. Torque Analysis

It is accurate to assert that the viscous torque is a measure of the resistance that a fluid offers to the rotational motion of the conic twin screws and it is a function of the viscosity of the fluid itself [31]. By completely filling the micro compounder chamber (6 g), it can be observed that the higher the torque value, the higher the viscosity of the polymer. In Figure 1, the torque/time curves were recorded at 190 °C for PLA and R-PLA, over the extrusion time; the trend observed showed that for PLA, at time zero, the torque value turned out to be slightly higher than that of R-PLA, thus demonstrating the liability of the shorter chains after cleavages induced by the compounding, thermoforming and the second extrusion process suffered by R-PLA. Nevertheless, the difference tapers off as the mixing process advances, reaching similar torque values for PLA and R-PLA after 60 s. While the virgin PLA meets a continuous decrease in torque, associated with the chains breakage that is occurring for the first time as it not was not processed previously. R-PLA, instead, displayed a non-constant decrease, indeed after 40 s there is the presence of a plateau and, thus, the stabilization of the process conditions. The appropriate extrusion time to obtain the molten material to be transferred into the injection press lasted 60 s, however, we wanted to push the mixing time up to 100 s, in order to understand the torque trend, confirming that the torque of R-PLA remained stable up to 100 s, thus it was not going to encounter further cleaving or degradation; on the contrary, PLA continued to decrease its torque value for almost the entire test time.

The shape of the trend is mirrored in Figure 2 at 280 °C, regarding PC and R-PC. At their congenial extrusion temperature, they showed an almost continuous and constant decrease, concerning the virgin PC, basically matching what was reported by Chiu et al. [32], while the achievement of the torque stabilization was registered for R-PC, which started, as did R-PLA, from a relatively lower value than the virgin one, but it stabilized after 25 s and recorded a plateau. In the past, the decrease in the molecular weight of melt processed polycarbonate, was evidenced in different papers [33,34]. Similar papers were published about the PLA processing, proving the occurrence of a chain scission and the consequent decrease in the molecular weight [35,36]; indeed the rheological properties and the solution viscosity are very sensitive to the molecular weight changes and the correlations between the molecular weight and viscosity [37,38].

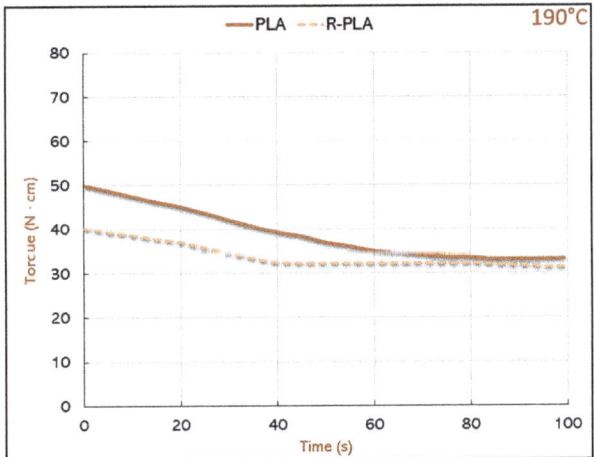

Figure 1. Torque trend for PLA and R-PLA.

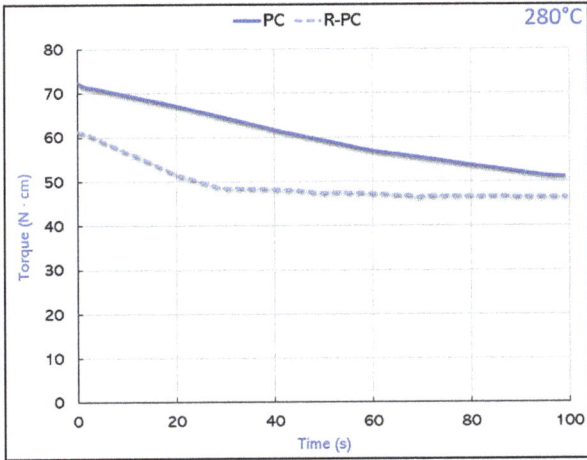

Figure 2. Torque trend for PC and R-PC.

3.1.2. Mechanical Properties

Starting from the PLA/R-PLA comparison, from the point of view of the tensile properties (two representative stress/strain graphs in Figure 3), except for the elongation at break, there are no substantial differences. These results can probably be ascribed to the struggle between some phenomena that occur contemporarily during the reprocessing, as the decrease of viscosity (registered by the torque decrement) is probably due to the molecular chain scission [39,40]. This is crucial feedback, because it allows for the reuse of the material for subsequent processing, as it has similar properties. Definitely, the PLA matrix blends are fragile and show a low elongation at break, whereas the bi-continuous blends, including a continuous PC phase, are better and show an improved elongation at break.

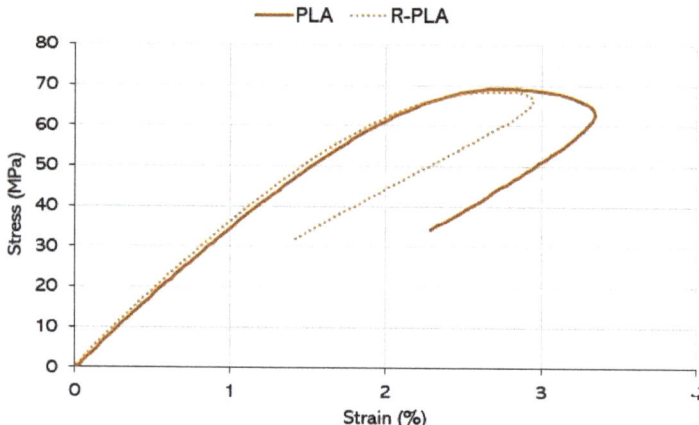

Figure 3. Stress/strain curves for PLA and R-PLA.

Furthermore, regarding the PC/R-PC comparison (where two representative curves are depicted in Figure 4), there is a slight lowering of the stress at break and elongation at break, again caused by reprocessing. A 3% degree decrease of the mechanical strength and stiffness of polycarbonate, after two reprocessing steps, is in line with what was found by Perez et al. [41] and more recently, by Reich et al. [42], in their studies on the mechanical property variations of polycarbonate after successive stages of extrusion/molding.

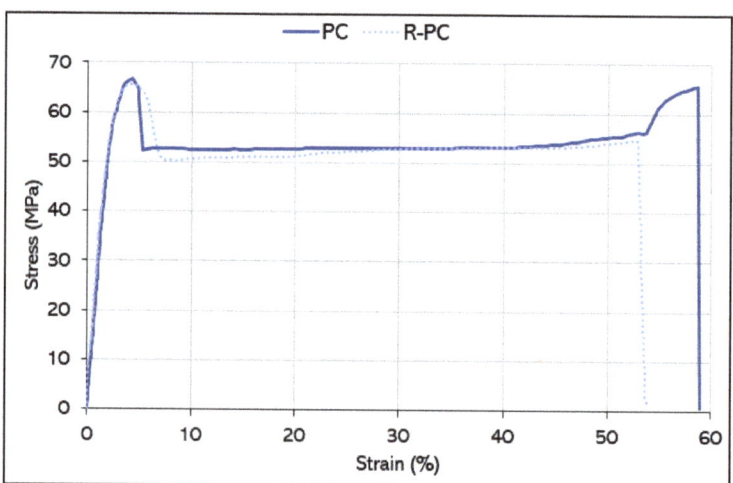

Figure 4. Stress/strain curves for PC and R-PC.

The results of the Charpy tests, shown in Table 2, reveal that the impact strength of the virgin PLA is slightly higher than the impact resistance of the recycled polymer; a reduction of around 5 percent in this property after the subsequent rework is what was also evaluated by Aguero et al. [43], who reprocessed PLA up to six times, at which point the lowering of the impact properties reached almost 50%. This progressive decrease in the energy absorption capacity can be linked with the degradation process on PLA. A 15% decrease has been evaluated for the impact resistance of R-PC, with respect to PC; but the value still turns out to be very useful for blending it with R-PLA and improving its impact resistance.

Table 2. Resumé of the mechanical properties for virgin and recycled polymers.

Polymer	Elastic Modulus (GPa)	Yield Stress (MPa)	Stress at Break (MPa)	Elongation at Break (%)	Charpy Impact Strength (kJ/m^2)
PLA	3.5 ± 0.3	-	67.3 ± 3.5	2.5 ± 0.4	2.9 ± 0.2
R-PLA	3.6 ± 0.2	-	65.1 ± 3.2	3.0 ± 0.6	2.7 ± 0.3
PC	2.2 ± 0.1	67.1 ± 0.8	63.1 ± 3.8	60.6 ± 3.2	10.9 ± 0.2
R-PC	2.3 ± 0.2	65.7 ± 1.6	54.8 ± 3.9	55.7 ± 4.5	9.4 ± 0.5

3.2. Blend Characterizations

3.2.1. Torque Analysis

Tests for the torque evaluation were performed on the studied blends at the same extrusion temperature (Figure 5). As for the pure polymers, a decrease in viscosity is shown for all blends with the increase of the dwell time in the micro compounder recirculation chamber. The trend decreased abruptly in the first 20 s, then a stabilization of the fluidity was registered.

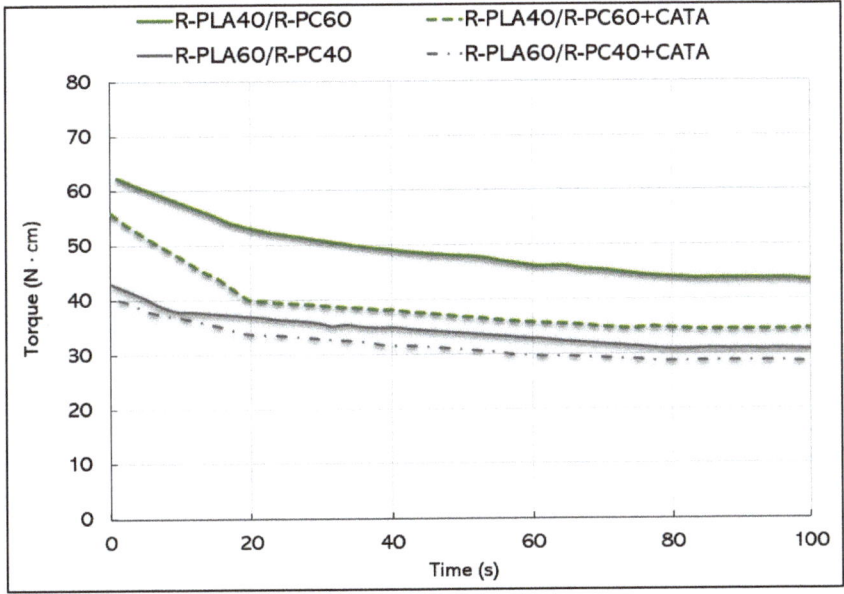

Figure 5. Torque trend for the R-PLA/R-PC blends.

The significant data concerning all of the materials, is that the addition of the catalytic system decreases the torque value with respect to the relative counterparts without TA and TBATPB; the reason is to be found in the typical process of the chain scission in the presence of a catalyst that polymers undergo during the transesterification reaction in the melt, with a decrease in the molecular weight and, consequently, of the torque value [44]. This difference is more pronounced for blends in which there is a higher amount of PC. These considerations should be related to the morphology of the blends that are analyzed in the following section.

3.2.2. Morphological Structure and the Mechanical Results

The fluidity of the polymeric melts, assessed indirectly by means of the torque measurement, necessarily goes to influence the morphological structures of these blends, which, in turn, influence the mechanical properties that are closely related. The micrographs presented in Figure 6a–d are explanatory: it can be seen in the 4000× magnifications, that the R-PLA60/R-PC40 blend (Figure 6a) is characterized by a co-continuity of phases, while the corresponding blend with the addition of TA and TBATPB (Figure 6c) shows rarefied areas of bi-continuity, but especially areas where the presence of the deformed ellipsoidal particles of PC, as dispersed phase in the PLA, is observed.

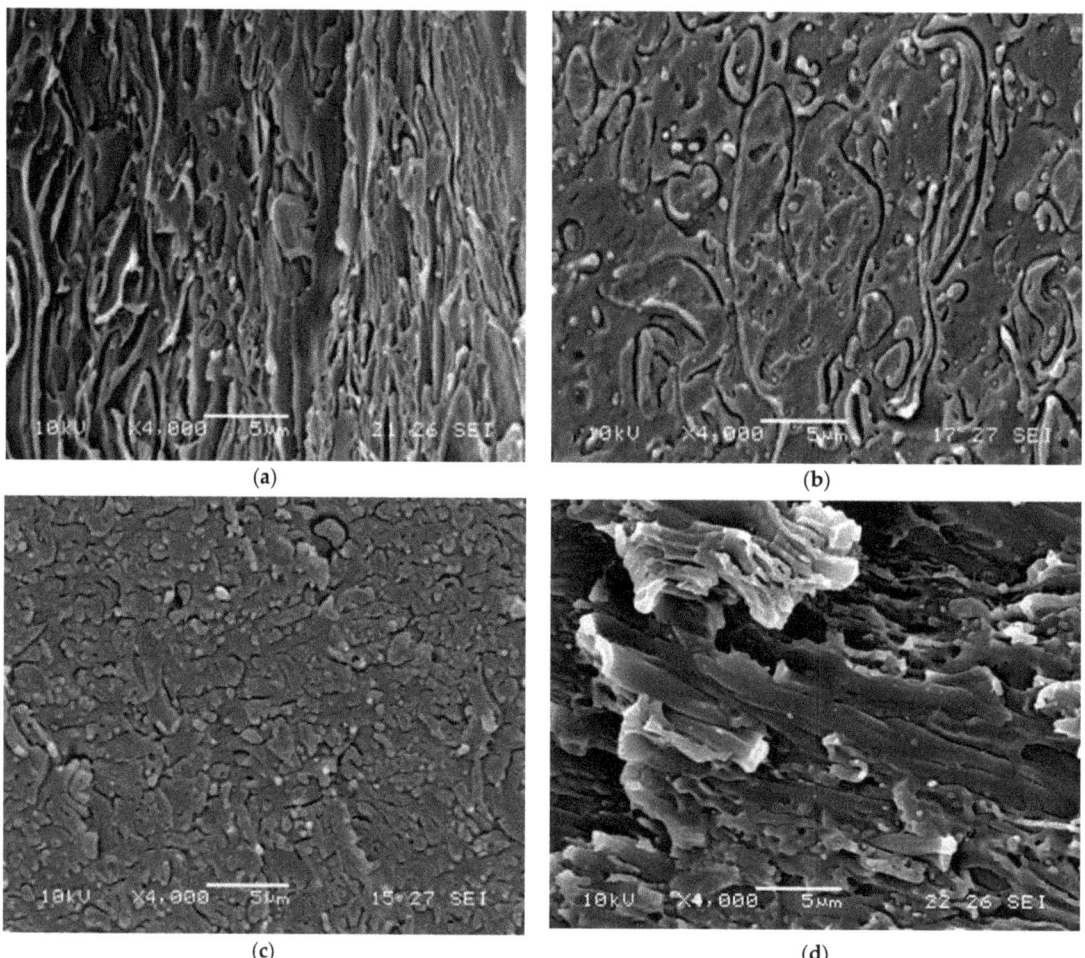

Figure 6. SEM Micrographs of (**a**) R-PLA60/R-PC40, (**b**) R-PLA40/R-PC60, (**c**) R-PLA60/R-PC40 + CATA, (**d**) R-PLA40/R-PC60 + CATA.

In contrast, in the case of the R-PLA40/R-PC60 blend, a PC matrix structure with spheroidal inclusions of PLA, is seen (Figure 6b), nevertheless the addition of the catalytic system caused the formation of a bi-continuous morphological structure (Figure 6d).

In the literature, it is well known that immiscible polymers, such as the PLA/PC systems, are characterized by heterogeneous morphologies achievable during the melting. The types and dimensions of the morphology determine the properties of the blend, depending on the interfacial tension, viscosities and compatibilizers [45,46]. The co-continuous structures can be considered as the coexistence of at least two adjacent structures within the same volume. The mixtures with a co-continuous structure can favorably combine the properties of both components [47] and the concept of the phase inversion must be taken into account.

Phase inversion is a phenomenon that occurs when within a mixture, as the composition changes, the polymer that had the continuous phase changes to a dispersed phase, and vice versa [48], but why is such a morphology obtained at different percentages of PLA with and without a catalyst? According to Avgeroupolos et al. [49], phase inversion occurs when the ratios of the torques and volume fractions of the components of a blend are equal. For blends without CATA, in this paper, therefore, the phase inversion point is reached for larger quantities of PLA, than for the compatibilized blends. The motivation we propose is that the lowering of the viscosity generated by the catalyst flattening the torque values to similar values, identifies the phase inversion around 50/50 between R-PLA and R-PC, for such systems causing the different behaviors, in response to the tensile stress.

In this context Veenstra et al. [50] stated that the co-continuous morphology improves the characteristics of both polymer components, with respect to all possible morphologies. This assumption is confirmed by the quasi-static tensile properties (Figure 7 and Table 3) in which the bi-continuous blends (R-PLA60/R-PC40 and R-PLA40/R-PC60 + CATA) exhibit a much higher ductility, with elongations at break, even exceeding those of the pure PC, without decreasing in tensile strength. In contrast, the other two blends show a comparable ultimate tensile strength but with an evident brittleness. What drives the achievement of the improved properties over the pure PLA is the accomplishment of a co-continuous morphology during the processing.

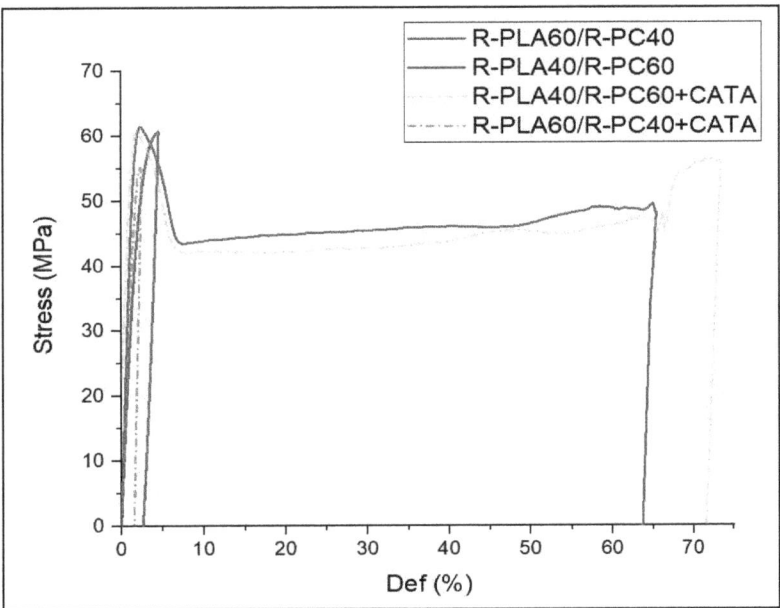

Figure 7. Stress-curves strain curves for the R-PLA/R-PC blends with and without CATA.

Table 3. Resumé of the mechanical properties for the blends with the recycled polymers.

Blend	Elastic Modulus (GPa)	Yield Stress (MPa)	Stress at Break (MPa)	Elongation at Break (%)	Charpy Impact Strength (kJ/m^2)
R-PLA60/R-PC40	2.7 ± 0.4	61.1 ± 1.1	57.2 ± 2.0	63.9 ± 8.7	7.5 ± 1.9
R-PLA60/R-PC40 + CATA	3.0 ± 0.2	-	60.1 ± 0.7	4.0 ± 0.6	6.3 ± 0.2
R-PLA40/R-PC60	3.2 ± 0.3	-	63.6 ± 2.8	3.5 ± 0.2	5.6 ± 0.8
R-PLA40/R-PC60 + CATA	3.4 ± 0.3	61.7 ± 1.5	66.4 ± 3.5	72.4 ± 9.5	4.6 ± 0.9

The elastic modulus exhibits higher values for the mixtures with CATA; this is due to the formation of bonds, due to the interchange reaction between the components. In a previous study [22], it was seen that, after the quasi-static tests, the elongations at break, even greater than 120%, were achieved, but always with polycarbonate amounts of at least 60%; in the present work, with the recycled polymers, it is possible to state that this range of bi-continuity is much wider, allowing even blends with 60% PLA to have a co-continuous morphology, resulting in its ductile behavior.

This strong relationship between the phase morphology and the elongation at break values for the studied blends, has been highlighted in Figure 8.

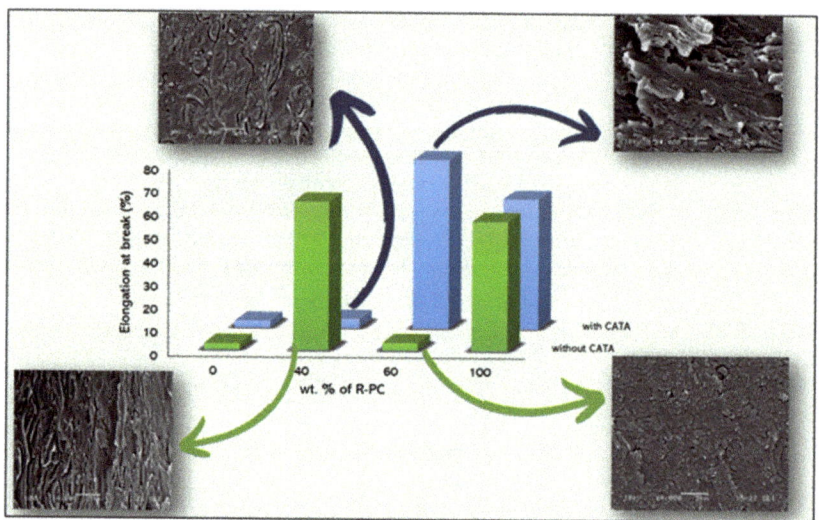

Figure 8. Morphology/elongation at break relationship.

Conversely, when these materials were examined at high-speeds, through the impact tests, it has been noticed that what increases the toughness is the amount of R-PC in the compound, rather than the morphology; in fact, the trend is almost linear for both blends without and with a catalyst (Figure 9). This difference, in response to the slow test, versus the fast test, has also been found in other polymer systems, such as PLA/PBAT [51] or PLA/POE-g-GMA [52].

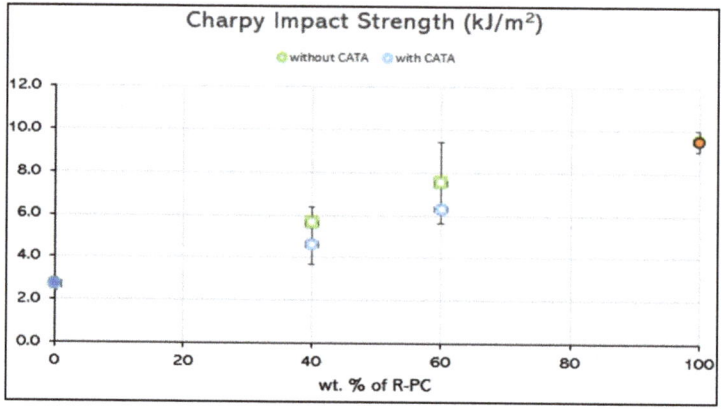

Figure 9. Trend of the Charpy impact strength over the R-PC quantity in the blend.

The catalytic system, operating through the chain rupture, did not favor the achievement of the impact strengths equivalent to the blends without a catalyst.

Definitely, as a result of the mechanical response in the slow test, specifically the tensile toughness is increased when a bi-continuous structure has been achieved during the processing, while in the fast test (impact test) it is the higher PC content that causes high value of energy absorbed before the crack propagation.

3.2.3. DMTA Analysis

The development of such blends, based on polylactic acid and polycarbonate, could have a higher impact if the studied catalytic system could form copolymers during the reactive extrusion. Even polymers suffered in the industrial recycling process. In this regard, the DMTA analysis guarantees the possibility, through the study of tan delta peaks generated by a tensile test carried out at a certain frequency and in temperature sweeps, to evaluate the formation of copolymers, as demonstrated by Liu et al. [18]

As shown in Figure 10, the energy released by the viscous motion of the polymer chains is reflected in the relaxation peak of tan δ, whose maximum can be considered an expression of Tg. Since the immiscible blends are those without CATA, a clear phase separation structure occurs, as revealed by two maxima in the tan δ curve. The small intermediate peaks of the blends without catalyst (red and black dots in Figure 10) are explained as the occurrence of the crystallization of the material; moreover, the more evident and significant ones (circled in yellow in Figure 10) that are present in the curves, concerning the blends with a catalyst are attributable to the formation of a copolymer that has, as Tg, a temperature intermediate between those of PLA and PC. This finding also confirms the data obtained with the virgin polymers, by Phuong et al. [22], namely that the presence of a new species is represented by the tan δ peak at around 110 °C.

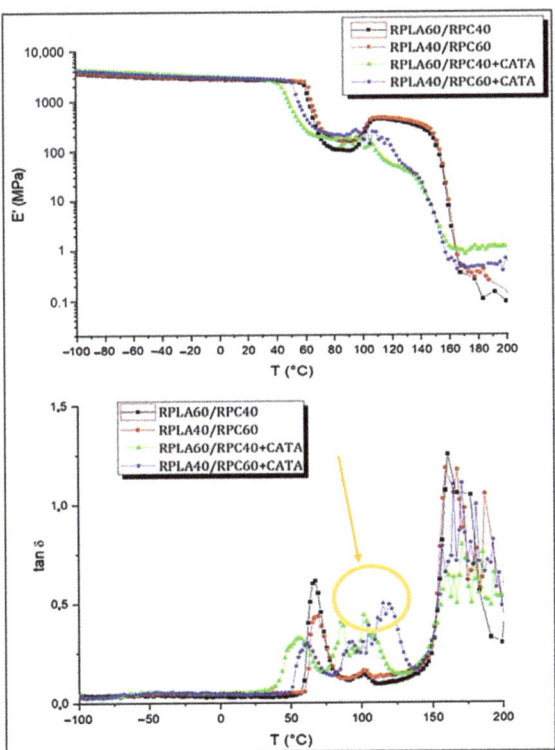

Figure 10. DMTA analysis to evaluate the presence of copolymers.

Moreover, it can be noticed that the blends with the catalyst system exhibit a glass transition that starts at lower temperatures; this is probably due to the viscosity decrease caused by TA. The modulus storage E′ is higher for the mixtures with CATA, when high temperatures are reached; this is due to the bond formation because of the interchange reaction between the components.

4. Conclusions

The development of materials that are environmentally friendly and recyclable, that also have good mechanical properties (comparable to benchmarks), compatible with affordable price and capable to replace petroleum derivatives, is the path currently pursued in polymer research.

In this paper, the melt viscosity and mechanical properties of the virgin PLA and PC were first compared with the corresponding materials that had undergone extrusion, thermoforming processes and were recovered as production scraps. Such processing was seen to decrease the properties, but always in an acceptable range of values that guaranteed their reuse and subsequent compounding.

For this purpose, the blends with 40 and 60% wt. of recycled PLA (R-PLA) were processed by studying whether a system of compatibilizers (successfully tested on the virgin polymers in a previous paper) would also work for the blends obtained from the recycled polymers. An interesting phenomenon was seen, as a result of the mechanical response in the slow test, specifically, the tensile toughness is increased when a bi-continuous structure has been achieved during the processing, while in the fast test (impact test), it is the higher PC content that causes the high value of energy absorbed before the crack propagation. The catalytic system, through the DMTA analysis, was seen to be able to induce the formation of the PLA-PC copolymers, since the presence of a peak of the tan delta at an intermediate temperature, with respect to PLA and PC, an α-transition has been registered.

The R-PLA/RPC blends with an improved ductility, with respect to the pure recycled PLA, were obtained in the present work, thanks to the achieved R-PC phase continuity, thus suggesting the methodologies to foster the use of recycled renewable polymers in a wider range of durable applications, such as in automotive and electronic equipment, where recyclability is requested.

With respect to the blends produced by using the virgin PLA and PC [22] where 60 wt.% of PC was necessary to observe blends with a good ductility, the use of recycled polymers allowed to obtain blends with an improved ductility, using only 40% of R-PC, thus allowing to increase the renewable content (and thus the carbon storage potential) of the developed secondary material). Such compounds, therefore, represent a great opportunity because they combine a good technical potential, a high renewable content, eco-sustainability, recyclability and can be a viable solution to post-consumer disposal problems, which are increasingly burdensome in both economic and environmental terms. The challenge that needs to be addressed concerns logistics, i.e., ensuring the suitable recycling lines for materials of this type and the widespread possibility of implementing the circular-by-design concept.

Author Contributions: Conceptualization, A.L. and M.-B.C.; methodology, V.G and L.A.; validation, A.L., M.-B.C., V.G. and L.A.; investigation, V.G. and L.A.; resources, A.L. and M.-B.C.; data curation, V.G and L.A.; writing—original draft preparation, V.G and L.A.; writing—review and editing, A.L. and M.-B.C.; supervision, A.L. and M.-B.C. All authors have read and agreed to the published version of the manuscript.

Funding: The authors gratefully acknowledge the partial financial support of project EVOLUTION (The Electric Vehicle revOLUTION enabled by advanced materials highly hybridized into lightweight components for easy integration and dismantling providing a reduced life cycle cost logic) Contract No. 314744 FP7-2012-GC-MATERIALS, funded by the European Commission under the 7th Framework Programme (FP7).

Institutional Review Board Statement: Not applicable.

Informed Consent Statement: Not Applicable.

Data Availability Statement: Not applicable.

Acknowledgments: Romei s. r.l. company is thanked for providing the grounded R-PLA and R-PC.

Conflicts of Interest: The authors declare no conflict of interest.

References

1. Muñoz Meneses, R.A.; Cabrera-Papamija, G.; Machuca-Martínez, F.; Rodríguez, L.A.; Diosa, J.E.; Mosquera-Vargas, E. Plastic recycling and their use as raw material for the synthesis of carbonaceous materials. *Heliyon* **2022**, *8*, e09028. [CrossRef]
2. Mazur-Wierzbicka, E. Circular economy: Advancement of European Union countries. *Environ. Sci. Eur.* **2021**, *33*, 111. [CrossRef]
3. Kumar, R.; Verma, A.; Shome, A.; Sinha, R.; Sinha, S.; Jha, P.K.; Kumar, R.; Kumar, P.; Shubham; Das, S.; et al. Impacts of Plastic Pollution on Ecosystem Services, Sustainable Development Goals, and Need to Focus on Circular Economy and Policy Interventions. *Sustainability* **2021**, *13*, 9963. [CrossRef]
4. Farah, S.; Anderson, D.G.; Langer, R. Physical and mechanical properties of PLA, and their functions in widespread applications—A comprehensive review. *Adv. Drug Deliv. Rev.* **2016**, *107*, 367–392. [CrossRef] [PubMed]
5. Naser, A.Z.; Deiab, I.; Defersha, F.; Yang, S. Expanding Poly(lactic acid) (PLA) and Polyhydroxyalkanoates (PHAs) Applications: A Review on Modifications and Effects. *Polymers* **2021**, *13*, 4271. [CrossRef] [PubMed]
6. Yang, Y.; Zhang, L.; Xiong, Z.; Tang, Z.; Zhang, R.; Zhu, J. Research progress in the heat resistance, toughening and filling modification of PLA. *Sci. China Chem.* **2016**, *59*, 1355–1368. [CrossRef]
7. Zhao, X.; Hu, H.; Wang, X.; Yu, X.; Zhou, W.; Peng, S. Super tough poly(lactic acid) blends: A comprehensive review. *RSC Adv.* **2020**, *10*, 13316–13368. [CrossRef] [PubMed]
8. Liu, H.; Zhang, J. Research progress in toughening modification of poly(lactic acid). *J. Polym. Sci. Part B Polym. Phys.* **2011**, *49*, 1051–1083. [CrossRef]
9. Lin, L.; Deng, C.; Wang, Y. Improving the impact property and heat-resistance of PLA/PC blends through coupling molecular chains at the interface. *Polym. Adv. Technol.* **2015**, *26*, 1247–1258. [CrossRef]
10. Yuryev, Y.; Mohanty, A.K.; Misra, M. Novel biocomposites from biobased PC/PLA blend matrix system for durable applications. *Compos. Part B Eng.* **2017**, *130*, 158–166. [CrossRef]
11. Nagarajan, V.; Mohanty, A.K.; Misra, M. Perspective on Polylactic Acid (PLA) based Sustainable Materials for Durable Applications: Focus on Toughness and Heat Resistance. *ACS Sustain. Chem. Eng.* **2016**, *4*, 2899–2916. [CrossRef]
12. Zeng, J.-B.; Li, K.-A.; Du, A.-K. Compatibilization strategies in poly(lactic acid)-based blends. *RSC Adv.* **2015**, *5*, 32546–32565. [CrossRef]
13. Srithep, Y.; Rungseesantivanon, W.; Hararak, B.; Suchiva, K. Processing and characterization of poly(lactic acid) blended with polycarbonate and chain extender. *J. Polym. Eng.* **2014**, *34*, 665–672. [CrossRef]
14. Wang, Y.; Chiao, S.M.; Lai, M.-T.; Yang, S.-Y. The role of polycarbonate molecular weight in the poly(L-lactide) blends compatibilized with poly(butylene succinate-co-L-lactate). *Polym. Eng. Sci.* **2013**, *53*, 1171–1180. [CrossRef]
15. Chen, Y.; Zeng, G.S.; Jiang, P.; Lu, W.; Huang, W.L. Study on the Thermal and Rheological Properties of Reactive Blending PC/PLA/EVA Blends. *Appl. Mech. Mater.* **2011**, *79*, 282–287. [CrossRef]
16. Ikehara, T.; Nishikawa, Y.; Nishi, T. Evidence for the formation of interpenetrated spherulites in poly(butylene succinate-co-butylene carbonate)/poly(l-lactic acid) blends investigated by atomic force microscopy. *Polymer* **2003**, *44*, 6657–6661. [CrossRef]
17. Wang, Y.; Chiao, S.M.; Hung, T.-F.; Yang, S.-Y. Improvement in toughness and heat resistance of poly(lactic acid)/polycarbonate blend through twin-screw blending: Influence of compatibilizer type. *J. Appl. Polym. Sci.* **2012**, *125*, E402–E412. [CrossRef]
18. Liu, C.; Lin, S.; Zhou, C.; Yu, W. Influence of catalyst on transesterification between poly(lactic acid) and polycarbonate under flow field. *Polymer* **2013**, *54*, 310–319. [CrossRef]
19. Tripathi, N.; Misra, M.; Mohanty, A.K. Durable Polylactic Acid (PLA)-Based Sustainable Engineered Blends and Biocomposites: Recent Developments, Challenges, and Opportunities. *ACS Eng. Au* **2021**, *1*, 7–38. [CrossRef]
20. Pompe, G.; Häußler, L. Investigations of transesterification in PC/PBT melt blends and the proof of immiscibility of PC and PBT at completely suppressed transesterification. *J. Polym. Sci. Part B Polym. Phys.* **1997**, *35*, 2161–2168. [CrossRef]
21. Ignatov, V.N.; Carraro, C.; Tartari, V.; Pippa, R.; Scapin, M.; Pilati, F.; Berti, C.; Toselli, M.; Fiorini, M. PET/PC blends and copolymers by one-step extrusion: 2. Influence of the initial polymer composition and type of catalyst. *Polymer* **1997**, *38*, 201–205. [CrossRef]
22. Phuong, V.T.; Coltelli, M.-B.B.; Cinelli, P.; Cifelli, M.; Verstichel, S.; Lazzeri, A. Compatibilization and property enhancement of poly(lactic acid)/polycarbonate blends through triacetin-mediated interchange reactions in the melt. *Polymer* **2014**, *55*, 4498–4513. [CrossRef]
23. Phuong, V.T.; Gigante, V.; Aliotta, L.; Coltelli, M.B.; Cinelli, P.; Lazzeri, A. Reactively extruded ecocomposites based on poly(lactic acid)/bisphenol A polycarbonate blends reinforced with regenerated cellulose microfibers. *Compos. Sci. Technol.* **2017**, *139*, 127–137. [CrossRef]
24. Aryan, V.; Maga, D.; Majgaonkar, P.; Hanich, R. Valorisation of polylactic acid (PLA) waste: A comparative life cycle assessment of various solvent-based chemical recycling technologies. *Resour. Conserv. Recycl.* **2021**, *172*, 105670. [CrossRef]

25. Yarahmadi, N.; Jakubowicz, I.; Enebro, J. Polylactic acid and its blends with petroleum-based resins: Effects of reprocessing and recycling on properties. *J. Appl. Polym. Sci.* **2016**, *133*, 43196. [CrossRef]
26. Freeland, B.; McCarthy, E.; Balakrishnan, R.; Fahy, S.; Boland, A.; Rochfort, K.D.; Dabros, M.; Marti, R.; Kelleher, S.M.; Gaughran, J. A Review of Polylactic Acid as a Replacement Material for Single-Use Laboratory Components. *Materials* **2022**, *15*, 2989. [CrossRef]
27. Penco, M.; Lazzeri, A.; Phuong, V.T.; Cinelli, P. Copolymers Based on Polyester and Aromatic Polycarbonate. U.S. Patent 20140128540A1, 8 May 2014.
28. Tejada-Oliveros, R.; Gomez-Caturla, J.; Sanchez-Nacher, L.; Montanes, N.; Quiles-Carrillo, L. Improved Toughness of Polylactide by Binary Blends with Polycarbonate with Glycidyl and Maleic Anhydride-Based Compatibilizers. *Macromol. Mater. Eng.* **2021**, *306*, 2100480. [CrossRef]
29. Guessoum, M.; Chelghoum, N.; Haddaoui, N. Reactive Compatibilization of Poly(lactic acid) and Polycarbonate Blends Through Catalytic Transesterification Reactions. *Int. J. Comput. Exp. Sci. Eng.* **2019**, *5*, 65–68. [CrossRef]
30. You, W.; Yu, W. Control of the dispersed-to-continuous transition in polymer blends by viscoelastic asymmetry. *Polymer* **2018**, *134*, 254–262. [CrossRef]
31. Santi, C.R.; Hage, E.; Correa, C.A.; Vlachopoulos, J. Torque viscometry of molten polymers and composites. *Appl. Rheol.* **2009**, *19*, 13148. [CrossRef]
32. Chiu, H.T.; Huang, J.K.; Kuo, M.T.; Huang, J.H. Characterisation of PC/ABS blend during 20 reprocessing cycles and subsequent functionality recovery by virgin additives. *J. Polym. Res.* **2018**, *25*, 124. [CrossRef]
33. Abbås, K.B. Thermal degradation of bisphenol A polycarbonate. *Polymer* **1980**, *21*, 936–940. [CrossRef]
34. Abbås, K.B. Degradational effects on bisphenol A polycarbonate extruded at high shear stresses. *Polymer* **1981**, *22*, 836–841. [CrossRef]
35. Cuadri, A.A.; Martín-Alfonso, J.E. Thermal, thermo-oxidative and thermomechanical degradation of PLA: A comparative study based on rheological, chemical and thermal properties. *Polym. Degrad. Stab.* **2018**, *150*, 37–45. [CrossRef]
36. Khademi, S.M.H.; Hemmati, F.; Aroon, M.A. An insight into different phenomena involved in continuous extrusion foaming of biodegradable poly(lactic acid)/expanded graphite nanocomposites. *Int. J. Biol. Macromol.* **2020**, *157*, 470–483. [CrossRef]
37. Polidar, M.; Metzsch-Zilligen, E.; Pfaendner, R. Controlled and Accelerated Hydrolysis of Polylactide (PLA) through Pentaerythritol Phosphites with Acid Scavengers. *Polymers* **2022**, *14*, 4237. [CrossRef] [PubMed]
38. Signori, F.; Coltelli, M.-B.; Bronco, S. Thermal degradation of poly(lactic acid) (PLA) and poly(butylene adipate-co-terephthalate) (PBAT) and their blends upon melt processing. *Polym. Degrad. Stab.* **2009**, *94*, 74–82. [CrossRef]
39. Coltelli, M.-B.; Mallegni, N.; Rizzo, S.; Cinelli, P.; Lazzeri, A. Improved Impact Properties in Poly(lactic acid) (PLA) Blends Containing Cellulose Acetate (CA) Prepared by Reactive Extrusion. *Materials* **2019**, *12*, 270. [CrossRef]
40. Coltelli, M.-B.; Toncelli, C.; Ciardelli, F.; Bronco, S. Compatible blends of biorelated polyesters through catalytic transesterification in the melt. *Polym. Degrad. Stab.* **2011**, *96*, 982–990. [CrossRef]
41. Pérez, J.M.; Vilas, J.L.; Laza, J.M.; Arnáiz, S.; Mijangos, F.; Bilbao, E.; Rodríguez, M.; León, L.M. Effect of reprocessing and accelerated ageing on thermal and mechanical polycarbonate properties. *J. Mater. Process. Technol.* **2010**, *210*, 727–733. [CrossRef]
42. Reich, M.J.; Woern, A.L.; Tanikella, N.G.; Pearce, J.M. Mechanical Properties and Applications of Recycled Polycarbonate Particle Material Extrusion-Based Additive Manufacturing. *Materials* **2019**, *12*, 1642. [CrossRef]
43. Agüero, A.; Morcillo, M.d.C.; Quiles-Carrillo, L.; Balart, R.; Boronat, T.; Lascano, D.; Torres-Giner, S.; Fenollar, O. Study of the Influence of the Reprocessing Cycles on the Final Properties of Polylactide Pieces Obtained by Injection Molding. *Polymers* **2019**, *11*, 1908. [CrossRef]
44. Bi, F.-L.; Xi, Z.-H.; Zhao, L. Reaction Mechanisms and Kinetics of the Melt Transesterification of Bisphenol-A and Diphenyl Carbonate. *Int. J. Chem. Kinet.* **2018**, *50*, 188–203. [CrossRef]
45. Sadiku-Agboola, O.; Sadiku, E.R.; Adegbola, A.T.; Biotidara, O.F. Rheological Properties of Polymers: Structure and Morphology of Molten Polymer Blends. *Mater. Sci. Appl.* **2011**, *02*, 30–41. [CrossRef]
46. Willemse, R.C.; Posthuma de Boer, A.; van Dam, J.; Gotsis, A.D. Co-continuous morphologies in polymer blends: The influence of the interfacial tension. *Polymer* **1999**, *40*, 827–834. [CrossRef]
47. Pötschke, P.; Paul, D.R. Formation of Co-continuous Structures in Melt-Mixed Immiscible Polymer Blends. *J. Macromol. Sci. Part C* **2003**, *43*, 87–141. [CrossRef]
48. Yeo, L.Y.; Matar, O.K.; de Ortiz, E.S.P.; Hewitt, G.F. Phase Inversion and associated phenomena. *Multiph. Sci. Technol.* **2000**, *12*, 66. [CrossRef]
49. Avgeropoulos, G.N.; Weissert, F.C.; Biddison, P.H.; Böhm, G.G.A. Heterogeneous Blends of Polymers. Rheology and Morphology. *Rubber Chem. Technol.* **1976**, *49*, 93–104. [CrossRef]
50. Veenstra, H.; Verkooijen, P.C.J.; van Lent, B.J.J.; van Dam, J.; de Boer, A.P.; Nijhof, A.P.H.J. On the mechanical properties of co-continuous polymer blends: Experimental and modelling. *Polymer* **2000**, *41*, 1817–1826. [CrossRef]
51. Gigante, V.; Canesi, I.; Cinelli, P.; Coltelli, M.B.; Lazzeri, A. Rubber toughening of Polylactic acid (PLA) with Poly(butylene adipate-co- terephthalate) (PBAT): Mechanical properties, fracture mechanics and analysis of brittle—Ductile behavior while varying temperature and test speed. *Eur. Polym. J.* **2019**, *115*, 125–137. [CrossRef]
52. Aliotta, L.; Gigante, V.; Acucella, O.; Signori, F.; Lazzeri, A. Thermal, mechanical and micromechanical analysis of PLA/PBAT/POE-g-GMA extruded ternary blends. *Front. Mater.* **2020**, *7*, 130. [CrossRef]

Article

Curing Kinetics of Bioderived Furan-Based Epoxy Resins: Study on the Effect of the Epoxy Monomer/Hardener Ratio

Angela Marotta [1,*], Noemi Faggio [1,2] and Cosimo Brondi [1]

1. Department of Chemical, Materials and Production Engineering, University of Naples Federico II, Piazzale Vincenzo Tecchio, 80, 80125 Naples, Italy
2. Institute for Polymers, Composites and Biomaterials (IPCB)—CNR, Via Campi Flegrei 34, 80078 Pozzuoli, Italy
* Correspondence: angela.marotta@unina.it

Abstract: The potential of furan-based epoxy thermosets as a greener alternative to diglycidyl ether of Bisphenol A (DGEBA)-based resins has been demonstrated in recent literature. Therefore, a deep investigation of the curing behaviour of these systems may allow their use for industrial applications. In this work, the curing mechanism of 2,5-bis[(oxiran-2-ylmethoxy)methyl]furan (BOMF) with methyl nadic anhydride (MNA) in the presence of 2-methylimidazole as a catalyst is analyzed. In particular, three systems characterized by different epoxy/anhydride molar ratios are investigated. The curing kinetics are studied through differential scanning calorimetry, both in isothermal and non-isothermal modes. The total heat of reaction of the epoxy resin as well as its activation energy are estimated by the non-isothermal measurements, while the fitting of isothermal data with Kamal's autocatalytic model provides the kinetic parameters. The results are discussed as a function of the resin composition. The global activation energy for the curing process of BOMF/MNA resins is in the range 72–79 kJ/mol, depending on both the model used and the sample composition; higher values are experienced by the system with balanced stoichiometry. By the fitting of the isothermal analysis, it emerged that the order of reaction is not only dependent on the temperature, but also on the composition, even though the values range between 0.31 and 1.24.

Keywords: epoxy resins; anhydride curing agent; curing kinetics; calorimetry; bio-based epoxy; furan-based epoxy

1. Introduction

Epoxy thermoset resins are polymeric materials widely used in the industrial field thanks to their excellent mechanical and thermal properties, as well as their chemical resistance [1]. Specific applications require remarkable structural characteristics of epoxy resins, which in turn are dictated by the nature of the epoxy monomer and hardener, as well as the network feature developed upon curing [2]. The curing reaction proceeds through different mechanisms depending on the nature of the crosslinking agent and the catalyst, as well as on the curing conditions [2].

In particular, the latter have a significant impact on the network structure, and as a result, the deeper the understanding of the curing mechanism and kinetics of an epoxy resin, the greater the control on the final properties of the material [3]. Several experimental techniques and methods are used to assess chemical reaction kinetics, such as Fourier transform infrared spectroscopy (FTIR) [4,5], nuclear magnetic resonance (NMR) [6,7], and dynamic mechanical analysis [8,9]. In this sense, one of the most widely used techniques to identify the kinetic parameters, such as the rate constant, activation energy, and reaction order, associated with a curing reaction is differential scanning calorimetry (DSC), used [10] under both isothermal and dynamic conditions [11]. The widespread use of the DSC technique is due to its higher feasibility compared to FTIR and NMR. In fact, NMR analysis requires highly expensive and difficult-to-use apparatus. FTIR instead, despite the easiness of use, is a less accurate technique due to the possible overlapping of the characteristic peaks associated with the reacting groups [12]. In fact, Shnawa [13] investigated the structure of a tannin-based epoxy resin (TER) by means of FTIR. Although this equipment provided some insights about the chemical architecture of this compound, it was possible to investigate the formation of the crosslinked network by DSC only. In particular, it was observed that the TER was successfully cured with an amine-based curing agent. Moreover, DSC analysis confirmed that TER blended to commercial epoxy resin could act as a curing accelerator by monitoring the exothermicity of the curing peaks and the completion of the curing temperatures. In another work, Yang and co-authors [14] demonstrated that the addition of a phosphorous/imidazole-containing compound to epoxy resins significantly decreases the curing time.

In recent studies carried out by the authors, the curing behavior of 2,5-bis[(oxiran-2-ylmethoxy)methyl]furan (BOMF) with methyl nadic anhydride (MNA) has been already reported [12]. This resin has been applied as a tinplate coating, showing excellent adhesion and chemical resistance, thus being a good substitute for DGEBA-based products in the field of metal packaging [15]. An interesting aspect of such systems relies on the possibility of tuning their properties by changing the epoxy/anhydride ratio, the nature of the catalyst, or by including an opportunely selected inorganic filler.

Considering the already well-established use of carbohydrate derivatives concerning the category of thermosetting epoxy resins, furanic compounds attracted an increased interest as a potential alternative in the replacement of oil-based ones. Under this perspective, furan-based epoxy resins have been studied in the last years but have not been fully exploited yet. To assess the potential use of these compounds, a study on the reaction kinetics by FTIR [12] and the coating properties [15] was already performed. To our knowledge, a deeper study on the reaction mechanisms and the kinetic parameters of BOMF/MNA has not been performed yet.

Taking all the above considerations into account, the goal of this work is not only to provide insight into the curing kinetics of the biobased BOMF/MNA resin, but also to evaluate the effects of different epoxy/anhydride combinations in the reacting mixture. To this aim, DSC has been selected as a tool to carry out these studies. In particular, non-isothermal and isothermal DSC analyses were conducted, and different thermokinetic methods were used to calculate the characteristic kinetic parameters of the BOMF/MNA curing process as function of the resin composition.

2. Materials and Methods

2.1. Materials

In this work 2,5-bis[(oxiran-2-ylmethoxy)methyl]furan (BOMF) was used as a biobased epoxy monomer and methyl-5-norbornene-2,3-dicarboxylic anhydride (methyl nadic anhydride, MNA, 90%), purchased from Sigma-Aldrich (Darmstadt, Germany), was used as a curing agent. To ensure the obtainment of a crosslinked material, 2-methylimidazole (2-MI, 99%), purchased from Acros Organics (Geel, Belgium), was used as an initiator. MNA and 2-MI were used as received without further purification, while BOMF was synthesized following the procedure present in the literature [12].

2.2. Sample Preparation

Epoxy/anhydride mixtures were prepared by mixing at room temperature the epoxy monomer and the curing agent in the proper amounts and the initiator to the extent of 0.5 wt.% of the total amount of the epoxy monomer and anhydride. The weights of BOMF (the epoxy monomer) and MNA (the anhydride curing agent) were calculated to obtain different molar epoxy-to-anhydride ratios, namely 1/1 (stoichiometric ratio), 1/0.8 (epoxy excess), and 0.8/1 (anhydride ratio).

2.3. Differential Scanning Calorimetry

Differential scanning calorimetric analyses (DSC) were performed by means of a DSC Q2000 (TA Instrument, New Castle, DE, USA) equipped with a refrigerator cooling system (RCS). The analyses were performed on samples of about 7–8 mg sealed in aluminum pans immediately after the preparation of the mixtures. The samples were stored at −30 °C for a maximum of 2 days. Measurements under inert nitrogen flux (50 mL/min) in both non-isothermal conditions, heating from 25 to 200 °C at heating rates of 1, 1.5, 3, 5, and 10 °C/min, and isothermal conditions, at temperatures of 90, 100, 110, 120, and 130 °C, were performed.

3. Modeling

The degree of cure (α) corresponds to the integrated area of a specific DSC peak and can range from 0 to 1. At the current time (t), α can be measured as reported [16–18]:

$$\alpha(t) = \frac{1}{\Delta H_{tot}} \int_0^t dH = \frac{\Delta H(t)}{\Delta H_{tot}} \quad (1)$$

where $\Delta H(t) = \int_0^t dH$ is the reaction enthalpy at the time t and ΔH_{tot} is the total reaction enthalpy [19,20]. As commonly performed in kinetic analyses conducted by DSC [16], it is assumed that there is a direct proportionality among the heat flow (dQ/dt) detected during the analysis and the conversion rate ($d\alpha/dt$) [21] according to the following equation:

$$\frac{dQ}{dt} = Q_{tot} \frac{d\alpha}{dt} \quad (2)$$

In the case of the isothermal mode, the functional dependencies of the conversion rate can be regarded as the product of two basic functions, respectively, related to the dependency on the temperature and the degree of cure:

$$\frac{d\alpha}{dt} = k(T) f(\alpha) \quad (3)$$

where $k(T)$ is the rate constant and $f(\alpha)$ is the reaction model. According to the transition-state theory [22], the activation energy represents the difference in the energy amount between the reactant molecule at its initial configuration and the corresponding transition-state molecule in its activated configuration. In the context of a chemical reaction, this quantity can be formally regarded as the minimum amount of energy required to activate the reactant molecules so that they can undergo chemical transformations [23]. Under this perspective, the activation energy can be enclosed in a rate constant that is dependent on the temperature, not the concentration of the reactants, and can be formally expressed by the Arrhenius equation:

$$k(T) = A \, e^{\left(\frac{-E_a}{RT}\right)} \quad (4)$$

where A is a preexponential factor, R is the gas constant, T is the temperature, and E_a is the activation energy. The reaction model $f(\alpha)$ reflects the reaction mechanism and, in the specific case of an epoxy curing reaction, it generally takes the form of a nth-order kinetic model (Equation (5)) or an autocatalytic process (Kamal's model, Equation (6)) [23–25]:

$$f(\alpha) = (1 - \alpha)^n \quad (5)$$

$$f(\alpha) = \alpha^m (1 - \alpha)^n \quad (6)$$

In general, the nth-order and the autocatalytic model provide a suitable description of the behavior of the thermosetting materials [26,27]. In particular, the nth-order kinetics report a reaction rate proportional to the n exponent of the reactant concentration. In this case, the model assumes that only one reaction occurs during the curing process and, therefore, can lead to some limitations since more simultaneous reactions can take place during the whole process. With reference to the autocatalytic model, reaction products also participate in the reaction, leading to a reaction rate increase during the initial stage, reaching a maximum, and then decreasing [16,28].

The abovementioned models imply that a system achieves a full extent of cure, that is, $\alpha = 1$. As a complete cure is not always observed in epoxy curing systems, Equations (5) and (6) can be manipulated so that the kinetic models are also applicable to incomplete cure, and Equations (7) and (8) are, respectively, obtained:

$$f(\alpha) = (\alpha_{max} - \alpha)^n \tag{7}$$

$$f(\alpha) = \alpha^m (\alpha_{max} - \alpha)^n \tag{8}$$

where α_{max} is the maximum degree of conversion (experimentally measured) and must be implemented to cope with the incomplete curing reaction. As these concepts are generally valid in the simple case of one-step chemical reactions, the aforementioned considerations and the formalism introduced for the activation energy can be extended to complex multistep reactions, taking into account the rate-controlling step. In our case, for multistep reaction mechanisms, the reaction rate can be expressed as a function of more than one rate constant. Indeed, Kamal's model can be rearranged and expressed as reported in Equation (9) [29]:

$$\frac{d\alpha}{dt} = (k_1 + k_2 \alpha^m)(\alpha_{max} - \alpha)^n \tag{9}$$

where k_1 is the rate constant of a catalyzed n-order reaction and k_2 is the rate constant of an autocatalytic m-order reaction.

When dealing with the non-isothermal mode, the formalism implied for the expression of the conversion rate in the case of the isothermal analysis (Equation (3)) can be adopted in the same way to express the functional dependencies. As heating experiments are performed at constant heating rates $\beta = dT/dt > 0$, the conversion rate can be expressed as the following:

$$\frac{d\alpha}{dt} = \frac{1}{\beta} k(T) f(\alpha) \tag{10}$$

Typically, the rate of reaction at a constant heating rate as function of temperature exhibits a relative maximum within the temperature interval under study. Moreover, the rate constant can be expressed by the Arrhenius equation also in the case of the non-isothermal analysis. In our case, the Kissinger and the Flynn–Wall–Ozawa models have been applied for the determination of the energy of activation of the overall reaction.

The Kissinger model [30,31] retrieved the E_a at several peak temperatures (T_p) due to the different applied heating rates β, assuming that the reaction rate in a constant conversion is solely dependent on the temperature. In this case, the following linear relationship is reported:

$$\ln\left(\frac{\beta}{T_p^2}\right) = -\frac{E_a}{RT_p} + \ln\left(\frac{AR}{E_a}\right) \tag{11}$$

Under these conditions, Kissinger [30–32] reported that, for a series of non-isothermal tests, the following dependency of the heating rate on the inverse of the peak temperature holds true:

$$\frac{d \ln\left(\beta/T_p^2\right)}{d(1/T_p)} = -\frac{E_a}{R} \tag{12}$$

In the case of the Flynn–Wall–Ozawa model, the Arrhenius rate law was integrated and then the Doyle approximation [33,34] was applied to obtain Equation (13). According to ASTM E698, [35] also in this case, the analysis was performed on several heating rates, with T_p being independent of the heating rate:

$$\ln(\beta) = -1.052 \frac{E_a}{RT_p} - const. + \ln\left(\frac{A\,E_a}{R\,f(\alpha)}\right) \quad (13)$$

where *const.* stands for a constant term depending on the integration method applied to the Arrhenius law, while $f(\alpha)$ is the reaction model. As well in this case, for a series of non-isothermal tests, the following dependency of the heating rate on the inverse of the peak temperature holds true: [36,37]

$$\frac{d\,\ln(\beta)}{d(1/T_p)} = -1.052 \frac{E_a}{R} \quad (14)$$

Both Equations (12) and (14) suggest two alternative ways to estimate the activation energy. Through the linear fitting of data derived by dynamic DSC measurements, from a plot of $\ln(\beta/T_p^2)$ [or $\ln(\beta)$] versus $1/T_p$, it is possible to determine the apparent activation energy value with further insight on the preexponential factor A in the case of the Kissinger method.

4. Results

Differential scanning calorimetry was used to investigate the curing behavior of epoxy/anhydride mixtures with different molar ratios in both dynamic and isothermal conditions. DSC thermograms collected at different heating rates (from 1 to 10 °C/min) for the three formulations (stoichiometric, epoxy excess, and anhydride excess) are respectively reported in Figure 1a,c,e with their respective peak temperature (T_p) values listed in Table 1. The absence of any residual post-cure reactions was certified by the absence of a further exotherm peak during the second heating ramp performed on all the studied samples. As no further peak was detected, samples were assumed as fully cured during the dynamic run. Therefore, the total heat of reaction ΔH_{tot} was evaluated using the DSC analysis conducted at 10 °C/min for each sample. The calculated values of ΔH_{tot} are 353.7, 301.1, and 377.2 J/g for the samples at a stoichiometric epoxy/anhydride ratio, epoxy excess, and anhydride excess, respectively. The thermal stability of the same three epoxy/anhydride systems at different molar ratios was investigated by means of thermogravimetric analysis (TGA) and reported in the supplementary information of a previous work [12]. The systems did not experience any thermal degradation in the range of temperatures utilized in the DSC analyses.

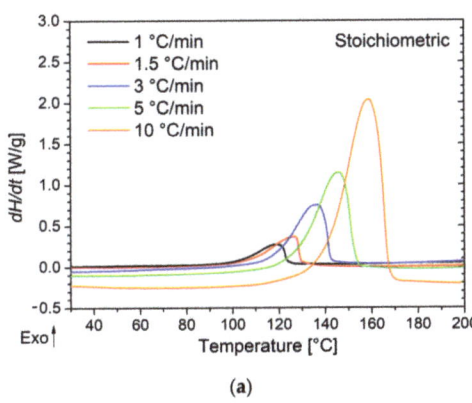
(a)

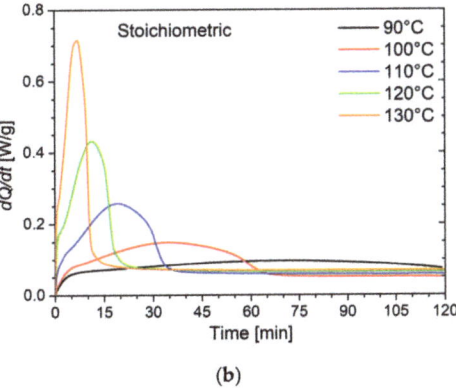
(b)

Figure 1. *Cont.*

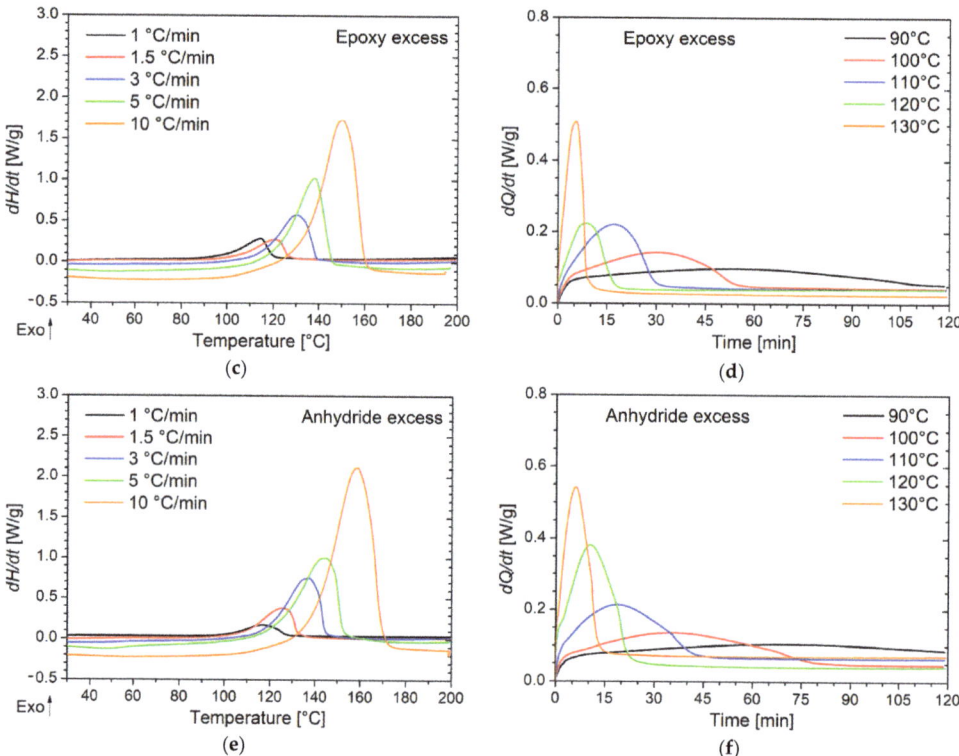

Figure 1. Non-isothermal DSC thermograms of BOMF/MNA resins with (**a**) a stoichiometric epoxy/anhydride ratio, (**c**) epoxy excess, and (**e**) anhydride excess. Dynamic DSC analyses were conducted at five heating rates (i.e., 1 °C/min, 1.5 °C/min, 3 °C/min, 5 °C/min, and 10 °C/min) for each sample. Isothermal DSC thermograms of BOMF/MNA resins with (**b**) a stoichiometric epoxy/anhydride ratio, (**d**) epoxy excess, and (**f**) anhydride excess at five different temperatures (i.e., 90, 100, 110, 120, and 130 °C) for each sample.

Table 1. Peak temperature (T_p) values for the non-isothermal DSC analyses at different heating rates for mixtures with balanced and unbalanced epoxy/anhydride ratios.

	10 °C/min	5 °C/min	3 °C/min	1.5 °C/min	1 °C/min
Stoichiometric	158.4 °C	145.6 °C	136.7 °C	126.1 °C	119.0 °C
Epoxy Excess	153.1 °C	140.7 °C	130.8 °C	120.9 °C	114.5 °C
Anhydride Excess	158.3 °C	144.9 °C	137.3 °C	125.9 °C	116.7 °C

All the systems exhibit a single exothermic peak with a T_p, which shifts to higher temperatures by increasing the heating rate regardless of the epoxy/anhydride ratio. Furthermore, when an epoxy excess is used, at each heating rate, T_p is about 5 °C lower than that of other samples. It has been reported that a lower T_p appears to be an indicator of higher curing reactivity [38]; however, this occurrence will be confuted by considerations regarding the activation energy of the curing reaction, reported further in this section. This trend can be, instead, explained by the increased mobility of the growing macromolecular units associated with the decrease in the mixture viscosity when the epoxy monomer is added in stoichiometric excess [12].

As both the Kissinger and Ozawa approaches are based on fitting procedures applied at various heating rates [30–32,35–37] and rely on the assumption that the extent of reaction α at the peak temperature (T_p) is constant and independent of such heating rates, these methods were applied to calculate the activation energy as solely a function of the epoxy/anhydride ratio and provide a better elucidation on the stoichiometry effects. Figure 2a reports the values of $\ln(\beta/T_p^2)$ as function of $1/T_p$; these data were linearly fitted, and the obtained parameters were used to evaluate the activation energy as well as the pre-exponential factor A for the overall reaction, according to Equation (11). Similarly, Figure 2b reports the linear plot of $\ln(\beta)$ versus $1/T_p$ according to the Ozawa model (Equation (13)), and its linear fitting can be observed. The calculated values of the kinetic parameters are summarized in Table 2.

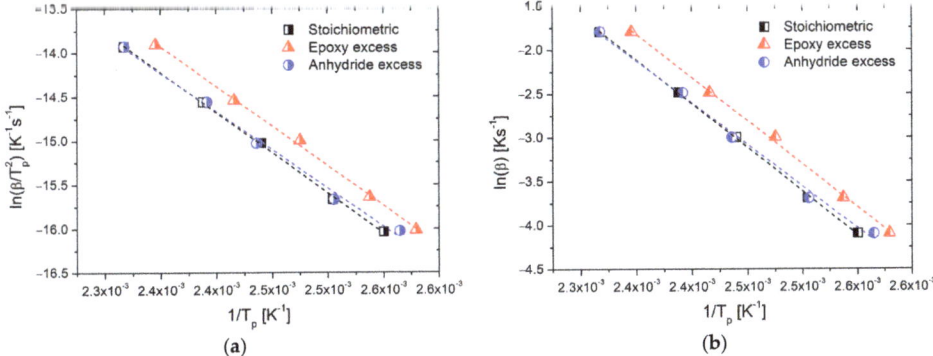

Figure 2. Plots of the heating rate versus $1/T_p$. Dash lines represent the linear fit according to (**a**) Kissinger (Equation (11)) and (**b**) Ozawa (Equation (13)).

Table 2. Values of the global activation energy (E_a^K) and the preexponential factor (A^K) determined by the Kissinger method and the global activation energy (E_a^O) determined by the Ozawa method.

	Kissinger		Ozawa
	E_a^K [kJ/mol]	A^K [s^{-1}]	E_a^O [kJ/mol]
Stoichiometric	76.2	1.37×10^7	78.9
Epoxy Excess	75.2	1.40×10^7	77.9
Anhydride Excess	72.4	4.51×10^6	75.3

Both the Kissinger and Ozawa models return similar values of activation energy for systems with a stoichiometric ratio and an excess of epoxide, indicating the same global reactivity of the two systems. In particular, it can be observed that the lower peak temperatures when epoxides are added in larger quantities, compared to those of mixtures at a stoichiometric ratio, may suggest an enhanced reactivity of the samples at epoxy excess. However, we remark that this observation may lead to an erroneous evaluation of the tendency of reaction of the system. On the other hand, for the system with an excess of anhydride, the E_a is lower, and this occurrence indicates an inhibited reactivity of the system despite the fact that T_p is comparable with the stoichiometric system. The E_a values evaluated with the Kissinger method are systematically lower than those evaluated with the Ozawa method, as expected by the comparison with similar data reported in the literature [30–32,35–37,39–41].

In the case of the isothermal DSC tests conducted on the BOMF/MNA resins at different epoxy/anhydride ratios, to prevent vitrification as much as possible, five temperature values lower than the onset temperature value of the cure peak detected in the dynamic DSC analyses were chosen. In Figure 1b,d,f, the thermograms of the isothermal analyses are reported in which the heat flows as a function of time (directly related to the rate of conversion according to Equation (2)) are depicted. As can be seen, when decreasing the curing temperature, a longer reaction time is needed for the system to complete the cure and to reach the maximum conversion rate. This behavior is characteristic of autocatalytic reaction kinetics [18,42]. As already observed, reactions characterized by n-order kinetics show a peak in isothermal scans, eventually approaching zero for longer times regardless of the temperature. In our case, all systems cured at 90 °C did not fully react during the analysis time (two hours). It was proved by performing a heating scan on the samples that underwent the isothermal analysis in which a residual cure was detected. In addition, lower peak temperatures and higher values of the conversion rate are associated with an increase in the curing temperature. For all the samples cured above 90 °C, no post-cure reactions were detected by dynamic scans performed on the samples that underwent isothermal tests.

The kinetic parameters of the curing reaction were calculated using the analyses of thermograms obtained from the isothermal DSC experiments. To this regard, an accurate baseline construction was required to obtain consistent results relevant to the kinetic analysis. In this work, for all the samples, a baseline was constructed, according to the procedure reported by Barton et al. [10], by performing an isothermal scan on the post-cured samples. The degree of cure as a function of time for each temperature (Figure 3a,c,e) was determined relating the reaction enthalpy obtained by isothermal analysis to the ΔH_{tot} value derived from the dynamic measurements previously reported, following Equation (1).

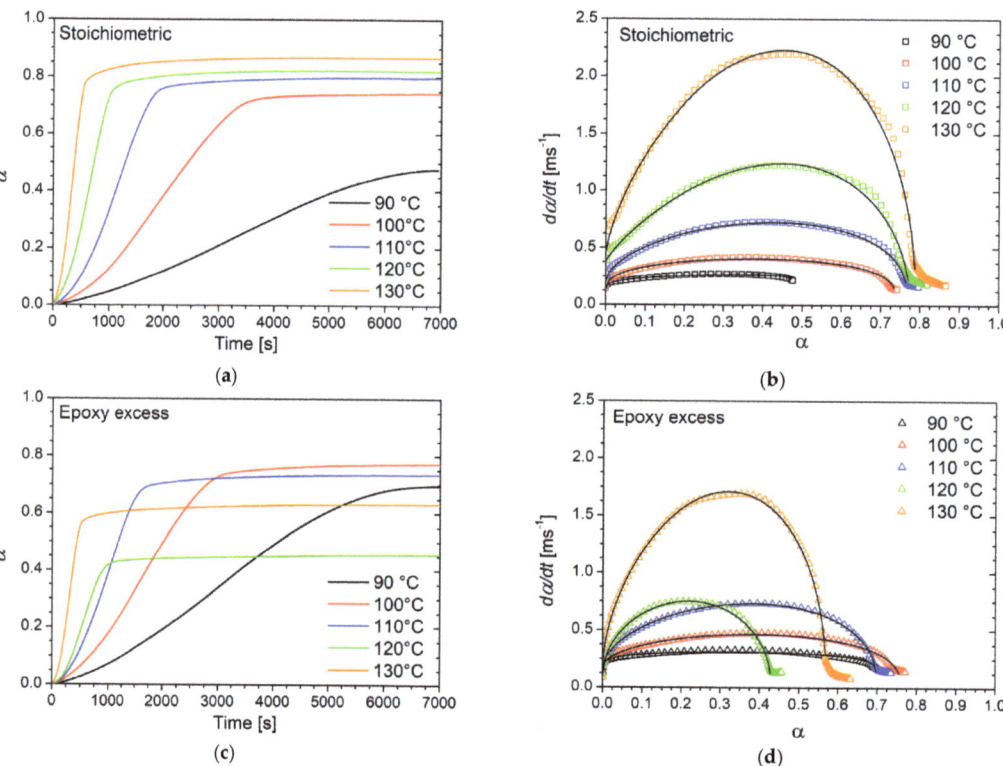

Figure 3. *Cont.*

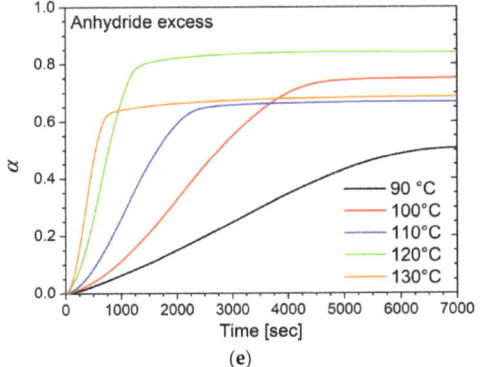
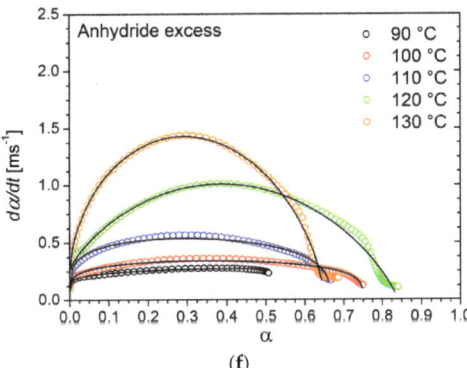

Figure 3. Plots of the degree of conversion (α) as a function of time and conversion rate ($d\alpha/dt$) as function of conversion for epoxy/anhydride mixtures with (**a**,**b**) a stoichiometric ratio, (**c**,**d**) epoxy excess, and (**e**,**f**) anhydride excess, respectively. Solid lines in (**b**,**d**,**f**) represent the fitting curves of the experimental data by the Kamal model (Equation (9)).

The sigmoidal shape of the α–time curves, especially for low temperatures, is characteristic of autocatalytic processes for which the reaction rate is relatively low at the beginning of the process, and the conversion degree only slowly increase due to the initial reaction steps. As new functional groups, which can catalyze the curing process, are created by the first reaction steps, the cure degree rapidly increases until a high degree of cure is achieved. At this time, the reaction rate drastically decreases, and α reaches a plateau. Nevertheless, no sample reaches the maximum degree of cure (Table 3). In particular, a linear increase in α_{max} with temperature can be noticed for the system with an equimolar epoxy/anhydride ratio, while systems with unbalanced stoichiometry exhibited a non-linear behavior.

Table 3. Values of the maximum degree of cure α_{max} reached after 2 h of isothermal DSC analysis for samples with different epoxy/anhydride ratios.

	α_{max}				
	90 °C	100 °C	110 °C	120 °C	130 °C
Stoichiometric	0.48	0.74	0.77	0.78	0.79
Epoxy Excess	0.70	0.76	0.70	0.43	0.57
Anhydride Excess	0.51	0.75	0.67	0.84	0.65

Reaction rate data were fitted to the Kamal model (Equation (9)) by minimizing the least square error function to evaluate the k_1, k_2, m, and n kinetic parameters with no prior assumption for the total reaction order $m + n$. The best-fitting values obtained for the kinetic parameters at various curing temperatures are listed in Table 4. In Figure 3b,d,f the experimental data (symbols) are compared to the fitted data (black solid lines). Overall, the curing degree rate is adequately described by autocatalytic behavior.

The total reaction order $m + n$ as well as the rate constants show an increasing trend with the curing temperature at each epoxy/anhydride ratio.

Further analysis can be carried out on the obtained values for the rate constants k_1 and k_2, which express, respectively, the influence of the initiation step on the overall reaction and the dependence of the reaction rate on the species formed during the reaction steps (that also act as catalysts in autocatalytic processes) [18,42]. As the rate constants comply with the Arrhenius law (Equation (4)), by plotting their logarithm against the inverse of the cure temperatures (Figure 4), it is possible to gather values of the preexponential factor A and the activation energy E_a for the single-reaction processes (Table 5).

Table 4. Kinetic parameters obtained by least square fitting of the curing degree rate $d\alpha/dt$ versus α by the Kamal method for systems with different epoxy/anhydride ratios as function of curing temperature (T_c).

	T_c [°C]	k_1 [s^{-1}]	k_2 [s^{-1}]	m	n	$m+n$
Stoichiometric	90	1.31×10^{-4}	2.27×10^{-4}	0.29	0.06	0.35
	100	1.40×10^{-4}	5.87×10^{-4}	0.40	0.28	0.68
	110	2.18×10^{-4}	1.41×10^{-3}	0.58	0.37	0.95
	120	4.29×10^{-4}	3.49×10^{-3}	0.86	0.50	1.36
	130	5.09×10^{-4}	5.92×10^{-3}	0.75	0.49	1.24
Epoxy Excess	90	1.21×10^{-4}	2.86×10^{-4}	0.18	0.13	0.31
	100	1.60×10^{-4}	7.72×10^{-4}	0.51	0.36	0.87
	110	1.89×10^{-4}	1.67×10^{-3}	0.57	0.40	0.97
	120	2.21×10^{-4}	3.23×10^{-3}	0.56	0.47	1.03
	130	2.82×10^{-4}	6.18×10^{-3}	0.65	0.46	1.11
Anhydride Excess	90	1.19×10^{-4}	2.39×10^{-4}	0.28	0.05	0.33
	100	1.45×10^{-4}	3.76×10^{-4}	0.27	0.25	0.52
	110	1.77×10^{-4}	7.49×10^{-4}	0.27	0.31	0.58
	120	3.02×10^{-4}	3.35×10^{-4}	0.79	0.78	1.57
	130	3.28×10^{-4}	5.03×10^{-4}	0.59	0.64	1.23

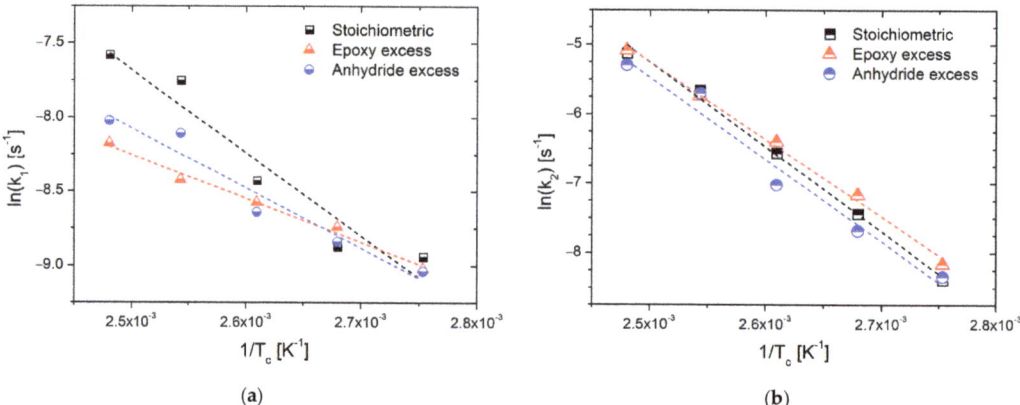

Figure 4. Linear plots of the logarithm of the rate constants, (a) $\ln(k_1)$ and (b) $\ln(k_2)$), obtained by fitting the isothermal data with the Kamal model (Equation (9)) as function of the inverse of the cure temperature ($1/T_c$) at different epoxy/anhydride ratios.

Table 5. Values of the preexponential factor (A_1) and activation energy (E_{a1}) from the linear fitting of the rate constant k_1 and values of the preexponential factor (A_2) and activation energy (E_{a2}) from the linear fitting of the rate constant k_2 for systems at different epoxy/anhydride ratios.

	A_1 [s^{-1}]	E_{a1} [kJ/mol]	A_2 [s^{-1}]	E_{a2} [kJ/mol]
Stoichiometric	5.49×10^2	46.5	8.83×10^{10}	101.3
Epoxy excess	4.13×10^{-1}	24.5	6.22×10^9	92.4
Anhydride excess	7.64	33.6	2.85×10^{10}	98.2

5. Discussion

One of the key factors influencing the morphological changes in the physical structure during thermosetting processes is the curing kinetics. In particular, the epoxy/anhydride crosslinking mechanism is extremely complex, as it involves the simultaneous occurrence of competitive reactions, such as etherification and esterification reactions, which globally lead to the network formation [39,43,44].

When anhydrides are used as curing agents in presence of imidazole as an initiator, the first reaction step is the initiation, a nucleophilic attack of the imidazole nitrogen to the anhydride group (Scheme 1a) that generates a zwitterion intermediate containing a quaternary nitrogen cation and an active oxyanion. The latter reacts with the anhydride or the epoxy group, giving rise to a carboxylate anion (Scheme 1b), which initiates chain-wise polymerization [43–45]. The formed anion can further react both with an epoxy group (polyetherification, Scheme 1c) or with another anhydride (polyesterification, Scheme 1d), propagating the chain growth through an alternating copolymerization anionic mechanism.

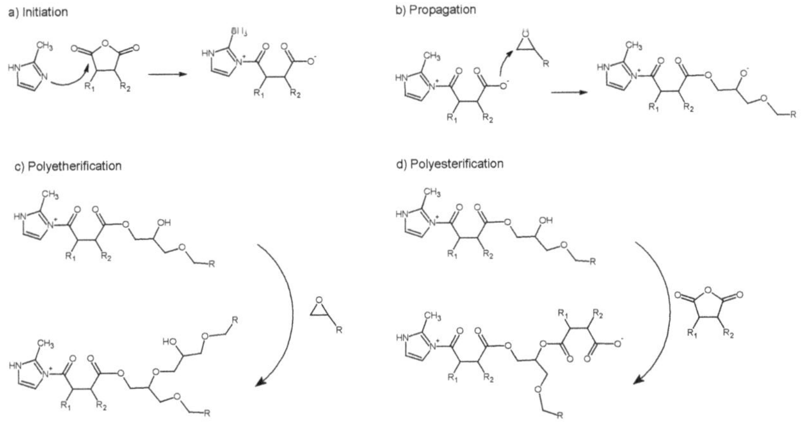

Scheme 1. Reaction steps of epoxy/anhydride cure: (**a**) initiation, (**b**) propagation, (**c**) polyetherification and (**d**) polyesterification.

Due to the complexity of the mechanism underlying the generation of the thermosetting network of epoxy/anhydride resins, a single-step reaction mechanism is unlikely to correctly fit the experimental data. Moreover, the continuous generation of reactive groups throughout the propagation steps suggests the autocatalytic behavior of the epoxy/anhydride reaction. In fact, the autocatalytic nature of the BOMF/MNA curing reaction clearly is suggested by the shape of the thermograms derived from isothermal DSC analysis: the presence of a maximum at different times, depending on the cure temperature, is particular to autocatalytic processes [46–48].

Non-isothermal DSC analyses were elaborated by means of two different methods to derive information about the activation energy of the overall curing process. As expected, the Kissinger and Ozawa methods did not return the same value of E_a, [40,41] but the latter are reasonable values for an epoxy/anhydride cure, in trend with the reported literature [39,49]. In the specific case of BOMF/MNA, if an excess of anhydride is used, even though the maximum reaction rate is reached at a temperature comparable with the systems with a stoichiometric epoxy/anhydride molar ratio, the system is more reactive (lower activation energy, Table 2). An accelerating effect is indicated by a decrease in the activation energy since less energy from the reacting components is required to complete the reaction [40]. This finding could be related to the higher number of initiation sites formed in the presence of higher amounts of anhydride [12]. Consequently, the number of propagating chains is increased, and so is the number of reactive sites, with an autocatalytic effect responsible for a globally faster reaction (reflected also by the lower value of A^k, Table 2). We remark that the activation energy only provides data on the reaction rate, not about the reaction mechanism, and that both the Kissinger and Ozawa methods provide a single value of the activation energy for the whole process despite the fact that, in complex systems such as epoxy/anhydride resins, this parameter fluctuates with the curing period.

The isothermal analysis of the curing process can, instead, give a major degree of detail about the reaction mechanisms. Isothermal DSC data were fitted to an autocatalytic model developed by Kamal, which well characterizes systems wherein the reaction rate increases due to the catalytic impact of the groups formed during the reaction itself, to define the relationship among conversion, time, and temperature. First of all, the conversion α as a function of time was evaluated at different temperatures for the three studied systems (Figure 3a,c,e). The lack of a linear dependance of α with temperature for the systems with epoxy excess and anhydride excess, together with the partial extent of cure experienced by all the samples, can be explained by the occurrence, to different extents, of vitrification. The reactivity of the molecules dominates the rate of reaction at the beginning of a curing reaction. As the reaction proceeds, the degree of cure increases and the system first undergoes gelation (i.e., the development of molecular branching), which is related to an increase in viscosity and a subsequent worsening of the molecular mobility. When diffusion-controlled processes overstep the mobility of the reacting groups, the reaction becomes diffusion-controlled and vitrification occurs [50]. By correlating this phenomenon to the conversion degree, we can observe that at low conversion values, the network is far from being formed and the reacting species, still present in high concentrations, can easily diffuse, and the reaction proceeds fast. Once high conversions are reached, long chains are formed, resulting in a viscosity increase which limits the chain mobility; at this point, the reaction becomes diffusion-controlled, and a plateau value for the conversion is reached. When the system vitrifies, increasing the temperature above the glass transition temperature (T_g), the viscosity decreases, and subsequently, the chain mobility is somehow improved, enabling the reaction of the remaining unreacted functional groups [51]. For the systems studied in this work, all the temperatures selected for the isothermal analysis are above T_g [12]. However, due to the narrowness of the cure peak (Figure 1a,c,e), even though the temperature of isothermal analysis is higher than T_g, this latter is still too close to the peak temperature, so vitrification is unavoidable. It is noteworthy to highlight the behavior of the system with an epoxy excess, for which strong vitrification is verified at temperatures of 120 and 130 °C. Using the non-isothermal DSC analysis, it emerged that when BOMF is present in stoichiometric excess, the curing reaction begins at a lower temperature (T_{onset}) compared to the other systems. In particular, when this system was heated at 10 °C/min to the temperature of 120 °C, the degree of conversion reached was higher than the α experienced by the other samples at the same temperature. This behavior was also reflected in the dependence of conversion on temperature resulting from the isothermal analysis.

The fitting of experimental data with the mechanistic Kamal method was also carried out. As can be seen from Figure 3b,d,f, the model fits well the experimental data, especially for the system with epoxy excess, regardless of the conversion degree. For both mixtures at a stoichiometric ratio and anhydride excess, instead, the fit is in good agreement with the model, particularly at low α values, and a lack of accuracy at higher conversions is noted, when the reaction is controlled by diffusion phenomena rather than by kinetic parameters [52].

The parameters obtained by the fitting of the experimental data, reported in Table 4, reflect, as far as k_1 and m are concerned, the contribution of non-autocatalytic phenomena (in this specific case of study, the initiation step). On the other hand, k_2 and n account for the contribution of autocatalytic phenomena (i.e., propagation in epoxy/anhydride cure). Regardless of the resin composition, all four parameters generally increase with temperature, even though with a different trend, and k_2 and m are systematically higher than k_1 and n. Specifically, when the stoichiometry is unbalanced, the order of reaction increases faster with temperature. For the stoichiometric system, instead, m increases much more than n with temperature, indicating a stronger temperature dependance of propagation compared to initiation. Noteworthy are the values of m and n for the system with an excess of anhydride: they are almost the same, with an exception made at 90 °C. This occurrence indicates that initiation and propagation proceed simultaneously. Generally speaking, the overall reaction order, $m + n$, is in the range of 0.35–1.23 regardless the composition.

Further analysis was conducted on the rate constants obtained by the fitting procedure, extrapolating the values of the activation energy (E_{a1} and E_{a2}) and the preexponential factor (A_1 and A_2) for k_1 and k_2, respectively. In particular, while the trend of k_2 with temperature is well fitted by a linear trend, the behavior of k_1 is less prominent of a linear behavior, although it is still well described by a linear fit. This is due to the fact that k_1 is only calculated using two or three of the first experimental datapoints, which might result in inaccurate estimated values [52]. As a general consideration, it emerges that the propagation step is much more kinetically disfavored.

All the considerations expressed here can be summarized into practical advice for the application of BOMF/MNA resins. As the curing degree of samples with a stoichiometric ratio among the reactants has an increasing trend with the curing temperature, this composition should be used for applications in which significant differences in curing temperatures might be experienced by different samples, or also at different points of the same sample (i.e., thick samples). If lower curing temperatures must be used, instead, a composition with an epoxy excess should be chosen, which ensures a higher α_{max} at 90 °C. However, for the cure of this sample, it is mandatory to control the curing temperature, as an uncontrolled increase in the temperature will have a detrimental effect on the degree of cure. On the other hand, when energy consumption is a limiting factor, the use of anhydride excess may be preferred due to the lower activation energy overall required for the curing process.

6. Conclusions

The cure kinetics of bioderived furan-based epoxy resins reacted with anhydride in the presence of 2-methylimidazole as an initiator were investigated by both non-isothermal and isothermal differential scanning calorimetry (DSC). The cure kinetics of BOMF/MNA resulted in being dependent on the epoxy/anhydride molar ratio. The main findings of this study can be summed up as follows:

When anhydride is in a molar excess, the promotion of the initiation step leads to an overall improved reactivity. In fact, it was observed that the activation energy of the curing process decreased from 76.2 to 72.4 kJ/mol (Kissinger method) and from 78.9 kJ/mol to 75.3 kJ/mol (Ozawa method) for stoichiometric and anhydride excess, respectively;

The high reactivity of the system, independent of the resin composition, leads to incomplete cure due to vitrification. This occurrence was observed even at the higher curing temperature (130 °C) from the maximum degree of conversion (0.79, 0.57, and 0.65 for stoichiometric, epoxy excess, and anhydride excess, respectively);

Both propagation and initiation are favored by increasing the temperature, but with a different dependence, as a function of the composition.

The autocatalytic Kamal model fits well the experimental data but did not succeed in fully describing the complex epoxy anhydride curing process. The global order of reaction ($m + n$) was in the range 0.35–1.24, 0.31–1.11, and 0.33–1.23 for stoichiometric, epoxy excess, and anhydride excess, respectively.

Overall, the observed results enlighten a promising use of these compounds, and a deeper understanding of the reaction kinetics and mechanisms of these systems could help in promoting the application of these biobased materials at an industrial level, addressing the global dependence on oil-derived polymers.

Author Contributions: Conceptualization, A.M.; methodology, A.M. and C.B.; investigation, A.M., N.F. and C.B.; data curation, A.M., N.F. and C.B.; writing—original draft preparation, A.M. and N.F.; writing—review and editing, A.M. and C.B. All authors have read and agreed to the published version of the manuscript.

Funding: This research received no external funding.

Institutional Review Board Statement: Not applicable.

Informed Consent Statement: Not applicable.

Data Availability Statement: The data presented in this study are available on request from the corresponding author.

Conflicts of Interest: The authors declare no conflict of interest.

References

1. Zhou, S.; Chen, Z.; Tusiime, R.; Cheng, C.; Sun, Z.; Xu, L.; Liu, Y.; Jiang, M.; Zhou, J.; Zhang, H.; et al. Highly improving the mechanical and thermal properties of epoxy resin via blending with polyetherketone cardo. *Compos. Commun.* **2019**, *13*, 80–84. [CrossRef]
2. Vidil, T.; Tournilhac, F.; Musso, S.; Robisson, A.; Leibler, L. Control of reactions and network structures of epoxy thermosets. *Prog. Polym. Sci.* **2016**, *62*, 126–179. [CrossRef]
3. Guo, Q. Use of Thermosets in the Building and Construction Industry. In *Thermosets Structure, Properties, and Applications*, 2nd ed.; Elsevier: Amsterdam, The Netherlands, 2017; pp. 279–300.
4. Musto, P.; Martuscelli, E.; Ragosta, G.; Russo, P.; Villano, P. Tetrafunctional epoxy resins: Modeling the curing kinetics based on FTIR spectroscopy data. *J. Appl. Polym. Sci.* **1999**, *74*, 532–540. [CrossRef]
5. Fernández-Francos, X.; Rybak, A.; Sekula, R.; Ramis, X.; Serra, A. Modification of epoxy–anhydride thermosets using a hyperbranched poly (ester-amide): I. Kinetic study. *Polym. Int.* **2012**, *61*, 1710–1725. [CrossRef]
6. Martin-Gallego, M.; González-Jiménez, A.; Verdejo, R.; Lopez-Manchado, M.A.; Valentin, J.L. Epoxy resin curing reaction studied by proton multiple-quantum NMR. *J. Polym. Sci. Part B Polym. Phys.* **2015**, *53*, 1324–1332. [CrossRef]
7. Buist, G.J.; Hamerton, I.; Howlin, B.J.; Jones, J.R.; Liu, S.; Barton, J.M. Comparative kinetic analyses for epoxy resins cured with imidazole–metal complexes. *J. Mater. Chem.* **1994**, *4*, 1793–1797. [CrossRef]
8. Nuñez, L.; Fraga, F.; Castro, A.; Fraga, L. Elastic Moduli and Activation Energies for an Epoxy/m-XDA System by DMA and DSC. *J. Therm. Anal.* **1998**, *52*, 1013–1022. [CrossRef]
9. Stark, W. Investigation of the curing behaviour of carbon fibre epoxy prepreg by Dynamic Mechanical Analysis DMA. *Polym. Test.* **2013**, *32*, 231–239. [CrossRef]
10. Barton, J.M. *The Application of Differential Scanning Calorimetry (DSC) to the Study of Epoxy Resin Curing Reactions*; Springer: Berlin, Germany, 1985; pp. 111–154. [CrossRef]
11. Vyazovkin, S.; Sbirrazzuoli, N. Kinetic methods to study isothermal and nonisothermal epoxy-anhydride cure. *Macromol. Chem. Phys.* **1999**, *200*, 2294–2303. [CrossRef]
12. Marotta, A.; Faggio, N.; Ambrogi, V.; Cerruti, P.; Gentile, G.; Mija, A. Curing Behavior and Properties of Sustainable Furan-Based Epoxy/Anhydride Resins. *Biomacromolecules* **2019**, *20*, 3831–3841. [CrossRef]
13. Shnawa, H.A. Curing and thermal properties of tannin-based epoxy and its blends with commercial epoxy resin. *Polym. Bull.* **2020**, *78*, 1925–1940. [CrossRef]
14. Yang, S.; Huo, S.; Wang, J.; Zhang, B.; Wang, J.; Ran, S.; Fang, Z.; Song, P.; Wang, H. A highly fire-safe and smoke-suppressive single-component epoxy resin with switchable curing temperature and rapid curing rate. *Compos. Part Eng.* **2020**, *207*, 108601. [CrossRef]
15. Marotta, A.; Faggio, N.; Ambrogi, V.; Mija, A.; Gentile, G.; Cerruti, P. Biobased furan-based epoxy/TiO2 nanocomposites for the preparation of coatings with improved chemical resistance. *Chem. Eng. J.* **2020**, *406*, 127107. [CrossRef]
16. Yousefi, A.; LaFleur, P.G.; Gauvin, R. Kinetic studies of thermoset cure reactions: A review. *Polym. Compos.* **1997**, *18*, 157–168. [CrossRef]
17. Hardis, R.; Jessop, J.L.; Peters, F.E.; Kessler, M.R. Cure kinetics characterization and monitoring of an epoxy resin using DSC, Raman spectroscopy, and DEA. *Compos. Part A Appl. Sci. Manuf.* **2013**, *49*, 100–108. [CrossRef]
18. Kenny, J.M. Determination of autocatalytic kinetic model parameters describing thermoset cure. *J. Appl. Polym. Sci.* **1994**, *51*, 761–764. [CrossRef]
19. Kessler, M.R.; White, S.R. Cure kinetics of the ring-opening metathesis polymerization of dicyclopentadiene. *J. Polym. Sci. Part A Polym. Chem.* **2002**, *40*, 2373–2383. [CrossRef]
20. Roudsari, G.M.; Mohanty, A.K.; Misra, M. Study of the Curing Kinetics of Epoxy Resins with Biobased Hardener and Epoxidized Soybean Oil. *ACS Sustain. Chem. Eng.* **2014**, *2*, 2111–2116. [CrossRef]
21. Vyazovkin, S.; Chrissafis, K.; Di Lorenzo, M.L.; Koga, N.; Pijolat, M.; Roduit, B.; Sbirrazzuoli, N.; Suñol, J.J. ICTAC Kinetics Committee recommendations for collecting experimental thermal analysis data for kinetic computations. *Thermochim. Acta* **2014**, *590*, 1–23. [CrossRef]
22. Laidler, K.J.; King, M.C. Development of transition-state theory. *J. Phys. Chem.* **1983**, *87*, 2657–2664. [CrossRef]
23. Anslyn, E.V.; Dougherty, D.A. Transition State Theory and Related Topics. In *Modern Physical Organic Chemistry*; University Science Books: Melville, NY, USA, 2006; pp. 365–373.
24. Ghaemy, M.; Barghamadi, M.; Behmadi, H. Cure kinetics of epoxy resin and aromatic diamines. *J. Appl. Polym. Sci.* **2004**, *94*, 1049–1056. [CrossRef]
25. Li, C.; Fan, H.; Hu, J.; Li, B. Novel silicone aliphatic amine curing agent for epoxy resin: 1,3-Bis(2-aminoethylaminomethyl) tetramethyldisiloxane. 2. Isothermal cure, and dynamic mechanical property. *Thermochim. Acta* **2012**, *549*, 132–139. [CrossRef]

26. Ferdosian, F.; Zhang, Y.; Yuan, Z.; Anderson, M.; Xu, C. Curing kinetics and mechanical properties of bio-based epoxy composites comprising lignin-based epoxy resins. *Eur. Polym. J.* **2016**, *82*, 153–165. [CrossRef]
27. Wan, J.; Gan, B.; Li, C.; Molina-Aldareguia, J.; Kalali, E.N.; Wang, X.; Wang, D.-Y. A sustainable, eugenol-derived epoxy resin with high biobased content, modulus, hardness and low flammability: Synthesis, curing kinetics and structure–property relationship. *Chem. Eng. J.* **2016**, *284*, 1080–1093. [CrossRef]
28. Hu, J.; Shan, J.; Wen, D.; Liu, X.; Zhao, J.; Tong, Z. Flame retardant, mechanical properties and curing kinetics of DOPO-based epoxy resins. *Polym. Degrad. Stab.* **2014**, *109*, 218–225. [CrossRef]
29. Kamal, M.R. Thermoset characterization for moldability analysis. *Polym. Eng. Sci.* **1974**, *14*, 231–239. [CrossRef]
30. Kissinger, H.E. Reaction Kinetics in Differential Thermal Analysis. *Anal. Chem.* **1957**, *29*, 1702–1706. [CrossRef]
31. Boey, F.; Qiang, W. Experimental modeling of the cure kinetics of an epoxy-hexaanhydro-4-methylphthalicanhydride (MHHPA) system. *Polymer* **2000**, *41*, 2081–2094. [CrossRef]
32. Barros, J.J.P.; Silva, I.D.D.S.; Jaques, N.G.; Wellen, R.M.R. Approaches on the non-isothermal curing kinetics of epoxy/PCL blends. *J. Mater. Res. Technol.* **2020**, *9*, 13539–13554. [CrossRef]
33. Doyle, C.D. Kinetic analysis of thermogravimetric data. *J. Appl. Polym. Sci.* **1961**, *5*, 285–292. [CrossRef]
34. Doyle, C.D. Estimating isothermal life from thermogravimetric data. *J. Appl. Polym. Sci.* **1962**, *6*, 639–642. [CrossRef]
35. *ASTM E698*; Standard Test Method for Kinetic Parameters for Thermally Unstable Materials Using Differential Scanning Calorimetry and the Flynn/Wall/Ozawa Method. ASTM International: West Conshohocken, PA, USA, 2016.
36. Ozawa, T. A New Method of Analyzing Thermogravimetric Data. *Bull. Chem. Soc. Jpn.* **1965**, *38*, 1881–1886. [CrossRef]
37. Flynn, J.H. A General Differential Technique for the Determination of Parameters for d (α)/dt= f (α) A exp (− E/RT) Energy of Activation, Preexponential Factor and Order of Reaction (when Applicable). *J. Therm. Anal. Calorim.* **1991**, *37*, 293–305. [CrossRef]
38. Deng, J.; Liu, X.; Li, C.; Jiang, Y.; Zhu, J. Synthesis and properties of a bio-based epoxy resin from 2,5-furandicarboxylic acid (FDCA). *RSC Adv.* **2015**, *5*, 15930–15939. [CrossRef]
39. Kumar, S.; Samal, S.K.; Mohanty, S.; Nayak, S.K. Study of curing kinetics of anhydride cured petroleum-based (DGEBA) epoxy resin and renewable resource based epoxidized soybean oil (ESO) systems catalyzed by 2-methylimidazole. *Thermochim. Acta* **2017**, *654*, 112–120. [CrossRef]
40. Harsch, M.; Karger-Kocsis, J.; Holst, M. Influence of fillers and additives on the cure kinetics of an epoxy/anhydride resin. *Eur. Polym. J.* **2007**, *43*, 1168–1178. [CrossRef]
41. Sbirrazzuoli, N.; Girault, Y.; Elégant, L. Simulations for evaluation of kinetic methods in differential scanning calorimetry. Part 3—Peak maximum evolution methods and isoconversional methods. *Thermochim. Acta* **1997**, *293*, 25–37. [CrossRef]
42. Barghamadi, M. Kinetics and Thermodynamics of Isothermal Curing Reaction of Epoxy-4, 40-Diaminoazobenzene Reinforced with Nanosilica and Nanoclay Particles. *Polym. Compos.* **2010**, *31*, 1442–1448. [CrossRef]
43. Pin, J.-M.; Sbirrazzuoli, N.; Mija, A. From Epoxidized Linseed Oil to Bioresin: An Overall Approach of Epoxy/Anhydride Cross-Linking. *ChemSusChem* **2015**, *8*, 1232–1243. [CrossRef]
44. Fernàndez-Francos, X.; Ramis, X.; Serra, À. From curing kinetics to network structure: A novel approach to the modeling of the network buildup of epoxy-anhydride thermosets. *J. Polym. Sci. Part A Polym. Chem.* **2014**, *52*, 61–75. [CrossRef]
45. Kumar, S.; Samal, S.K.; Mohanty, S.; Nayak, S.K. Epoxidized Soybean Oil-Based Epoxy Blend Cured with Anhydride-Based Cross-Linker: Thermal and Mechanical Characterization. *Ind. Eng. Chem. Res.* **2017**, *56*, 687–698. [CrossRef]
46. Paramarta, A.; Webster, D.C. Bio-based high performance epoxy-anhydride thermosets for structural composites: The effect of composition variables. *React. Funct. Polym.* **2016**, *105*, 140–149. [CrossRef]
47. Pan, X.; Sengupta, P.; Webster, D.C. High Biobased Content Epoxy–Anhydride Thermosets from Epoxidized Sucrose Esters of Fatty Acids. *Biomacromolecules* **2011**, *12*, 2416–2428. [CrossRef] [PubMed]
48. Mahendran, A.R.; Wuzella, G.; Kandelbauer, A.; Aust, N. Thermal cure kinetics of epoxidized linseed oil with anhydride hardener. *J. Therm. Anal.* **2011**, *107*, 989–998. [CrossRef]
49. Sun, G.; Sun, H.; Liu, Y.; Zhao, B.; Zhu, N.; Hu, K. Comparative study on the curing kinetics and mechanism of a lignin-based-epoxy/anhydride resin system. *Polymer* **2006**, *48*, 330–337. [CrossRef]
50. Atarsia, A.; Boukhili, R. Relationship between isothermal and dynamic cure of thermosets via the isoconversion representation. *Polym. Eng. Sci.* **2000**, *40*, 607–620. [CrossRef]
51. Zhao, L.; Hu, X. Autocatalytic curing kinetics of thermosetting polymers: A new model based on temperature dependent reaction orders. *Polymer* **2010**, *51*, 3814–3820. [CrossRef]
52. Thomas, R.; Durix, S.; Sinturel, C.; Omonov, T.; Goossens, S.; Groeninckx, G.; Moldenaers, P.; Thomas, S. Cure kinetics, morphology and miscibility of modified DGEBA-based epoxy resin—Effects of a liquid rubber inclusion. *Polymers* **2007**, *48*, 1695–1710. [CrossRef]

MDPI
St. Alban-Anlage 66
4052 Basel
Switzerland
www.mdpi.com

Polymers Editorial Office
E-mail: polymers@mdpi.com
www.mdpi.com/journal/polymers

Disclaimer/Publisher's Note: The statements, opinions and data contained in all publications are solely those of the individual author(s) and contributor(s) and not of MDPI and/or the editor(s). MDPI and/or the editor(s) disclaim responsibility for any injury to people or property resulting from any ideas, methods, instructions or products referred to in the content.

www.ingramcontent.com/pod-product-compliance
Lightning Source LLC
LaVergne TN
LVHW070744100526
838202LV00013B/1301